Bioenergy Systems: Assessment and Analysis

Bioenergy Systems: Assessment and Analysis

Edited by Alice Wheeler

SYRAWOOD
PUBLISHING HOUSE

New York

Published by Syrawood Publishing House,
750 Third Avenue, 9th Floor,
New York, NY 10017, USA
www.syrawoodpublishinghouse.com

Bioenergy Systems: Assessment and Analysis
Edited by Alice Wheeler

International Standard Book Number: 978-1-68286-691-7 (Hardback)

Cataloging-in-Publication Data

Bioenergy systems : assessment and analysis / edited by Alice Wheeler.
    p. cm.
Includes bibliographical references and index.
ISBN 978-1-68286-691-7
1. Biomass energy. 2. Energy conversion. 3. Energy crops. I. Wheeler, Alice.
TP339 .B55 2019
662.88--dc23

TABLE OF CONTENTS

Permissions

List of Contributors

Index

PREFACE

Every book is initially just a concept; it takes months of research and hard work to give it the final shape in which the readers receive it. In its early stages, this book also went through rigorous reviewing. The notable contributions made by experts from across the globe were first molded into patterned chapters and then arranged in a sensibly sequential manner to bring out the best results.

Bioenergy is generated from renewable biological sources of energy like bioethanol, biodiesel and biocrops like sugarcane or corn. Bioresources like agriculture, animal, household and forestry waste are treated with techniques of bioproducts engineering to develop energy efficient products. Bioenergy systems also delve into applications of contemporary biological processes for the purposes of electricity generation. This book provides significant information about this discipline to help develop a good understanding of bioenergy and related fields. It brings forth some of the most innovative concepts and elucidates the unexplored aspects of this discipline. Environmentalists, engineers, researchers, experts and students in search of information to further their knowledge will be greatly assisted by this book.

It has been my immense pleasure to be a part of this project and to contribute my years of learning in such a meaningful form. I would like to take this opportunity to thank all the people who have been associated with the completion of this book at any step.

Editor

Assessing the hydrologic and water quality impacts of biofuel-induced changes in land use and management

YONG CHEN[1,2], SRINIVASULU ALE[1,3] (iD), NITHYA RAJAN[2] and CLYDE MUNSTER[3]

[1]Texas A&M AgriLife Research (Texas A&M University System), PO Box 1658, 11708 Highway 70S, Vernon, TX 76384, USA, [2]Department of Soil and Crop Sciences, Texas A&M University, 370 Olsen Blvd, TAMU MS 2474, College Station, TX 77843, USA, [3]Department of Biological and Agricultural Engineering, Texas A&M University, TAMU MS 2117, College Station, TX 77843, USA

Abstract

The Southern High Plains (SHP) of Texas, where cotton (Gossypium hirsutum L.) is grown in vast acreage, and the Texas Rolling Plains (TRP), which is dominated by an invasive brush, honey mesquite (Prosopis glandulosa) have the potential for biofuel production for meeting the U.S. bioenergy target of 2022. However, a shift in land use from cotton to perennial grasses and a change in land management such as the harvesting of mesquite for biofuel production can significantly affect regional hydrology and water quality. In this study, APEX and SWAT models were integrated to assess the impacts of replacing cotton with Alamo switchgrass (Panicum virgatum L.) and Miscanthus × giganteus in the upstream subwatershed and harvesting mesquite in the downstream subwatershed on water and nitrogen balances in the Double Mountain Fork Brazos watershed in the SHP and TRP regions. Simulated average (1994–2009) annual surface runoff from the baseline cotton areas decreased significantly ($P < 0.05$) by 88%, and percolation increased by 28% under the perennial grasses scenario compared to the baseline cotton scenario. The soil water content enhanced significantly under the irrigated switchgrass scenario compared to the baseline irrigated cotton scenario from January to April and August to October. However, the soil water content was depleted significantly under the dryland Miscanthus scenario from April to July relative to the baseline dryland cotton scenario. The nitrate-nitrogen (NO_3-N) and organic-N loads in surface runoff and NO_3-N leaching to groundwater reduced significantly by 86%, 98%, and 100%, respectively, under the perennial grasses scenario. Similarly, surface runoff, and NO_3-N and organic-N loads through surface runoff reduced significantly by 98.9%, 99.9%, and 99.5%, respectively, under the post-mesquite-harvest scenario. Perennial grasses exhibited superior ethanol production potential compared to mesquite. However, mesquite is an appropriate supplementary bioenergy source in the TRP region because of its standing biomass and rapid regrowth characteristics.

Keywords: Agricultural Policy/Environmental eXtender, biomass, honey mesquite, Miscanthus, Soil and Water Assessment Tool, Southern High Plains of Texas, switchgrass, Texas Rolling Plains

Introduction

The U.S. agriculture is facing an unprecedented challenge in securing the nation's energy future in addition to meeting the traditional goal of food security. According to the current Renewable Fuels Standard Program (RFS2), the volume of renewable fuel required to be blended into transportation fuel will be 136 million m[3] by 2022 (U.S. Department of Agriculture; USDA, 2010). As one of the world's largest food producer, exporter, and donor, the U.S. plays a vital role in addressing these challenges. Further increase in crop production or change in land use will be needed for meeting these challenges in the coming years. Since the industrial revolution, human actions (including agriculture) have become a major driving factor for global environmental change, land and water degradation, and biodiversity loss (Rockström et al., 2009; Foley et al., 2011). As a result, agriculture must address its environmental consequences as it seeks to meet the aforementioned food security and renewable fuel challenges.

There are two general types of renewable biofuels. First-generation biofuels are usually produced through intensive agricultural activities, which are similar to those used in growing primary food crops such as maize (Zea mays L.) and grain sorghum (Sorghum bicolor (L.) Moench). Production of first-generation biofuels potentially competes with the production of food. Thus, if the production of first-generation biofuels rises to certain levels, there can be detrimental social consequences

Correspondence: Srinivasulu Ale
e-mails: sriniale@ag.tamu.edu; Srinivasulu.Ale@gmail.com

in the form of reduced food supplies and associated increases in commodity prices that can be passed on to consumers. As intensive agricultural activities typically utilize more resources (prime farmland, irrigation water, fertilizers and pesticides, and fuel for farming operations), there can be increased negative environmental effects associated with the production of first-generation biofuels. To address these concerns, the USDA recommended that of the targeted production of 136 million m^3 of biofuels by 2022, 76 million m^3 should be produced from cellulosic and other advanced biofuel feedstocks (USDA, 2010). Cellulosic biofuels, which are also called second-generation biofuels, are primarily made from the by-products of intensive agricultural activities or from less-intensive agricultural activities performed on nonfood croplands using substantially reduced resource inputs.

The Southern High Plains (SHP) of Texas in the United States is one of the most intensively managed cotton-growing regions in the world. The cotton planting area in the SHP accounted for approximately 31% of the entire U.S. cotton acreage in 2015 (National Agricultural Statistics Service; NASS, 2015). The Ogallala Aquifer is the primary source of irrigation water for this region. Intensive agricultural production in the SHP since 1950s has resulted in a continuous decline of groundwater levels and deterioration of groundwater quality, mainly due to high concentrations of nitrate nitrogen (NO_3-N) (Chaudhuri & Ale, 2014a,b). The land use change from high water- and N-consuming crops such as cotton to more water- and nitrogen-use-efficient perennial grasses such as Alamo switchgrass (*Panicum virgatum* L.) and *Miscanthus* × *giganteus* may benefit this region by prolonging the availability of groundwater and improving groundwater quality. Using the Soil and Water Assessment Tool (SWAT), Cibin *et al.* (2016) simulated a reduction in annual surface runoff by about 12% and 15% under the *Miscanthus* and switchgrass land use scenarios, respectively, compared to the baseline corn/soybean land use in the Wildcat Creek watershed in Indiana. In another SWAT simulation study, Ng *et al.* (2010) showed that a 10% change in land use from cropland to *Miscanthus* would decrease the NO_3-N load in streamflow by about 6.4% at the outlet of the Salt Creek watershed in Illinois. Sarkar & Miller (2014) also predicted from a SWAT modeling study that the loss of N to surface runoff from switchgrass systems was approximately 73% lower than that from cotton systems in the Black Creek watershed in South Carolina. Through a GIS-based approach, Rao & Yang (2010) predicted that the increase in the extent of grassland could significantly increase groundwater recharge and thereby decrease the groundwater level decline rates, especially in the environmentally sensitive Texas High Plains (THP) region.

Honey mesquite (*Prosopis glandulosa*) is a polymorphic woody legume that invaded grasslands and rangelands in the Southwestern United States, and it is spread over 21 million ha in Texas alone (SCS, 1988; Asner *et al.*, 2003). It has been recognized as a bioenergy feedstock (Padron & Navarro, 2004; Singh *et al.*, 2007; Ansley *et al.*, 2010; Wang *et al.*, 2014), and it is grown under a vast acreage in the Texas Rolling Plains (TRP), which is adjacent to the SHP. The invasion of honey mesquite on grasslands of the TRP caused several negative impacts such as increasing the extent of bare ground and thereby increasing erosion potential, and reducing herbaceous production, which is harmful to the livestock industry and grassland ecosystems (Teague & Dowhower, 2003; Ansley *et al.*, 2010; Wang *et al.*, 2014). Mesquite harvest may not only supply feedstock for biofuel production, but also help in the recovery of grassland functions. Park *et al.* (2012) also reported that honey mesquite has the potential for use as bioenergy feedstock given its high density and presence in large extent of area in the TRP.

The SWAT model (Arnold *et al.*, 1998) and the Agricultural Policy/Environmental eXtender (APEX) model (Williams, 1995), which are widely used across the world, have demonstrated potential to satisfactorily predict long-term impacts of land use change and land management practices on hydrologic processes and water quality in complex watersheds (Ko *et al.*, 2009; Gassman *et al.*, 2010; Ghaffari *et al.*, 2010; Srinivasan *et al.*, 2010; Tuppad *et al.*, 2010; Powers *et al.*, 2011; Wu & Liu, 2012). Specifically, the APEX model has the capability to accurately predict hydrology and water quality in intensively managed agricultural watersheds with large extents of irrigated areas (Saleh & Gallego, 2007; Wang *et al.*, 2011; Jung *et al.*, 2014). The auto-irrigation function included in the APEX model simulates irrigation water as the precipitation, which results in a realistic simulation of percolation during the irrigation process. In contrast, auto-irrigation feature in the SWAT model applies irrigation water until the soil moisture content reaches the field capacity level, and hence, the model, in general, simulates negligible percolation during irrigation events. In addition, the APEX model includes detailed cotton growth parameters, which are very useful for accurate prediction of cotton growth. Furthermore, APEX outputs both cotton lint and seed yields, and it permits specifying disease severity and plant population. As for the SWAT model, it provides reasonable crop management functions for satisfactorily simulating range grass and honey mesquite land uses. For example, the SWAT model allows users to input initial biomass for honey mesquite (tree crop in the crop database), which eliminates the need to grow honey mesquite from seed at the beginning of the simulation.

The SWAT model also simulates reservoir operations accurately using four different methods when compared to the APEX model. While the SWAT model allows users to input daily observed reservoir releases, the APEX model allows inputting an average annual reservoir release only. Therefore, the APEX and SWAT models were integrated in this study (hereafter referred as 'Integrated APEX-SWAT model') to make use of the strengths of both models.

A majority of the published biofuel-induced water quantity and quality studies were conducted in the watersheds located in the humid regions of the United States such as the Upper Mississippi River Basin (Daloğlu *et al.*, 2012; Demissie *et al.*, 2012; Scherer *et al.*, 2015). However, such assessments are limited in the semi-arid SHP and TRP regions. The objectives of this study were: (1) to assess the impacts of biofuel-induced land use change from cotton to perennial bioenergy crops such as switchgrass and *Miscanthus*, and the harvest of mesquite for biofuel use on hydrology and water quality in the semi-arid Double Mountain Fork Brazos watershed that spans across the SHP and TRP regions using the Integrated APEX-SWAT model; (2) to estimate the biomass and biofuel production potential of three bioenergy crops considered in this study; and (3) to compare and contrast the effects of the proposed changes in land use on water and N balances under irrigated and dryland conditions.

Materials and methods

Study watershed

The delineated area of the Double Mountain Fork Brazos watershed is about 6000 km² (Fig. S1). The areas of the upstream subwatershed (upstream of Gauge I) and the downstream subwatershed (downstream of Gauge I and upstream of Gauge II) are about 3297 and 2703 km², respectively. The upstream subwatershed is located in the Hockley, Lynn, and Garza counties (Fig. S1), where cotton is the dominant land use (Fig. 1). The downstream subwatershed, which is primarily composed of rangelands (Fig. 1), is situated in Scurry, Kent, and Stonewall counties. The long-term (1981–2010) average annual rainfall across the watershed varies between 457 and 559 mm, and the long-term average annual maximum and minimum temperatures are about 24 °C and 9 °C, respectively. The topography of the watershed is relatively flat. The major soil types in the watershed are classified as Amarillo sandy loam, Acuff sandy clay loam, and Olton clay loam (Soil Survey Staff, 2010).

Description of SWAT and APEX models

SWAT is a continuous-time, semidistributed, process-based, river basin scale model (Arnold *et al.*, 2012). It divides a

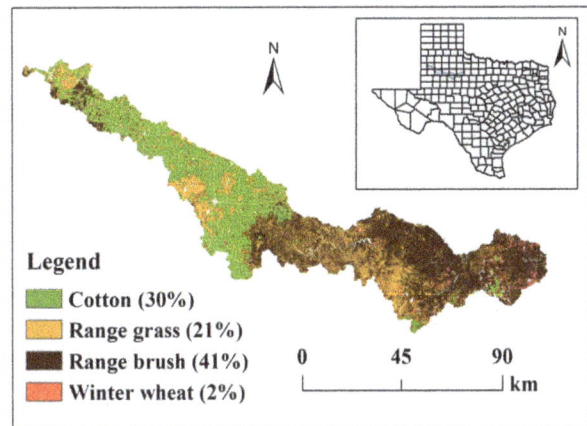

Fig. 1 Major land uses in the study watershed according to the 2008 National Agricultural Statistics Service (NASS) Cropland Data Layer (CDL).

watershed into a number of subwatersheds, which are further divided into several hydrologic response units (HRUs). HRUs are the basic building blocks of the SWAT model from which all landscape processes are computed. They consist of homogeneous land use, soil characteristics, and soil slope. SWAT is operated on a daily time step, and it is widely proven to be a feasible tool to predict the impact of land use management on water, sediment, and agricultural chemical yields (Gassman *et al.*, 2014). A large number of input parameters are needed for SWAT to evaluate the effects of land use change or management practices on hydrology and water quality. The primary model components in SWAT relate to hydrology, water quality, and crop growth (Knisel, 1980; Leonard *et al.*, 1987; Williams *et al.*, 2008). Major model inputs are related to hydrography, terrain, land use, soil, tile drainage, weather, and management practices (Srinivasan *et al.*, 2010). Additional details about the SWAT model can be found in Arnold *et al.* (2012).

In this study, ArcSWAT 2012.10_2.16 (Revision 627) for Arc-GIS 10.2.2 platform was used. The SWAT Calibration and Uncertainty Procedures (SWAT-CUP) tool (Abbaspour *et al.*, 2007) was used for the sensitivity analysis, and calibration and validation of the SWAT model. The Sequential Uncertainty Fitting version-2 (SUFI-2) procedure (Abbaspour *et al.*, 2007) available in SWAT-CUP 2012 was used to estimate various SWAT parameters related to streamflow and N load.

APEX model is a flexible and dynamic tool that is used for simulating management and land use change impacts on hydrology and water quality for whole farms and small watersheds (Williams, 1995; Williams *et al.*, 2008). The APEX model divides a watershed into a number of subareas, which have the same function as the SWAT HRUs. The APEX model consists of 12 major components: climate, hydrology, crop growth, pesticide fate, nutrient cycling, erosion–sedimentation, carbon cycling, management practices, soil temperature, plant environment control, economic budgets, and subarea/routing (Golmohammadi *et al.*, 2014). More details about the APEX model components can be found in Williams (1995), Tuppad *et al.* (2009), and Wang *et al.* (2012). The ArcAPEX model (version

0806), which was interfaced with ArcGIS 10.2.2 platform, was used in this study.

SWAT and APEX models' setup and integration

The DEM (30 × 30 m) of the study watershed was downloaded from the U.S. Geological Survey (http://viewer.nationa lmap.gov/viewer/#) and input to the Integrated APEX-SWAT model. The 2008 NASS Cropland Data Layer (CDL) (http://na ssgeodata.gmu.edu/CropScape/) was used to represent the prevalent land use conditions during the period of model simulations (1994 to 2009). The dominant agricultural land use in the watershed in 2008 was cotton, which occupied about 30% of the entire watershed area, and about 52% of the upstream subwatershed (Fig. 1). About 41% and 21% of the entire watershed area were covered by range brush and range grass, respectively. The soil data were obtained from the Soil Survey Geographic Database (SSURGO) (Soil Survey Staff, 2014), which was compatible with the Integrated APEX-SWAT model. Four soil slopes were considered: ≤1%, 1–3%, 3–5%, and >5%. Daily weather data from a total of seven weather stations for the period from 1992 to 2009 were obtained from the National Climatic Data Center (NCDC) and used in this study (Fig. S1) (NOAA-NCDC, 2014). The missing weather data for a weather station were filled with the average value of weather parameter for two adjacent weather stations (Ale et al., 2009). More detailed information for the model setup can be found in Chen et al. (2016a,b). For the HRU and subarea definitions, thresholds of 5%, 5%, and 10% were used for land use, soil type, and slope, respectively. A total of 25 APEX subareas were delineated in the upstream subwatershed, and 35 SWAT subbasins and 1417 HRUs were identified in the downstream subwatershed.

The Alan Henry Reservoir (storage capacity: 4882 × 10⁴ m³) exists in the downstream subwatershed. The SWAT parameters related to the operation of this reservoir were obtained from the Texas Water Development Board's report on 'Volumetric Survey of Alan Henry Reservoir' (Texas Water Development Board, 2005). The 'Measured Daily Outflow' method available in the SWAT model was used to estimate the reservoir discharge based on the reservoir storage levels recorded by the USGS gauge. More details about the parameters used for reservoir simulation are provided in Chen et al. (2016a).

As described earlier, the APEX model is capable of simulating croplands better when compared to the SWAT model, and on the other hand, SWAT performs better in simulating noncroplands and transport of flow, sediment, and nutrients through detailed in-stream channel and reservoir processes (Santhi et al., 2014). To take advantage of the strengths of APEX and SWAT models, the APEX model was integrated with SWAT model in this study. Initially, the APEX model was set up for the upstream subwatershed (Fig. 2), where cotton was the dominant land use. The SWAT model was then set up for the entire watershed, and the upstream subwatershed was simulated as one subbasin in the SWAT model. The APEX-simulated net flows, and sediment and nutrient loads were then input as a point source to the SWAT model at Gauge I (outlet of the upstream subwatershed). The downstream subwatershed, which is dominated by the range land use and contains the Alan Henry Reservoir, was therefore essentially modeled using the SWAT model. Although measured streamflow data were available at both Gauge I and Gauge II in the watershed, measured data on N concentration in streamflow were available at Gauge II (watershed outlet) only, and hence, this integration of two models enabled assessment of water quality effects of proposed land use change and mesquite harvest from the entire 6000-km² watershed.

Management practices of crops in the study watershed

The management-related parameters for cotton were specified based on the locally followed practices. Spring tillage was implemented for cotton (Table S1). About 138 and 69 kg N ha⁻¹ were applied to the irrigated and dryland cotton, respectively. According to the NASS county-wise cotton acreage estimates over the period from 1994 to 2009 (NASS, 2015), about 39% of the cotton acreage in the watershed was irrigated. Remote sensing images were used to identify irrigated subareas. Subareas that contain large number of circular fields, which represent center pivot irrigated areas, were considered as irrigated subareas. It was also made sure that the total extent of irrigated cotton area was about 39% of the entire

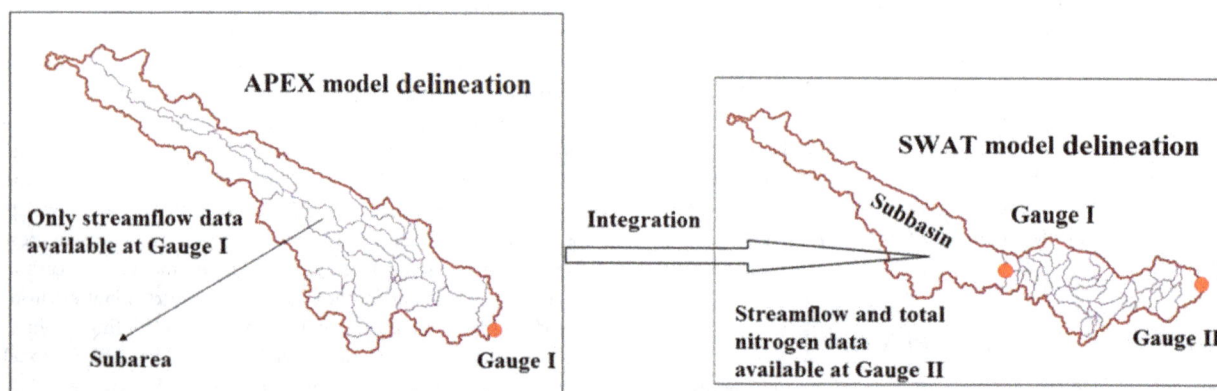

APEX model delineation

Only streamflow data available at Gauge I

Subarea

Gauge I

Integration

SWAT model delineation

Subbasin

Gauge I

Streamflow and total nitrogen data available at Gauge II

Gauge II

Fig. 2 Illustration showing the APEX model integration with the SWAT model.

cotton-growing areas in the watershed. Auto-irrigation was therefore simulated in about 39% of cotton planting area in the watershed based on plant water stress.

The land uses of range grass and range brush were simulated as Southwestern U.S. range and honey mesquite, respectively. The most commonly adopted heavy continuous grazing management practice was simulated on the range grassland (Park *et al.*, 2017). The detailed management-related parameters for the range grass were set up according to Park *et al.* (2017) (Table S1). Biomass of honey mesquite at the beginning of the simulation was assumed as 19.4 Mg ha^{-1} (Whisenant & Burzlaff, 1978) (Table S1).

Observed streamflow, cotton lint yield, and water quality data used for model calibration

Observed daily streamflow recorded at Gauge I and Gauge II during the period from 1994 to 2009 was obtained from the USGS National Water Information System (http://waterdata.usgs.gov/nwis/sw). The observed dryland and irrigated cotton lint yield data over the period from 1994 to 2009 for Lynn County, the county with the highest cotton acreage in the study area, were obtained from the NASS reports (http://quicksta ts.nass.usda.gov/). The instantaneous total nitrogen (TN) concentration data measured at the watershed outlet at Gauge II on some specific days (from a total of 39 grab samples) (Fig. S2) were used for model water quality calibration. These concentrations were used to estimate continuous daily TN loads using the USGS Load Estimator (LOADEST) regression model (Runkel *et al.*, 2004). A detailed description of LOADEST can be found in Jha *et al.* (2007). The estimated daily TN load data were distributed over 1995–2000 period.

Integrated APEX-SWAT model calibration

The APEX model was initially calibrated against observed streamflow and cotton lint yield data for the upstream subwatershed (Chen *et al.*, 2016b). As the observed N load data were not available for this upstream subwatershed, the APEX model was integrated with the SWAT model and the net flow, and sediment and nutrient loads from the upstream subwatershed were input as a point source to the downstream subwatershed at Gauge I (Fig. 2). The Integrated APEX-SWAT model was then calibrated against the observed streamflow data at Gauge II by solely adjusting SWAT model parameters in the downstream subwatershed. The calibration and validation periods considered for streamflow prediction were 1994–2001 and 2002–2009, respectively. After achieving a satisfactory streamflow calibration, the Integrated APEX-SWAT model was calibrated for the TN load prediction by changing the water quality parameters of both APEX (in the upstream subwatershed) and the SWAT (in the downstream subwatershed) models. Based on the available data, 1995–1997 and 1998–2000 periods were considered as the calibration and validation periods for TN, respectively. The calibrated Integrated APEX-SWAT model was then used to simulate the impacts of land use change from cotton to perennial grasses, and mesquite harvest on water and N balances. The values of calibrated

parameters related to hydrology, crop growth, and water quality are shown in Table S2.

The performance of the Integrated APEX-SWAT model in predicting streamflow and water quality during the calibration and validation periods was evaluated using three different statistical measures: square of Pearson's product-moment correlation coefficient (R^2) (Legates & McCabe, 1999), Nash–Sutcliffe efficiency (*NSE*) (Nash & Sutcliffe, 1970), and percent bias (*PBIAS*). The goal of calibration for monthly streamflow and water quality predictions was to achieve all three objective functions: minimize *PBIAS*, maximize *NSE*, and maximize R^2. We aimed to achieve $NSE \geq 0.60$, $R^2 \geq 0.65$, and *PBIAS* within $\pm15\%$ in monthly streamflow, and $NSE \geq 0.60$, $R^2 \geq 0.65$, and *PBIAS* within $\pm40\%$ in monthly TN load. The model performance in cotton lint yield prediction was assessed using R^2 and *PBIAS* only, and we aimed to achieve a *PBIAS* within $\pm10\%$ in average annual cotton lint yield under both irrigated and dryland conditions. Statistical analyses of the scenario analysis results were carried out using the Statistical Package for Social Science (SPSS 19.0). Analysis of variance (ANOVA) was used to test the difference with significance levels set at $P < 0.05$ or $P < 0.1$. Microsoft Excel 2013 was used for other data analysis.

Scenario analysis

In this study, switchgrass and *Miscanthus*, which were identified as ideal bioenergy grasses for this study region (Chen *et al.*, 2016a), were selected to hypothetically replace irrigated and dryland cotton areas, respectively. Honey mesquite, which was dominant in the range brush areas, was also considered as the bioenergy crop. Although honey mesquite harvest was recommended at a ten-year interval (Wang *et al.*, 2014), a nine-year harvest interval was assumed in this study so that it could be harvested twice (in 2000 and 2009) over the total simulation period of 18 years. In addition, standing honey mesquite biomass of 19.4 Mg ha^{-1} was harvested in 1992, at the beginning of the simulation period.

The Integrated APEX-SWAT model simulations were run from 1992 to 2009, and the 1992–1993 period was considered as the model warm-up period. The impacts of hypothetical biofuel-induced land use change and mesquite harvest on hydrology and water quality under simulated scenarios were evaluated over the remaining simulation period from 1994 to 2009. The land use change effects under both irrigated and dryland conditions were compared and contrasted.

Perennial grasses were planted on May 15, 1992, and harvested once every year on November 15 (Table S3). Irrigated switchgrass was assigned the same irrigation management practices as irrigated cotton. A recommended fertilizer application rate of about 124 kg N ha^{-1} was applied for irrigated switchgrass (Yimam *et al.*, 2014). About 98 kg N ha^{-1} was applied to dryland *Miscanthus* (Lewandowski & Schmidt, 2006; Danalatos *et al.*, 2007) (Table S3). No fertilizer was applied to honey mesquite. Tillage was not simulated under any of these hypothetical scenarios. Heat units to maturity of all simulated crops were estimated using the SWAT Potential Heat Unit (SWAT-PHU) program (http://swat.tamu.edu/software/pote

ntial-heat-unit-program/) (Tables S1 and S3). As crop growth parameter values for *Miscanthus* were not available in the models' crop database, the values from Trybula *et al.* (2015) field study were adopted.

Results

Integrated APEX-SWAT model calibration and validation results

The simulated monthly streamflow at the watershed outlet (Gauge II) during the calibration (1994–2001) and validation (2002–2009) periods closely matched with the observed streamflow (Fig. S3). The *NSE*, R^2, and *PBIAS* values for monthly predictions of streamflow were 0.64, 0.67, and 10.7%, respectively, during the calibration period, and they were 0.60, 0.65, and −9.3%, respectively, during the validation period. These values demonstrate a 'satisfactory' agreement between the simulated and observed streamflow according to the Moriasi *et al.* (2007) criteria. The *PBIAS* in predicting dryland and irrigated cotton lint yields over the entire simulation period (1994–2009) was about 0.1% and 0.7%, respectively, indicating a good overall match between the simulated and observed yields.

The simulated monthly TN load and the LOADEST estimated load during the calibration (1995–1997) and validation (1998–2000) periods also matched well as shown in Fig. S4. The *NSE* for monthly TN load prediction was 0.70 and 0.65 during the calibration and validation periods, respectively. The *PBIAS* in predicting TN load was −20.4% and 34.8% for the calibration and validation periods, respectively. The model performance ratings for the monthly TN load predictions were considered as 'satisfactory' for both the calibration and validation periods based on the *NSE* and *PBIAS* values, according to Moriasi *et al.* (2007) and Wang *et al.* (2012) criteria.

Simulated water and N mass balances in the upstream subwatershed under the baseline cotton scenario

The primary component of water balance in the upstream subwatershed, which is dominated by cropland, is the evapotranspiration (ET). Results showed that approximately 89% and 95% of the average annual (1994–2009) input water (precipitation + irrigation) was lost due to ET in the irrigated and dryland conditions, respectively, under the baseline cotton scenario (Table 1). Less than 1% of the input water yielded as surface runoff under the baseline cotton scenario in both the irrigated and dryland conditions (Table 1). Average annual percolation accounted for approximately 10% and 4% of the total water input under the baseline

Table 1 Hydrologic and water quality impacts of land use change from irrigated cotton to irrigated switchgrass, and dryland cotton to dryland *Miscanthus* in the upstream subwatershed

Irrigated and dryland areas combined	Cotton (baseline)	Perennial grasses	Change (%)
Precipitation (mm)	490.4	490.4	–
Irrigation (mm)	189.0	198.9	5.2
ET (mm)	624.1	626.0	0.3
Surface runoff (mm)	6.1	0.7	−88.1**
Percolation (mm)	47.0	60.3	28.1
NO_3-N load in surface runoff (kg ha^{-1})	0.11	0.01	−86.3**
Organic-N load in surface runoff (kg ha^{-1})	0.21	0.003	−98.4**
NO_3-N leaching (kg ha^{-1})	9.43	0.05	−99.5**

Irrigated areas	Irrigated cotton	Irrigated switchgrass	Change (%)
Precipitation (mm)	481.2	481.2	–
Irrigation (mm)	484.5	509.9	5.2
ET (mm)	863.7	841.3	−2.6
Surface runoff (mm)	8.2	1.0	−87.3**
Percolation (mm)	92.3	147.6	59.9*
NO_3-N load in surface runoff (kg ha^{-1})	0.16	0.04	−77.5**
Organic-N load in surface runoff (kg ha^{-1})	0.26	0.001	−99.5**
NO_3-N leaching (kg ha^{-1})	21.03	0.12	−99.4**

Dryland areas	Dryland cotton	Dryland *Miscanthus*	Change (%)
Precipitation (mm)	496.3	496.3	–
ET (mm)	472.5	489.6	3.6
Surface runoff (mm)	4.7	0.5	−88.9**
Percolation (mm)	18.4	4.9	−73.1
NO_3-N load in surface runoff (kg ha^{-1})	0.08	0.002	−97.7**
Organic-N load in surface runoff (kg ha^{-1})	0.18	0.005	−97.4**
NO_3-N leaching (kg ha^{-1})	2.08	0.01	−99.7*

**A significant difference at $P < 0.05$; *A significant difference at $P < 0.1$.

cotton scenario in the irrigated and dryland conditions, respectively (Table 1).

The simulated N mass balance under the baseline cotton scenario is shown in Table 2. On average, approximately 48% of the N inputs remained in soil under the baseline irrigated cotton scenario with 27% of N inputs taken up by the harvested portion of cotton and 13% leached to groundwater. Bronson *et al.* (2004) reported that the TN content of surface soil (0 to 10 cm) in some fields within our study watershed in 2001 was about 479 kg ha^{-1} under the long-term irrigated cotton land

Table 2 Simulated average (1994–2009) annual nitrogen mass balances (kg N ha^{-1}) of the upstream subwatershed under cotton and perennial grass land uses

| Land use | Nitrogen (N) inputs | | Nitrogen outputs | | | | | | | |
	N fertilizer	N in rainfall	N in Runoff	N in return flow	N in sediment	N leaching	Denitrification	Volatilization	N uptake by harvested portion	Change in soil N
Irrigated cotton	138	4.0	0.15 (0.11)*	6.2 (4.4)	0.25 (0.2)	18.8 (13.2)	0.49 (0.3)	9.36 (6.6)	38.34 (27.0)	68.28 (48.1)
Dryland cotton	69	4.0	0.08 (0.11)	0.62 (0.8)	0.18 (0.2)	1.9 (2.6)	0.09 (0.1)	5.32 (7.3)	18.18 (24.9)	46.37 (63.5)
Irrigated switchgrass	124	4.0	0.03 (0.02)	0.09 (0.1)	0.001 (0.001)	0.28 (0.2)	2.50 (2.0)	5.98 (4.7)	122.14 (95.4)	−3.02 (−2.4)
Dryland Miscanthus	98	4.0	0.002 (0.002)	0.009 (0.004)	0.004 (0.004)	0.026 (0.03)	0.74 (0.7)	4.57 (4.5)	58.52 (57.4)	38.13 (37.4)

*The numbers in the parentheses indicate the percentages of total nitrogen inputs that were either lost in different pathways or accumulated in soil.

use. At the same sampling locations, Zobeck *et al.* (2007) further documented that the soil TN content in 0–10 cm soil profile in 2003 was about 590 kg ha^{-1} under the long-term irrigated cotton land use. They found that about 56 kg ha^{-1} TN was accumulated in the soil each year under the irrigated cotton production. The simulated annual soil TN accumulation under the baseline irrigated cotton scenario in our study (about 68 kg ha^{-1}; Table 2) was comparable to the value reported in the above studies. The simulated average annual total N uptake by the irrigated cotton in our study (about 189 kg ha^{-1}; Table 2) was also comparable to the measured N uptake of 168 kg ha^{-1} by cotton, which was irrigated at 75% ET replacement in a field study by Li & Lascano (2011) in the SHP. Under the baseline dryland cotton scenario, approximately 64%, 25%, and 3% of N inputs were accumulated in soil, taken up by the harvested portion of cotton, and leached to groundwater, respectively (Table 2). The simulated total N uptake during the period of simulation (1994–2009) ranged from 56 to 152 kg ha^{-1} under the baseline dryland cotton scenario. Mullins & Burmester (1990) also documented that the total N uptake by dryland cotton ranged from 127 to 155 kg ha^{-1} in their field experiments conducted in Alabama in 1986 and 1987.

Biomass and biofuel availability from the changes in watershed land use and management

The simulated average annual harvestable biomass under the irrigated switchgrass, dryland *Miscanthus*, and honey mesquite scenarios was 17.3, 15.7, and 4.0 Mg ha^{-1}, respectively (Table 3). The APEX-simulated annual irrigated switchgrass and dryland *Miscanthus* biomass yields in this study were similar to those simulated by the SWAT model (17.5 and 15.6 Mg ha^{-1} biomass of irrigated switchgrass and dryland *Miscanthus*, respectively) in this watershed (Chen *et al.*, 2016a). The predicted annual *Miscanthus* biomass yield under the dryland conditions in this study was also within the range of reported *Miscanthus* biomass yield (9.8–17.8 Mg ha^{-1}) in the dryland production conditions in the U.K. (Christian *et al.*, 2008).

The crop database of honey mesquite was adjusted based on the studies of Kiniry (1998) and Ansley *et al.* (2010) in Texas to match the simulated biomass with the observed biomass in the TRP region (Table S2). The simulated total tree biomass of a 30-year-old honey mesquite plant in this study was about 40 Mg ha^{-1}, which was comparable to the measured biomass of 43 Mg ha^{-1} in a field study in the TRP (Ansley *et al.*, 2010). In addition, the predicted total tree biomass of a nine-year-old regrown honey mesquite was the same (28 Mg ha^{-1}) as that reported in Ansley *et al.* (2010).

Table 3 Average (1994–2009) annual biomass and biofuel production of irrigated switchgrass, dryland *Miscanthus*, and honey mesquite

Annual production	Harvestable biomass (Mg ha^{-1})	Biofuel conversion efficiency (liters ethanol Mg^{-1} biomass)*	Biofuel production (liters ethanol ha^{-1})
Irrigated switchgrass	17.3	366	6332
Dryland *Miscanthus*	15.7	366	5746
Honey mesquite	4.0	309	1246

*The theoretical ethanol yield data was taken from http://www.afdc.energy.gov/fuels/ethanol_feedstocks.html.

The simulated annual regrowth rate was about 2.4 Mg ha^{-1} for honey mesquite in this study. Ansley *et al.* (2010) and Wang *et al.* (2014) also documented a similar annual production rate of honey mesquite of 2.2 Mg ha^{-1}. According to the suggested theoretical ethanol yield (http://www.afdc.energy.gov/fuels/ethanolfeedstocks.html), the estimated average annual ethanol that could be produced with the simulated biomass of irrigated switchgrass, dryland *Miscanthus*, and honey mesquite was 6332, 5746, and 1246 L ha^{-1}, respectively (Table 3).

Impacts of biofuel-induced land use change and mesquite harvest on hydrology

The average (1994–2009) annual surface runoff from the upstream subwatershed decreased significantly ($P < 0.05$) by 88% under the perennial grasses scenario (i.e., irrigated cotton replaced by switchgrass and dryland cotton replaced by *Miscanthus*) compared to the baseline cotton scenario (Table 1). The annual percolation increased significantly ($P < 0.1$) by approximately 60% under the irrigated switchgrass scenario, and it decreased by approximately 73% under the dryland *Miscanthus* scenario relative to the baseline cotton scenario (Table 1). Overall, under the perennial grasses scenario, the average annual percolation in the upstream subwatershed increased by 28% relative to the baseline cotton scenario. However, this trend was not statistically significant at the $P < 0.1$ level. The average annual ET of irrigated switchgrass decreased by approximately 2.6% (not significant at the $P < 0.1$ level) when compared to the irrigated cotton. In contrast, the annual ET increased by about 3.6% under the dryland *Miscanthus* scenario compared to that under the baseline dryland cotton scenario. Overall, there was a statistically insignificant small increase in annual ET (0.3%) under the perennial grasses scenario relative to the baseline cotton scenario.

In the case of downstream subwatershed, the simulated average annual ET increased by 8.4% under the post-mesquite-harvest scenario relative to the baseline mesquite scenario (Table 4). However, this difference was not statistically significant. Also, a considerable interannual variability was found in this trend. For example, under the post-mesquite-harvest scenario, the average annual ET increased by 11% during the normal and wet years (rainfall > 500 mm) and reduced by about 1% during the dry years when compared to the baseline mesquite scenario. The increase in ET in wet years might have been caused by a much higher increase in evaporation when compared to reduction in transpiration after the mesquite harvest. The increase in average annual ET caused a significant ($P < 0.05$)

Table 4 The changes in hydrologic and water quality variables due to the harvest of honey mesquite from the downstream watershed for bioenergy purposes

Variable	Mesquite (baseline)	Post-mesquite-harvest
Hydrologic variables		
ET (mm)	458.0	496.3 (8.4)†
Surface runoff (mm)	20.8	0.23 (−98.9**)
Percolation (mm)	43.4	27.0 (−37.7)
Water quality variables		
NO$_3$-N load in surface runoff (kg ha^{-1})	0.009	0.000007 (−99.9**)
Organic-N load in surface runoff (kg ha^{-1})	0.06	0.0009 (−99.5**)
NO$_3$-N leaching (kg ha^{-1})	0.12	0.18 (56.1)

†The numbers in the parentheses indicate the percent changes between the post-mesquite-harvest and baseline mesquite scenarios.
**Significant differences between the post-mesquite-harvest and baseline mesquite scenarios at $P < 0.05$.

decrease in surface runoff by about 98.9% and an insignificant (at $P < 0.1$ level) 37.7% reduction in percolation under the post-mesquite-harvest scenario when compared to the baseline mesquite scenario (Table 4).

A hypothetical change in land use from cotton to perennial grasses altered average (1994–2009) monthly ET and soil water content significantly ($P < 0.05$) under the irrigated switchgrass and dryland *Miscanthus* scenarios (Fig. 3a,b,g,h). Monthly ET under the irrigated switchgrass scenario was significantly ($P < 0.05$) higher than that under the baseline irrigated cotton scenario in the months of May to July and November (Fig. 3a). However, it was significantly ($P < 0.05$) lower relative to the baseline irrigated cotton scenario in other months. Under the dryland *Miscanthus* scenario, monthly ET increased significantly ($P < 0.05$) in April, May, and November, but it decreased significantly ($P < 0.05$) in months of January, February, July, and December compared to the baseline dryland cotton scenario.

The simulated soil water content enhanced significantly ($P < 0.05$) under the irrigated switchgrass scenario from January to April and from August to October compared to the baseline irrigated cotton scenario. A significant ($P < 0.05$) decrease in soil water content was found under the irrigated switchgrass scenario in May and June when the simulated ET was significantly ($P < 0.05$) higher under the irrigated switchgrass scenario than the baseline irrigated cotton scenario (Fig. 3a,g). The soil water content depleted significantly ($P < 0.05$) under the dryland *Miscanthus* scenario from April to July relative to the baseline dryland cotton scenario due to the increase in simulated ET during those months. In general, soil water content was enhanced under the irrigated switchgrass scenario relative to the baseline irrigated cotton scenario, while it was reduced under the dryland *Miscanthus* scenario when compared to the dryland cotton scenario.

Negligible surface runoff was generated under the perennial grass scenarios relative to the baseline cotton scenario (Fig. 3c,d). The surface runoff under the irrigated switchgrass scenario decreased significantly ($P < 0.05$) in March and August compared to the baseline irrigated cotton scenario. The percolation under the irrigated switchgrass scenario increased from February to May and from July to November relative to the baseline irrigated cotton scenario (Fig. 3e). However, this trend was not statistically significant (at the $P < 0.05$ level). The percolation was negligible under the dryland *Miscanthus* scenario, and it also corresponded well with the depletion of soil water content under this scenario (Fig. 3f,h). Even under the baseline dryland cotton scenario, notable percolation was simulated only in May and June when the precipitation was relatively high and the simulated ET was relatively low.

Effects of biofuel-induced land use change and mesquite harvest on N losses

The average (1994–2009) annual NO_3-N and organic-N loads in the surface runoff and the NO_3-N leaching to the groundwater decreased significantly ($P < 0.05$) by about 86%, 98%, and 100%, respectively, under the perennial grasses scenario compared to the baseline cotton scenario (Table 1). The N lost through leaching was much higher compared to that lost through surface runoff (Table 1). For example, the NO_3-N leaching was about 20 times higher than the N lost through surface runoff under the baseline cotton land use. However, in the case of perennial grasses, NO_3-N leaching was only about four times higher than the N lost through surface runoff because of the higher N use efficiency of perennial grasses (Table 1).

A close look at the average annual N balances in irrigated conditions indicated that the NO_3-N and organic-N losses in surface runoff and NO_3-N leaching to groundwater were also significantly ($P < 0.05$) lower by about 78%, 100%, and 99%, respectively, under the irrigated switchgrass scenario than those under the baseline irrigated cotton scenario (Table 1). Although the average annual percolation increased by about 60% under the irrigated switchgrass scenario relative to the baseline irrigated cotton scenario (Table 1), the average annual NO_3-N leaching reduced by approximately 99% under the irrigated switchgrass scenario due to higher N uptake by harvested switchgrass (95% N uptake for irrigated switchgrass vs. 27% for irrigated cotton) (Table 2) and lower amount of N fertilizer application (124 kg N ha^{-1} for irrigated switchgrass vs. 138 kg N ha^{-1} for irrigated cotton) (Table S3). Similar results were found in case of dryland *Miscanthus* scenario. The annual NO_3-N load and organic-N load reduced significantly ($P < 0.05$) by about 98% and 97%, respectively, under the dryland *Miscanthus* scenario relative to the baseline dryland cotton scenario (Table 1). Ng et al. (2010) also predicted that a 50% change in land use from cropland to *Miscanthus* would result in a decrease in NO_3-N load in streamflow by about 30% at the watershed outlet in the Salt Creek watershed in Illinois. The average annual NO_3-N leaching also decreased significantly ($P < 0.1$) by about 100% under the dryland *Miscanthus* scenario compared to the baseline dryland cotton scenario. The N uptake by the dryland *Miscanthus* (167 kg N ha^{-1}) was also clearly higher than that of dryland cotton (92 kg N ha^{-1}) (Table 2).

The average annual NO_3-N leaching increased from 0.12 to 0.18 kg ha^{-1} under the post-mesquite-harvest scenario compared to the baseline mesquite scenario (Table 4), but these differences were not significant. The NO_3-N load and organic-N load through surface runoff

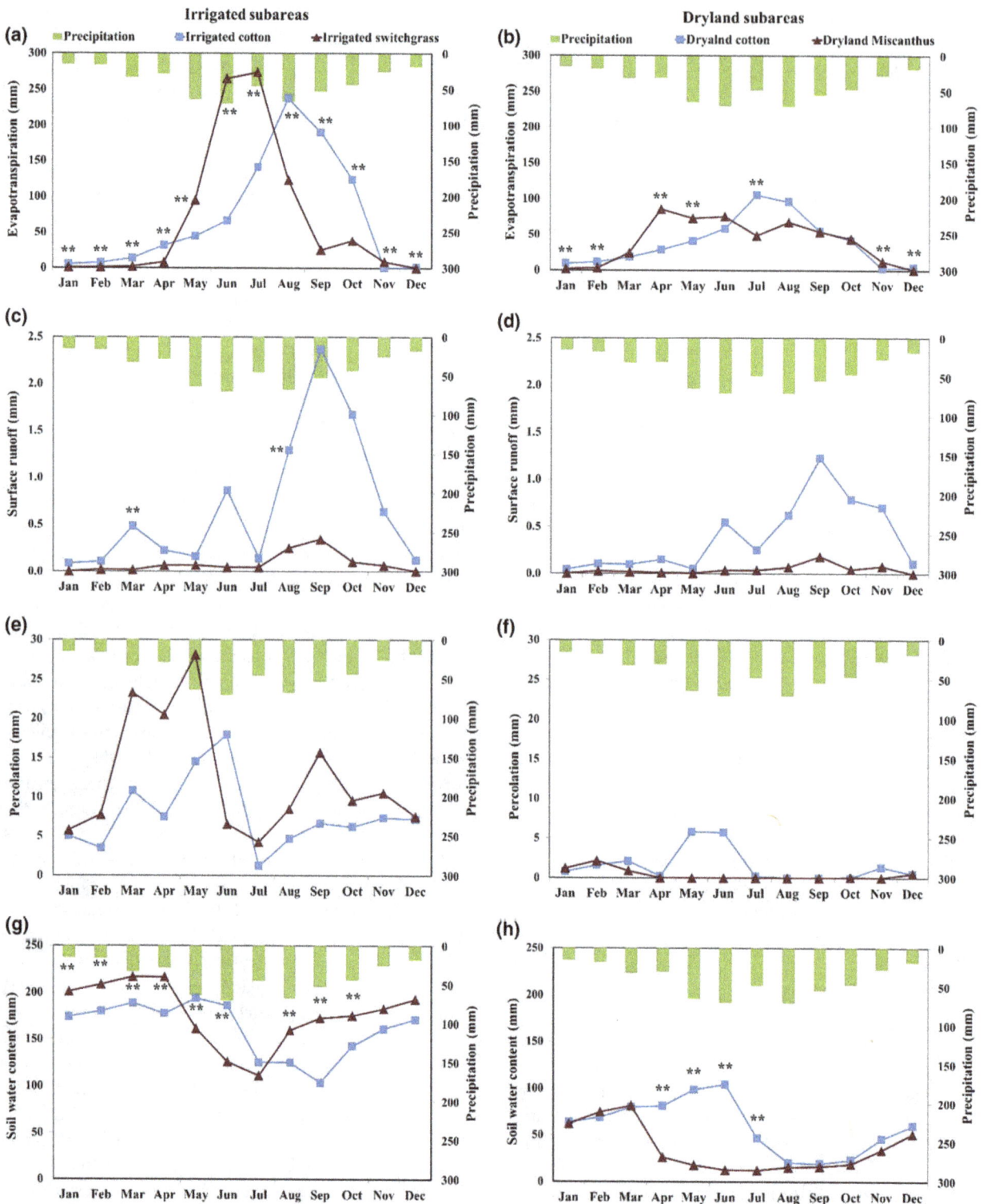

Fig. 3 Simulated average (1994–2009) monthly water fluxes in the irrigated and dryland areas under the baseline cotton and hypothetical perennial grass scenarios (** indicates a significant difference at $P < 0.05$).

were significantly reduced by about 99.9% and 99.5% under the post-mesquite-harvest scenario compared to the baseline mesquite scenario (at $P < 0.05$ level). The

significant decrease in surface runoff was the key reason for the associated significant reduction in N losses in surface runoff.

The average monthly NO_3-N and organic-N loads in surface runoff and NO_3-N leaching to groundwater were negligible under the perennial grass scenarios when compared to the baseline cotton scenario (Fig. 4). The NO_3-N load in the surface runoff decreased

significantly ($P < 0.05$) under the irrigated switchgrass scenario in March (100%), August (89%), September (93%), November (99%), and December (99%) relative to the baseline irrigated cotton scenario. Under the dryland *Miscanthus* scenario, NO_3-N load in surface runoff

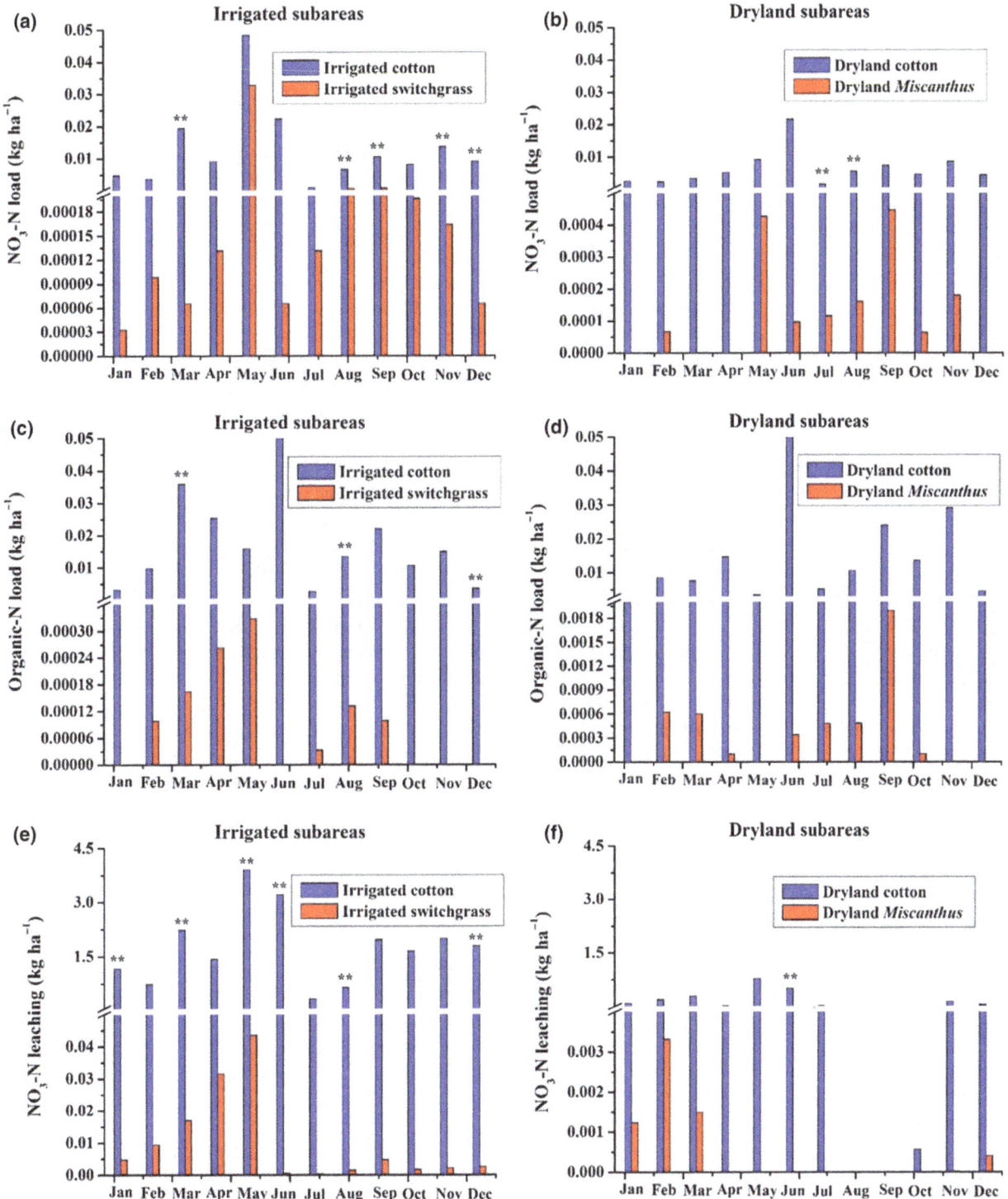

Fig. 4 Simulated average (1994–2009) monthly nitrogen loss through surface runoff and leaching under the irrigated and dryland areas under the baseline cotton and hypothetical perennial grass scenarios (** indicates a significant difference at $P < 0.05$).

decreased significantly ($P < 0.05$) in July (93%) and August (97%) compared to the baseline dryland cotton scenario. Using the APEX model, Feng *et al.* (2015) also found a significant ($P < 0.05$) reduction in N transported by surface water (91%) under the *Miscanthus* production scenario compared to the initial corn/soybean land use in the St. Joseph River watershed in Indiana. Generally, the organic-N load in surface runoff under the perennial grass scenarios was also much lower than that under the baseline cotton scenario (Fig. 4c,d). However, the differences in organic-N load between the dryland *Miscanthus* and baseline dryland cotton scenarios were not statistically significant (at $P < 0.1$ level). In contrast, the organic-N load in surface runoff under the irrigated switchgrass scenario reduced significantly ($P < 0.05$) in March (99.5%), August (99%), and December (100%) when compared to the baseline irrigated cotton scenario. The NO_3-N leaching to groundwater also decreased significantly ($P < 0.05$) under the irrigated switchgrass scenario in January (99.6%), March (99.2%), May (98.9%), June (100%), August (99.8%), and December (99.9%) when compared to the baseline irrigated cotton scenario. When the irrigated cotton land use was changed to irrigated switchgrass, the NO_3-N leaching decreased significantly ($P < 0.05$) during the high precipitation months such as May, June, and August. In this study, when compared to the baseline dryland cotton scenario, NO_3-N leaching under the dryland *Miscanthus* scenario decreased significantly ($P < 0.05$) by about 100% in June. The highest percolation under the baseline dryland cotton scenario and the lowest percolation under the dryland *Miscanthus* scenario were both predicted in June (Fig. 3h), and this might have contributed for the significant ($P < 0.05$) reduction in NO_3-N leaching under the dryland *Miscanthus* scenario in June compared to the baseline dryland cotton scenario.

Discussion

Impacts of biofuel-induced changes in land use and management on water and nitrogen balances

The average (1994–2009) annual surface runoff decreased significantly ($P < 0.05$) by 87% under the irrigated switchgrass scenario relative to the baseline irrigated cotton scenario in this study (Table 1). Using the SWAT model, Nelson *et al.* (2006) predicted a 55% decrease in surface runoff in the Delaware Basin in northeast Kansas due to the change in land use from the traditional corn–soybean cropping rotation to switchgrass. However, the average annual irrigation water requirement increased by approximately 5% (not significant at the $P < 0.1$ level) under the irrigated switchgrass

scenario when compared to the baseline irrigated cotton scenario (Table 1). It is interesting to find that the net groundwater use (irrigation water minus percolation) decreased by approximately 7.6% under the irrigated switchgrass scenario relative to the baseline irrigated cotton scenario. Monthly ET under the irrigated switchgrass scenario was significantly ($P < 0.05$) lower than that under the baseline irrigated cotton scenario in the months of January to April, August to October and December, while it was significantly higher in other months (Fig. 3a). This was primarily due to the simulated early initiation of regrowth of switchgrass in the study watershed in late April, and its late harvest in mid-November when compared to cotton. Cotton was planted in mid-May and harvested at the end of October. Yimam *et al.* (2015) also observed that the regrowth of switchgrass occurred around mid-April in a field experiment at Stillwater, Oklahoma. Also, the simulated peak ET occurred earlier in case of irrigated switchgrass when compared to irrigated cotton (Fig. 3a). As shown in Fig. 3a,b, the monthly ET pattern under the perennial grass scenarios shifted two to three months early relative to the baseline cotton scenario. This finding would be helpful in planning appropriate management strategies for growing perennial grasses in the SHP region.

A well-developed root system and better ground cover under the switchgrass scenario enhanced the soil water content significantly ($P < 0.05$) from January to April and from August to October compared to the baseline irrigated cotton scenario. However, the soil water content was reduced significantly from April to July under the dryland *Miscanthus* scenario when compared to the dryland cotton scenario. Several studies from the Midwestern United States have also reported reductions in soil water content under the *Miscanthus* land use when compared to that of maize (McIsaac *et al.*, 2010; VanLoocke *et al.*, 2010; Le *et al.*, 2011). The large leaf area index of *Miscanthus* (*Miscanthus* vs. switchgrass: 11 vs. 6) resulted in a very high ET which depleted the soil water content.

The average annual N loads through surface runoff decreased significantly ($P < 0.05$) by more than 86% under the perennial grasses scenario compared to the baseline cotton scenario (Table 1). Using the SWAT model, Sarkar & Miller (2014) also predicted that the N losses through surface runoff under switchgrass were approximately 73% lower than that under cotton in the Black Creek watershed in South Carolina. In another SWAT modeling study in the Arkansas–White–Red River Basin, Jager *et al.* (2015) predicted an 84% reduction in average annual NO_3-N load through surface runoff under the projected future (2022) switchgrass landscape compared to baseline no-cellulosic bioenergy grass scenario. Although the percent reductions in NO_3-

N losses in above studies were lower, absolute losses in their study were comparable to our results. The reduction in surface runoff and high N use efficiency were the main reasons for substantial reduction in N loads in surface runoff under the perennial grasses scenario when compared to baseline cotton scenario.

The NO_3-N leaching to groundwater was reduced by 99.5% under the perennial grasses scenario relative to the baseline cotton scenario (Table 1). From a four-year field experiment in central Illinois, McIsaac et al. (2010) also concluded that the average annual NO_3-N leaching under maize–soybean land use was much higher (about 40 kg N ha^{-1}) when compared to switchgrass (1.4 kg N ha^{-1}; 97% reduction) and Miscanthus (3 kg N ha^{-1}; 93% reduction) land uses. Approximately 95% and 57% of total N inputs (N in fertilizer and rainfall) were taken up by harvested portion of the irrigated switchgrass and dryland Miscanthus, respectively, whereas harvested portion of cotton used approximately 27% to 25% of total N inputs (Table 2). Powers et al. (2011) also reported that about 87% of the applied N was taken up by the harvested portion of the switchgrass in an APEX modeling study in eastern Iowa. Groundwater contamination by NO_3-N is a major concern in the THP region (Chaudhuri & Ale, 2014a,b), where groundwater is the major source of drinking water for >95% of rural population (Texas Water Development Board, 2007). Results from this study indicated that the land use change from cotton to perennial grasses could potentially reduce NO_3-N leaching to groundwater and thereby improve groundwater quality in this region in a long run.

It is also interesting to notice that the surface runoff, NO_3-N load, and organic-N load through surface runoff decreased significantly ($P < 0.05$) by about 98.9%, 99.9%, and 99.5% under the post-mesquite-harvest scenario compared to the baseline mesquite scenario (Table 4). The fast regrowth of the harvested mesquite showed higher simulated ET than the undisturbed mesquite, especially under the wet years. This was a major reason for the reduction in surface runoff and associated N losses. The harvest of honey mesquite could therefore not only benefit the water quality in the study watershed, but also supply biomass for biofuel production.

Biomass and biofuel production potential of irrigated switchgrass, dryland Miscanthus, and honey mesquite

The land use change from cotton to irrigated switchgrass and dryland Miscanthus exhibited superior biomass and ethanol production potential per ha compared to the honey mesquite harvest (Table 3). However, the vast extent of honey mesquite acreage in the Southern Great Plains (21 million ha in Texas alone) has the potential to supply abundant quantities of honey mesquite biomass for bioenergy purposes (Ansley et al., 2010; Park et al., 2012). For example, based on an estimated average standing biomass of a 10-year-old honey mesquite of 19.4 Mg ha^{-1} (about 1.94 Mg ha^{-1} biomass accumulation per year) reported in Whisenant & Burzlaff (1978), harvest of honey mesquite from the entire rangelands of Texas annually can provide the biomass required for producing 12.6 million m^3 of biofuels, which is approximately equal to 16.6% of the mandated 2022 U.S. biofuel target of 76 million m^3 of second-generation biofuel. In addition, honey mesquite has a high regrowth potential, and it does not require planting, irrigation, and fertilization costs (Park et al., 2012; Wang et al., 2014). These advantages make honey mesquite an appropriate supplementary bioenergy crop in the downstream subwatershed of this study and other similar areas in the TRP region.

Acknowledgements

This material is based on the work that is supported by the National Institute of Food and Agriculture, U.S. Department of Agriculture, under award number NIFA-2012-67009-19595. Any opinions, findings, conclusions, or recommendations expressed in this publication are those of the author(s) and do not necessarily reflect the view of the U.S. Department of Agriculture. We gratefully thank Dr. Raghavan Srinivasan, Texas A&M University, College Station, TX; Dr. Jongyoon Park, Texas A&M AgriLife Research, Vernon, TX; and Nancy Sammons, USDA-ARS Lab, Temple, TX, for their help. We gratefully thank the two anonymous reviewers for their valuable comments and suggestions for improving this manuscript.

References

Abbaspour KC, Vejdani M, Haghighat S (2007) SWAT-CUP calibration and uncertainty programs for SWAT. In: *Proc. Intl. Congress on Modelling and Simulation (MODSIM'07)* (eds Oxley L, Kulasiri D), pp. 1603–1609. Modelling and Simulation Society of Australia and New Zealand, Melbourne, Australia.

Ale S, Bowling LC, Brouder SM, Frankenberger JR, Youssef MA (2009a) Simulated effect of drainage water management operational strategy on hydrology and crop yield for Drummer soil in the Midwestern United States. *Agricultural Water Management*, **96**, 653–665.

Ansley RJ, Mirik M, Castellano MJ (2010) Structural biomass partitioning in regrowth and undisturbed mesquite (*Prosopis glandulosa*): implications for bioenergy uses. *Global Change Biology Bioenergy*, **2**, 26–36.

Arnold JG, Srinivasan R, Muttiah RS, Williams JR (1998) Large-area hydrologic modeling and assessment: Part I. Model development. *Journal of the American Water Resources Association*, **34**, 73–89.

Arnold JG, Moriasi DN, Gassman PW et al. (2012) SWAT: model use, calibration, and validation. *Transactions of the ASABE*, **55**, 1491–1508.

Asner GP, Archer SR, Hughes RF, Ansley RJ, Wessman CA (2003) Net changes in regional woody vegetation cover and carbon storage in Texas drylands, 1937-1999. *Global Change Biology*, **9**, 316–335.

Bronson KF, Zobeck TM, Chua TT, Acosta-Martinez V, van Pelt RS, Booker JD (2004) Carbon and nitrogen pools of Southern High Plains cropland and grassland soils. *Soil Science Society of America Journal*, **68**, 1695–1704.

Chaudhuri S, Ale S (2014a) Long-term (1930-2010) trends in groundwater levels in Texas: influences of soils, landcover and water use. *Science of the Total Environment*, **490**, 379–390.

Chaudhuri S, Ale S (2014b) Long-term (1960-2010) trends in groundwater contamination and salinization in the Ogallala aquifer in Texas, USA. *Journal of Hydrology*, **513**, 376–390.

Chen Y, Ale S, Rajan N, Morgan CLS, Park JY (2016a) Hydrological responses of land use change from cotton (*Gossypium hirsutum* L.) to cellulosic bioenergy crops in the Southern High Plains of Texas, USA. *Global Change Biology Bioenergy*, **8**, 981–999.

Chen Y, Ale S, Rajan N (2016b) Spatial variability of biofuel production potential and hydrologic fluxes of land use change from cotton (*Gossypium hirsutum* L.) to Alamo switchgrass (*Panicum virgatum* L.) in the Texas High Plains. *BioEnergy Research*, **9**, 1126–1141.

Christian D, Riche A, Yates N (2008) Growth, yield and mineral content of *Miscanthus × giganteus* grown as a biofuel for 14 successive harvests. *Industrial Crops and Products*, **28**, 320–732.

Cibin R, Trybula E, Chaubey I, Brouder S, Volenec JJ (2016) Watershed scale impacts of bioenergy crops on hydrology and water quality using improved SWAT model. *Global Change Biology Bioenergy*, **8**, 837–848.

Daloğlu I, Cho KH, Scavia D (2012) Evaluating causes of trends in long-term dissolved reactive phosphorus loads to Lake Erie. *Environmental Science & Technology*, **46**, 10660–10666.

Danalatos NG, Archontoulis SV, Mitsios I (2007) Potential growth and biomass productivity of Miscanthus×giganteus as affected by plant density and N-fertilization in central Greece. *Biomass and Bioenergy*, **31**, 145–152.

Demissie Y, Yan E, Wu M (2012) Assessing regional hydrology and water quality implications of large-scale regional biofuel feedstock production in the Upper Mississippi River Basin. *Environmental Science & Technology*, **46**, 9174–9182.

Feng QY, Chaubey I, Gu Her Y, Cibin R, Engel B, Volenec J, Wang XY (2015) Hydrologic and water quality impacts and biomass production potential on marginal land. *Environmental Modeling & Software*, **72**, 230–238.

Foley JA, Ramankutty N, Brauman KA et al. (2011) Solutions for a cultivated planet. *Nature*, **478**, 337–342.

Gassman PW, Williams JR, Wang X et al. (2010) The Agricultural Policy/Environmental Extender (APEX) model: An emerging tool for landscape and watershed environmental analyses. *Transactions of the ASABE*, **53**, 711–740.

Gassman PW, Sadeghi AM, Srinivasan R (2014) Applications of the SWAT model special section: overview and insights. *Journal of Environmental Quality*, **43**, 1–8.

Ghaffari G, Keesstra S, Ghodousi J, Ahmadi H (2010) SWAT-simulated hydrological impact of land-use change in the Zanjanrood Basin, Northwest Iran. *Hydrological Processes*, **24**, 892–903.

Golmohammadi G, Prasher S, Madani A, Rudra R (2014) Evaluating three hydrological distributed watershed models: MIKE-SHE, APEX, SWAT. *Hydrology*, **1**, 20–39.

Jager HI, Baskaran LM, Schweizer PE, Turhollow AF, Brandt CC, Srinivasan R (2015) Forecasting changes in water quality in rivers associated with growing biofuels in the Arkansas-White-Red river drainage, USA. *Global Change Biology Bioenergy*, **7**, 774–784.

Jha MK, Gassman PW, Arnold JG (2007) Water quality modeling for the Raccoon River watershed using SWAT. *Transactions of the ASABE*, **50**, 479–493.

Jung CG, Park JY, Kim SJ, Park GA (2014) The SRI (system of rice intensification) water management evaluation by SWAPP (SWAT-APEX Program) modeling in an agricultural watershed of South Korea. *Paddy and Water Environment*, **12**, 251–261.

Kiniry JR (1998) Biomass accumulation and radiation use efficiency in honey mesquite and eastern red cedar. *Biomass and Bioenergy*, **15**, 467–473.

Knisel WG (1980) *CREAMS: A Field-Scale Model for Chemicals, Runoff, and Erosion From Agricultural Management Systems*. Conservation Research Report No. 26. USDA National Resources Conservation Service, Washington, DC.

Ko JH, Piccinni G, Steglich E (2009) Using EPIC model to manage irrigated cotton and maize. *Agricultural Water Management*, **96**, 1323–1331.

Le PVV, Kumar P, Drewry DT (2011) Implications for the hydrologic cycle under climate change due to the expansion of bioenergy crops in the Midwestern United States. *PNAS*, **108**, 15085–15090.

Legates DR, McCabe GJ Jr (1999) Evaluating the use of ''goodness-of-fit'' measures in hydrologic and hydroclimatic model validation. *Water Resources Research*, **35**, 233–241.

Leonard RA, Knisel WG, Still DA (1987) GLEAMS: Groundwater loading effects on agricultural management systems. *Transactions of the ASABE*, **30**, 1403–1418.

Lewandowski I, Schmidt U (2006) Nitrogen, energy and land use efficiencies of miscanthus, reed canary grass and triticale as determined by the boundary line approach. *Biomass and Bioenergy*, **112**, 335–346.

Li H, Lascano RJ (2011) Deficit irrigation for enhancing sustainable water use: Comparison of cotton nitrogen uptake and prediction of lint yield in a multivariate

autoregressive state-space model. *Environmental and Experimental Botany*, **71**, 224–231.

McIsaac GF, David MB, Mitchell CA (2010) *Miscanthus* and switchgrass production in central Illinois: impacts on hydrology and inorganic nitrogen leaching. *Journal of Environmental Quality*, **39**, 1790–1799.

Moriasi DN, Arnold JG, Van Liew MW, Binger RL, Harmel RD, Veith T (2007) Model evaluation guidelines for systematic quantification of accuracy in watershed simulations. *Transactions of the ASABE*, **50**, 885–900.

Mullins GL, Burmester CH (1990) Dry matter, nitrogen, phosphorus, and potassium accumulation by four cotton varieties. *Agronomy Journal*, **82**, 729–736.

Nash JE, Sutcliffe JV (1970) River flow forecasting through conceptual models, Part I-a discussion of principles. *Journal of Hydrology*, **10**, 282–290.

NASS (National Agricultural Statistics Service NASS) (2015) Quick stats of cotton planting area. Available at: http://www.nass.usda.gov/Quick_Stats/Lite/#E0B0B59C-D3B5-38B3-B2DB-D7875C90B937 (accessed 15 March 2016).

Nelson RG, Ascough JC II, Langemeier MR (2006) Environmental and economic analysis of switchgrass production for water quality improvement in northeast Kansas. *Journal of Environmental Management*, **79**, 336–347.

Ng TZ, Eheart JW, Cai X, Miguez F (2010) Modeling miscanthus in the soil and water assessment tool (SWAT) to simulate its water quality effects as a bioenergy crop. *Environmental Science & Technology*, **44**, 7138–7144.

NOAA-NCDC (National Oceanic and Atmospheric Administration-National Climatic Data Center) (2014) Weather Data. Available at: http://gis.ncdc.noaa.gov/map/viewer/#app=cdo&cfg=cdo&theme=daily&layers=111&node=gis (accessed 18 August 2014)

Padron E, Navarro RM (2004) Estimation of above-ground biomass in naturally occurring populations of *Prosopis pallida* (H. & B. ex. Willd.) H.B.K. in the north of Peru. *Journal of Arid Environments*, **56**, 283–292.

Park SC, Ansley RJ, Mirik M, Maindrault MA (2012) Delivered biomass costs of honey mesquite (*prosopis glandulosa*) for bioenergy uses in the South Central USA. *BioEnergy Research*, **5**, 989–1001.

Park JY, Ale S, Teague WR, Dowhower SL (2017) Simulating hydrologic responses to alternate grazing management practices at the ranch and watershed scales. *Journal of Soil and Water Conservation*, **72**, 102–121.

Powers SE, Ascough JC II, Nelson RG, Larocque GR (2011) Modeling water and soil quality environmental impacts associated with bioenergy crop production and biomass removal in the Midwest USA. *Ecological Modelling*, **222**, 2430–2447.

Rao MN, Yang ZM (2010) Groundwater impacts due to conservation reserve program in Texas County, Oklahoma. *Applied Geography*, **30**, 317–328.

Rockström JW, Steffen K, Noone Å (2009) Planetary boundaries: exploring the safe operating space for humanity. *Ecology and Society*, **14**, 32.

Runkel RL, Crawford CG, Cohn TA (2004) Load Estimator (LOADEST): A FORTRAN program for estimating constituent loads in streams and rivers. USGS Techniques and Methods Book 4, Chapter A5. Reston, Va.: U.S. Geological Survey. Available at: http://pubs.er.usgs.gov/publication/tm4A5 (accessed 18 August 2015)

Saleh A, Gallego O (2007) Application of SWAT and APEX using the SWAPP (SWAT-APEX Program) for the upper North Bosque River watershed in Texas. *Transactions of the ASABE*, **50**, 1177–1187.

Santhi C, Kannan N, White M, Di Luzio M, Arnold JG, Wang X, Williams JR (2014) An integrated modeling approach for estimating the water quality benefits of conservation practices at river basin scale. *Journal of Environmental Quality*, **43**, 177–198.

Sarkar S, Miller SA (2014) Water quality impacts of converting intensively-managed agricultural lands to switchgrass. *Biomass and Bioenergy*, **68**, 32–43.

Scherer L, Venkatesh A, Karuppiah R, Pfister S (2015) Large-scale hydrological modeling for calculating water stress indices: implications of improved spatiotemporal resolution, surface-groundwater differentiation, and uncertainty characterization. *Environmental Science & Technology*, **49**, 4971–4979.

SCS (1988) *Texas Brush Inventory*. United States Department of Agriculture, Soil Conservation Service Misc. Report, Temple, TX.

Singh G, Mutha S, Bala N (2007) Effect of tree density on productivity of a *Prosopis cineraria* agroforestry system in north western India. *Journal of Arid Environments*, **70**, 152–163.

Soil Survey Staff (2010) *Keys to Soil Taxonomy* (11th edn). USDA-Natural Resources Conservation Service, Washington, DC.

Soil Survey Staff (2014) *Keys to Soil Taxonomy* (11th edn). USDA-Natural Resources Conservation Service, Washington, DC.

Srinivasan R, Zhang X, Arnold JG (2010) SWAT ungauged: hydrological budget and crop yield predictions in the upper Mississippi river basin. *Transactions of the ASABE*, **53**, 1533–1546.

Teague WR, Dowhower SL (2003) Patch dynamics under rotational and continuous grazing management in large, heterogeneous paddocks. *Journal of Arid Environments*, **53**, 211–229.

Texas Water Development Board (2005) Volumetric Survey of Alan Henry Reservoir. Available at: http://www.twdb.texas.gov/hydro_survey/alanhenry/ 2005-07/AlanHenry2005_FinalReport.pdf (accessed 5 July 2015)

Texas Water Development Board (2007) *Water for Texas II*. Texas Water Development Board, Austin, TX.

Trybula EM, Cibin R, Burks JL, Chaubey I, Brouder SM, Volenec JJ (2015) Perennial rhizomatous grasses as bioenergy feedstock in SWAT: parameter development and model improvement. *Global Change Biology Bioenergy*, **7**, 1185–1202.

Tuppad P, Winchell MF, Wang X, Srinivasan R, Williams JR (2009) ArcAPEX: Arc-GIS interface for Agricultural Policy Environmental Extender (APEX) hydrology/water quality model. *International Agricultural Engineering Journal*, **18**, 59–71.

Tuppad P, Santhi C, Wang X, Williams JR, Srinivasan R, Gowda PH (2010) Simulation of conservation practices using the APEX model. *Transactions of the ASABE*, **26**, 779–794.

USDA (2010) A USDA Regional roadmap to meeting the biofuels goals of the Renewable Fuels Standard by 2022. Biofuels strategic production report. USDA, Washington, DC. Available at: http://www.usda.gov/documents/USDA_Biofuels_Report_ 6232010.pdf (accessed 5 July 2015)

VanLoocke A, Bernacchi CJ, Twine TE (2010) The impacts of *Miscanthus×giganteus* production on the Midwest US hydrologic cycle. *Global Change Biology Bioenergy*, **2**, 180–191.

Wang X, Kannan N, Santhi C, Potter SR, Williams JR, Arnold JG (2011) Integrating APEX output for cultivated cropland with SWAT simulation for regional modeling. *Transactions of the ASABE*, **54**, 1281–1298.

Wang X, Williams JR, Gassman PW, Baffaut C, Izaurralde RC, Jeong J, Kiniry JR (2012) EPIC and APEX: Model use, calibration, and validation. *Transactions of the ASABE*, **55**, 1447–1462.

Wang T, Park SC, Ansley RJ, Amosson SH (2014) Economic and greenhouse gas efficiency of honey mesquite relative to other energy feedstocks for bioenergy uses in the Southern Great Plains. *BioEnergy Research*, **7**, 1493–1505.

Whisenant SG, Burzlaff DF (1978) Predicting green weight of mesquite (*Prosopis glandulosa* Torr.). *Journal of Range Management*, **31**, 396–397.

Williams JR (1995) The EPIC Model. In: *Computer Models of Watershed Hydrology* (ed. Singh VP), pp. 909–1000. Water Resources Publications, Highlands Ranch, CO.

Williams JR, Arnold JG, Kiniry JR, Gassman P, Green CH (2008) History of model development at Temple, Texas. *Hydrological Sciences Journal*, **53**, 948–960.

Wu YP, Liu SG (2012) Impacts of biofuels production alternatives on water quantity and quality in the Iowa River Basin. *Biomass and Bioenergy*, **36**, 182–191.

Yimam YT, Ochsner TE, Kakani VG, Warren JG (2014) Soil moisture dynamics and evapotranspiration under annual and perennial bioenergy crops. *Soil Science Society of America Journal*, **78**, 1584–1592.

Yimam YT, Ochsner TE, Kakani VG (2015) Evapotranspiration partitioning and water use efficiency of switchgrass and biomass sorghum managed for biofuel. *Agricultural Water Management*, **155**, 40–47.

Zobeck TM, Crownover J, Dollar M, van Pelt RS, Acosta-Martinez V, Bronson KF, Upchurch DR (2007) Investigation of soil conditioning index values for Southern High Plains agroecosystems. *Journal of Soil and Water Conservation*, **62**, 433–442.

2

Environmental impacts of bioenergy wood production from poplar short-rotation coppice grown at a marginal agricultural site in Germany

JANINE SCHWEIER[1], SAÚL MOLINA-HERRERA[2], ANDREA GHIRARDO[3], RÜDIGER GROTE[2], EUGENIO DÍAZ-PINÉS[2], JÜRGEN KREUZWIESER[4], EDWIN HAAS[2], KLAUS BUTTERBACH-BAHL[2], HEINZ RENNENBERG[4], JÖRG-PETER SCHNITZLER[3] and GERO BECKER[5]

[1]Chair of Forest Operations, Albert-Ludwigs-University Freiburg, Werthmannstraße 6, 79085 Freiburg, Germany, [2]Karlsruhe Institute of Technology (KIT), Institute of Meteorology and Climate Research, Atmospheric Environmental Research, Kreuzeckbahnstraße 19, 82467 Garmisch-Partenkirchen, Germany, [3]Helmholtz Zentrum München, Research Unit Environmental Simulation, Institute of Biochemical Pathology, Ingolstädter Landstraße 1, 85764 Neuherberg, Germany, [4]Chair of Tree Physiology, Albert-Ludwigs-University Freiburg, Georges-Köhler-Allee 53/54, 79110 Freiburg, Germany, [5]Chair of Forest Utilisation, Albert-Ludwigs-University Freiburg, Werthmannstraße 6, 79085 Freiburg, Germany

Abstract

For avoiding competition with food production, marginal land is economically and environmentally highly attractive for biomass production with short-rotation coppices (SRCs) of fast-growing tree species such as poplars. Herein, we evaluated the environmental impacts of technological, agronomic, and environmental aspects of bioenergy production from hybrid poplar SRC cultivation on marginal land in southern Germany. For this purpose, different management regimes were considered within a 21-year lifetime (combining measurements and modeling approaches) by means of a holistic Life Cycle Assessment (LCA). We analyzed two coppicing rotation lengths (7 × 3 and 3 × 7 years) and seven nitrogen fertilization rates and included all processes starting from site preparation, planting and coppicing, wood chipping, and heat production up to final stump removal. The 7-year rotation cycles clearly resulted in higher biomass yields and reduced environmental impacts such as nitrate (NO_3) leaching and soil nitrous oxide (N_2O) emissions. Fertilization rates were positively related to enhanced biomass accumulation, but these benefits did not counterbalance the negative impacts on the environment due to increased nitrate leaching and N_2O emissions. Greenhouse gas (GHG) emissions associated with the heat production from poplar SRC on marginal land ranged between 8 and 46 kg CO_2-eq. GJ^{-1} (or 11–57 Mg CO_2-eq. ha^{-1}). However, if the produced wood chips substitute oil heating, up to 123 Mg CO_2-eq. ha^{-1} can be saved, if produced in a 7-year rotation without fertilization. Dissecting the entire bioenergy production chain, our study shows that environmental impacts occurred mainly during combustion and storage of wood chips, while technological aspects of establishment, harvesting, and transportation played a negligible role.

Keywords: ammonium nitrate fertilization, ecosystem respiration, LandscapeDNDC, life cycle assessment, nitrate leaching, nitrous oxide, short-rotation coppices, technology and agronomy, wood chips, yield-scaled emissions

Introduction

Anthropogenic greenhouse gas (GHG) emissions need to decrease substantially in order to limit the global temperature rise to 2 °C compared to the pre-industrial period (UNFCCC, 2015) and to avoid that the global biosphere crosses irreversible tipping points (e.g., Ramanathan & Feng, 2008). In this context, the role of bioenergy production as a useful means to decrease GHG emissions from energy production is widely

Correspondence: Janine Schweier
e-mail: janine.schweier@foresteng.uni-freiburg.de

discussed. Currently, mankind already uses biomass with an annual gross calorific value of about 300 EJ (Haberl et al., 2007), but with the continuing rise in population and living standards, the demand for bioenergy is expected to increase further.

A promising option to increase lignocellulosic biomass production for energy use is the use of short-rotation coppices (SRCs) of fast-growing tree species. Such systems are considered as the most energy efficient carbon (C) conversion technology (Styles & Jones, 2007), which – if used for energetic purposes – can reduce the total GHG emissions by up to 90% compared to coal combustion (Djomo et al., 2010). In contrast to crops that can be used for food and energy (e.g., corn), SRCs are dedicated bioenergy crops only. However, due to their low nutritional demands and maintenance requirements, they can be cultivated on marginal lands, thus reducing the impacts on land availability for food and feed production (Butterbach-Bahl & Kiese, 2013; Dillen et al., 2013). Hybrid poplars have exceptional vegetative regeneration abilities (Aylott et al., 2008) and high biomass production rates and can be cultivated and adapted to a wide range of geographical conditions – especially in temperate climate (Fortier et al., 2015). Established as SRC on marginal agricultural sites, they further have the potential to increase soil C sequestration (Anderson-Teixeira et al., 2013), while reducing soil nitrate (Díaz-Pines et al., 2016).

The global environmental impact of hybrid poplar SRC cultivation is, however, not positive per se. Hybrid poplar SRCs are usually fertilized to increase biomass growth (Balasus et al., 2012), which can boost nitrogen (N) losses such as N_2O, a much more potent GHG than carbon dioxide (CO_2). Hence, the positive effect of C sequestration may be counterbalanced by N_2O emissions due to fertilization and also due to other processes during the plantations' lifetime. For example, technological processes such as storage and transport may cause high GHG emissions (Schweier et al., 2016). Therefore, a comprehensive evaluation of SRC cultivation focusing on the GHG balance of such systems together with other environmental impacts, for example, NO_3 leaching losses, needs to have a long-term perspective. Also, differences in management practices, in particular changing rotation cycle length, can have significant impacts on biomass yield and environmental effects such as soil C storage or soil N_2O emissions (e.g., Fang et al., 2007; Bacenetti et al., 2012).

Up to now, most analyzes addressing SRC cultivation and its environmental impacts have focused either on technological processes such as establishment, planting, and harvesting (Heller et al., 2003; Gasol et al., 2009; Nassi o di Nasso et al., 2010; Rödl, 2010; Bacenetti et al., 2012; Fiala & Bacenetti, 2012; Gabrielle et al., 2013; Manzone et al., 2014; Murphy et al., 2014; Quartucci et al., 2015; Schweier et al., 2016), or on agronomic aspects such as plant growth or N_2O fluxes (Pecenka et al., 2013; Rösch et al., 2013; Zona et al., 2013a,b; Walter et al., 2015; Brilli et al., 2016; Sabbatini et al., 2016). However, studies simultaneously addressing technological, agronomic as well as environmental aspects of SRC production are scarce. Moreover, they usually do not include long-term GHG emission balances for the full lifetime of a SRC, including a number of rotation cycles and the final removal of the remaining biomass.

In this study, we conducted an integrated analysis of the environmental impact categories Global Warming Potential (GWP) and the Eutrophication Potential (EP) related to energy produced from wood chips from a hybrid poplar SRC established on marginal land in southern Germany. We focused our analysis on these two categories, which are the primary criteria in numerous papers that deal with the cultivation and the use of biomass for energy production (Cherubini & Strømman, 2011), because they address different environmental spheres (air and soil) and are often found to show significant differences between management regimes (McBride et al., 2011). Our study addressed all phases of the technological and agronomic production of poplar wood chips, based on experimental (Díaz-Pines et al., 2016) and literature data (Burger, 2010) as well as data collections concerning technological activities (c.f. Schweier et al., 2016) and the use of a database (Ecoinvent, 2010) in combination with simulation estimates (for 21 years) performed with the process-based ecosystem model LandscapeDNDC (Haas et al., 2013) and Umberto, a software which supports ISO compliant LCAs (IFU, 2011). We hypothesize that the energy production from hybrid poplar SRC on marginal land (from cradle-to-site) results in a C sink due to C uptake during plant growth, while the overall production of energy out of SRC (from cradle-to-grave) results in a C source, however, being significantly lower compared to the use of fossil fuels.

Materials and methods

Life cycle assessment

To assess the environmental impacts of SRC wood chip production, the methodological framework of Life Cycle Assessment (LCA) was applied and 14 production chains were modeled using the software Umberto v5.6 (IFU, Hamburg, Germany).

Scope definition

All processes associated with the cultivation and growth of poplar SRC and the subsequent production of wood chips over a full rotation cycle were included, starting with the initial site preparation. This was followed by the cultivation and repeated harvesting, the chip production, and delivery of the chips at gate of the heating plant. The entire chain also included the final removal of the stems and stumps from the plantation site (Fig. 1) after 21 years of cultivation. To assess the impact of harvesting rotation cycle lengths within the 21-year plantation lifetime, we analyzed 2 different cycle lengths (7 × 3 years = seven rotation cycles: 7 harvests each 3 years and 3 × 7 years = 3 rotation cycles: 3 harvests each 7 years). In combination with this two management practices, we also analyzed seven different N fertilization rates (0/25/50/75/100/150/200 kg NH_4NO_3-N per hectare and rotation). Thus, in total, 14 production chains were assessed regarding their environmental impacts (Table 1).

1 Establishment & Maintenance

Production of plant cuttings.

- Application of herbicides.
- Ploughing, tillage (with rotary harrow and cultivator).
- Planting and weed control during first year.
- Rotary cultivation of the edge (yearly) and of the site (after each harvest).
- Application of herbicides after each harvest.

2 Fertilization

- One application of 0/25/50/75/100/150/200 kg NH_4NO_3-N per hectare and rotation period.

3 Field-Greenhouse gas emissions

- Photosynthesis & total ecosystem respiration.
- N_2O emissions.
- NO_3 leaching.

4 Harvesting

- Harvesting cycle of either 7 times in 21 years (=all 3 years) or 3 times in 21 years (=all 7 years).
- With modified forage harvester (400 kW) and use of accompanying tractor-trailer units.
- Interim storage in 2 km distance.

5 Transport

- Transportation of fresh wood chips to a heating plant in 50 km distance with trailer (capacity 80 loose m³).

6 Storage

- Storage and drying process of fresh wood to a water content (WC) of approximately 30%.

7 Removal

- Elimination of above- and belowground biomass at the end of the plantations' lifetime.

8 Combustion

- Combustion of wood chips in a heating plant (1.7 MW_{th}).
- Disposal of ashes (including carbon content)

Chips at heating plant

Upstream processes

Production of machines, fuels, lubricants, fertiliser, pesticides and other input material

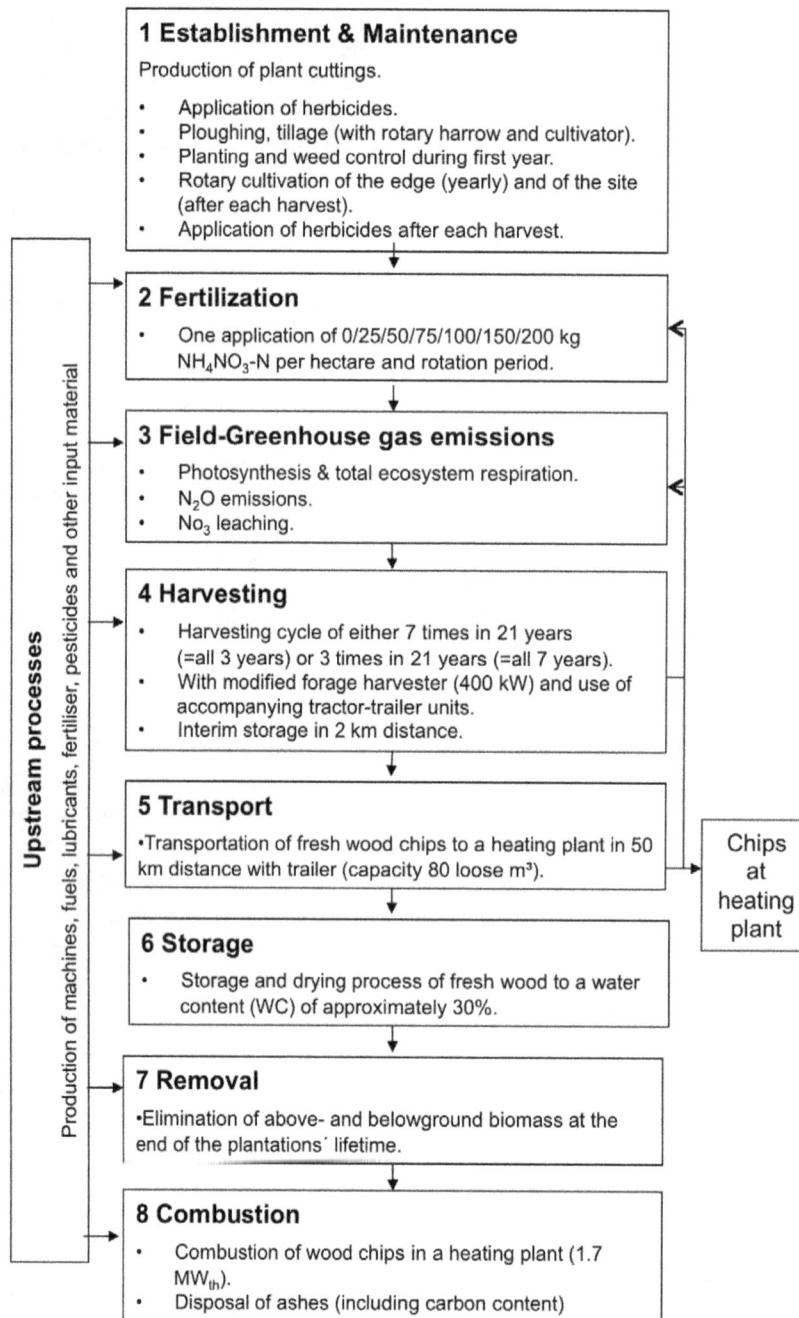

Fig. 1 System boundary of analyzed production chains of wood chips from hybrid poplar SRC. ammonium nitrate (NH_4NO_3), nitrous oxide (N_2O), nitrate (NO_3), megawatt (MW).

Site description

Most of the data that were required as inputs for the LCA have been collected on an experimental site in southern Germany. The site has a soil quality index (SQI) of 37 representing typical conditions for marginal agricultural land in the region (slope 10%, mean annual air temperature 7.2 °C and mean annual rainfall 790 mm yr^{-1} (May–September: 466 mm)). Thereby, the SQI is a numerical value that characterizes the quality and

production potential of cropland for annual crops. The scale of possible values ranges from 7 to 100 (c.f. Aust *et al.*, 2014). The 4.5 ha site was established in 2009 with two commercial hybrid poplar clones, that is, Max 4 (*Populus maximowiczii* A. Henry × *P. nigra* L.) and Monviso (*P.* × *generosa* A. Henry × *P. nigra* L.). It is located in the mountainous Swabian Alps region in southwest Germany (48°6'N/9°14'E; 650 m a.s.l.). Data on soil properties (including C and N contents, soil pH, bulk density, soil water-holding capacity, wilting point,

Table 1 Overview of the 14 analyzed production chains

Chain no.	Scenario name	Rotation cycle Year	Fertilization rate kg NH$_4$NO$_3$ ha^{-1} rotation^{-1}	Fertilization (in total) kg NH$_4$NO$_3$ ha^{-1}
1	3 yr/0 kgN	3-year: 7*3	0	0
2	3 yr/25 kgN	3-year: 7*3	25	175
3	3 yr/50 kgN	3-year: 7*3	50	350
4	3 yr/75 kgN	3-year: 7*3	75	525
5	3 yr/100 kgN	3-year: 7*3	100	700
6	3 yr/150 kgN	3-year: 7*3	150	1,050
7	3 yr/200 kgN	3-year: 7*3	200	1,400
8	7 yr/0 kgN	7-year: 3*7	0	0
9	7 yr/25 kgN	7-year: 3*7	25	75
10	7 yr/50 kgN	7-year: 3*7	50	150
11	7 yr/75 gN	7-year: 3*7	75	225
12	7 yr/100 kgN	7-year: 3*7	100	300
13	7 yr/150 kgN	7-year: 3*7	150	450
14	7 yr/200 kgN	7-year: 3*7	200	600

stone content, hydraulic conductivity, soil type, clay–silt, and sand contents), biomass production, gross primary production or photosynthesis, soil GHG fluxes, and nitrate leaching were obtained within four experimental years (Schnitzler et al., 2014; Díaz-Pines et al., 2016).

Simulation model

For providing comprehensive input data for *Umberto* regarding the biomass estimation during 21 years, the GHG exchange and nitrate leaching rates of poplar SRC cultivation, and the plant growth, we used the model *LandscapeDNDC* (Haas et al., 2013). *LandscapeDNDC* is an assembled modular modeling platform that integrates process-based models for describing C, N, and water fluxes within terrestrial ecosystems. It was initialized with data from the above-mentioned experimental site. The models' reliability has been shown in the previous studies evaluating C, N, and water balances (Holst et al., 2010; Grote et al., 2011a,b), plant growth for poplar plantations (Werner et al., 2012), GHG emissions under the influence of mean commodity crops and poplar plantations (Kim et al., 2014, 2015; Kraus et al., 2015; Molina-Herrera et al., 2015, 2016; Zhang et al., 2015; Díaz-Pines et al., 2016), and NO$_3$ leaching (Díaz-Pines et al., 2016; Dirnböck et al., 2016). For the present study, *LandscapeDNDC* was run with the physiological model 'PSIM' (Physiological Simulation Model) (Grote et al., 2011a), the soil biogeochemical model 'DNDC' (DeNitrification–DeComposition) (Li et al., 1992, 2000; Stange et al., 2000), the empirical microclimate model 'ECM' (Grote et al., 2009), and the hydrology module originating from 'DNDC' (Li et al., 1992). Several input data regarding soil, vegetation, climate, and air chemistry were required to run *LandscapeDNDC*. As stated, most of the input data were collected on the experimental site. The meteorological input data were obtained from the nearest German Weather Service meteorological station Sigmaringen (Deutscher Wetterdienst DWD, Offenbach, Germany), for the period 2009–2014 and then repeated until 2030 for the analysis of the LCA

in a long-term prospective. A constant atmospheric N deposition rate (15–20 kg N ha^{-1} yr^{-1}, estimated from regional values presented by Schaap et al., 2015) was applied along the 21 years for all cases. Physiological parameterization (e.g., RuBisCO (Ribulose-1.5-bisphosphate carboxylase/oxygenase) activity, water-use efficiency, respiration) has been derived from the literature and various previous experiments (Behnke et al., 2012; Schnitzler et al., 2014; Díaz-Pines et al., 2016). Additional parameters for clone-specific allometric relationships (e.g., maximum height: diameter ratio, crown width: diameter ratio) and final leaf area index became adjusted to the detailed measurements at the sites made throughout the first rotation phase and the beginning of the second (5 years). The ability to cover a wide range of site and climatic conditions has been shown by the representation of various poplar SRCs all over Europe (Werner et al., 2012).

To compute the total GHG balance, the results from *LandscapeDNDC*, such as net ecosystem C exchange (NEE), N$_2$O emissions, and NO$_3$ leaching, were combined with estimated indirect N$_2$O emissions due to soil nitrate leaching (calculated according to Denman et al., 2007), and measured soil CH$_4$ fluxes (based on a 4-year measurement campaign at the studied site; c.f. Díaz-Pines et al., 2016) were used as inputs in *Umberto*.

System boundaries

All 14 production chains (Fig. 1) comprise the following eight main process steps:
1. *Establishment and Maintenance*: We included the production of plant cuttings in a nursery, initial plowing, harrowing with a disk harrow, application of herbicides (5 l ha^{-1} Round up; Monsanto, St. Louis, MO, USA) with a boom sprayer, and mechanical weed control with a field cultivator. Planting of single rows (6350 cuttings ha^{-1}) was carried out with a professional planting machine owned by Probstdorfer Saatzucht GmbH (Vienna, Austria) (Fig. S11). GHG emissions due to these activities were based on data collected on site (Schweier, 2013; Schweier et al., 2016).

Table 2 Field operations and associated machinery data

Rotation length	Operation	Timeline	Operating rate (h ha^{-1})	Machine type	Power (kw)	Diesel consumption (kg ha^{-1})*	Implement *
3 years	Application of herbicides	Year 0, 1 and after each harvest	0.7	Tractor	83	94.5	Glyphosate (1.8 kg ha^{-1}) Dicamba (0.1 kg ha^{-1}) Pendimethalin (8 kg ha^{-1})
	Ploughing	Establishment	1.8	Tractor	102	23.2	
	Harrowing	Establishment	1.1	Tractor	83	13.5	
	Planting	Establishment	2.2	Tractor	83	21.9	6350 Cuttings
	Mechanical weed control	Year 0, 1 and after each harvest	0.8	Tractor	83	51.6	
	Application of fertilizer	1× per rotation	0.7	Tractor	83	0–1400 (Table 1)	Nitrogen
	Harvesting	1x per rotation	1.09–1.14 (Table S4)	Forager	400	444–464 (Table S4)	
	Removal	Year 21	9.0	Tractor	233	351.8	
7 years	Application of herbicides	Year 0, 1 & after each harvest	0.7	Tractor	83	52.5	Glyphosate (1.8 kg ha^{-1}) Dicamba (0.1 kg ha^{-1}) Pendimethalin (4 kg ha^{-1})
	Ploughing	Establishment	1.8	Tractor	102	23.2	
	Harrowing	Establishment	1.1	Tractor	83	13.5	
	Planting	Establishment	2.2	Tractor	83	21.9	6350 Cuttings
	Mechanical weed control	Year 0, 1 and after each harvest	0.8	Tractor	83	103.2	
	Application of fertilizer	1× per rotation	0.7	Tractor	83	0–600 (Table 1)	Nitrogen
	Harvesting	1× per rotation	1.52–1.53 (Table S4)	Forager	400	265–268 (Table S4)	
	Removal	Year 21	9.0	Tractor	233	351.8	

*Inputs refer to the overall lifetime of the plantation.

Information regarding machines and inputs is given in Table 2. Data regarding operating machines can be found in Table S3. Besides, it was assumed that after each harvesting cycle, a mechanical weed control was carried out with a field cultivator and herbicides were applied (2.5 l ha^{-1} Stomp SC; COMPO, Münster, Germany) with a boom sprayer. Respective emission data were taken from a database (Ecoinvent, 2010).

2. *Fertilization*: We considered one application of fertilizer in the first year of each rotation (Tables 1 and 2). Simulated fertilization rates were derived from past studies (Hellebrand *et al.*, 2008; Kavdir *et al.*, 2008; van den Driessche *et al.*, 2008; Kern *et al.*, 2010; Balasus *et al.*, 2012) and reflect common procedures for poplar SRC. Respective emission data were taken from *Ecoinvent* database, too (Ecoinvent, 2010). It should be noted that while liquid NPK fertilizer was given to the experimental site as fertigation, simulations only assumed the application of NH_4NO_3 because the model is not sensitive to P and K nutrition, implicitly assuming that differences between sites regarding these elements have no significant impact on plant development.

3. *Field-GHG*: We simulated the GHG emissions of this site with the *LandscapeDNDC* (as described in *simulation model*) and considered besides NEE (gross primary production minus autotrophic and heterotrophic respiration) also other components of the field-GHG balance, that is, soil N_2O and methane (CH_4) emissions as well as indirect N_2O emissions following NO_3 leaching.

4. *Harvesting*. We assumed harvesting cycles of either 7 times in 21 years (= each 3-years) or 3 times in 21 years (= each 7-years). Harvesting was carried out with a modified forage harvester (400 kW) (Fig. S12), cutting and chipping all stems and branches in one operation. The use of this machine in all rotation cycles was justified as the biomass simulation has shown that the stem diameters at ground level are unlikely to exceed the machines' capacity even after seven years of growth (Table S7). For all harvests, the accompanying tractor-trailer units were considered to transport the wood chips to an interim storage site at 2 km distance. Related data were collected from the first coppice after a 3-year cycle only, but detailed productivity figures of the machine were collected in an earlier study (Schweier & Becker, 2012). Thus, specific time and fuel consumptions were calculated for each harvesting operation within the 21-year lifetime (Table S4) depending on the amount of biomass per harvest.

5. *Transportation*: We included loading of fresh low-density wood chips (water content (WC) 55% (w/w)) at the interim storage site into trucks with a capacity of 80 loose m³, the

full loaded transport to a heating plant in 50 km distance, as well as the empty return of the trucks. GHG emissions due to transportation were taken from the database *Ecoinvent* (2010). Ton-kilometers were calculated per each harvest (Table S4).

6. *Storage*: We considered the drying process during storage of fresh wood chips to lower the water content (WC) down to ~30% (w/w), which is required before burning the biomass in small- and medium-sized heating plants. To quantify C losses in terms of CO_2 emissions (Table 3) from freshly harvested wood chips, around 60 kg biomass from the first harvest in 2012 was enclosed into 4 environmentally controlled chambers (temperature of 20 °C, relative air humidity of 40%, light intensity of ~50 μmol photons m^{-2} s^{-1}) at the phytotron facility at the Helmholtz Zentrum München (e.g., Vanzo *et al.*, 2015) and online measurements of trace gases (GHG and VOCs (volatile organic compounds)) were performed immediately after harvest and continuously for 6 weeks using infrared spectroscopy and online proton transfer reaction mass spectrometry (Ghirardo *et al.*, 2010, 2014; Vanzo *et al.*, 2015). GHG and VOC fluxes were calculated as previously described (Ghirardo *et al.*, 2011) and given per dried biomass.

7. *Removal*: We considered the removal of remaining above- and belowground biomass on site within 3 months after the last harvest at the end of the plantations' lifetime in year 21, thereby assuming that the disturbance effects have ceased during this time period (by Díaz-Pines *et al.*, 2016). The related C release is reported in Table S8. *LandscapeDNDC* did not consider any changes in soil properties caused by the extraction (e.g., changes in bulk density, redistribution of C contents, hydrological properties) or any priming associated with this process (Strömgren *et al.*, 2012). Data regarding machinery and fuel input of stump removal were

taken from the literature (Burger, 2010) and can be found in Table 2. The use of biomass from stump removal for energy production was not considered, as this is not a common practice in Germany.

8. *Combustion*: We considered the combustion of wood chips in a heating plant. In 2015, data from one year were collected in a modern medium-sized biomass heating plant (1.7 MWh a^{-1}, 90% efficiency, built in 2012) located in the Black Forest, Germany. The data included all technological processes and used inputs from takeover of wood chips until removal of ashes. As chips were dried before, it was assumed that the energy density of the chips is 11.84 GJ per ton wood chips at a WC of 31.8% (Hartmann, 2009). Resulting amounts of energy per hectare are shown in Table S6. The system boundary is when the product heat (GJ) is leaving the plant (water at 100 °C in winter, 75 °C in summer). Collected data refer to a mixed input of hardwood and softwood. However, to calculate the amount of required wood chips per year, we assumed that the heating plant was fed with poplar wood chips from SRC only.

Others: Following the LCA approach, we considered also CO_2 emissions caused by upstream processes, for example, due to the production and use of machineries or fuels. Inputs were calculated according to Nemecek & Kägi (2007), and related emission data were gathered from the commercial database *Ecoinvent* (Ecoinvent, 2010).

Functional Units

Emissions refer to the cultivated surface in hectares. In addition, we calculated all GHG emissions referring to dry matter in megagram (Mg$_{dm}$) of produced wood chips and to gigajoule

Table 3 *Global Warming Potential* for the production of poplar wood chips from SRC in 21 years, shown per process step and for all 14 production chains [in kg CO_2-eq. GJ^{-1}]. An overview of the 14 analyzed production chains can be found in Table 1. Results are reported per process step (EstMain = Establishment and Maintenance; Fert = Fertilization; Field-GHG = Field-Greenhouse gases; Har = Harvesting; Tra = Transport; Rem = Removal; Comb = Combustion). Negative signs indicate CO_2 sinks while positive signs indicate CO_2 sources

Chain	Process step							
	EstMain	Fert	Field-GHG	Har	Trans	Stor	Rem	Comb
1: 3 yr/0 kgN	+0.34	+0.00	−150.04	+0.55	+2.19	+28.02	+4.49	+139.20
2: 3 yr/25 kgN	+0.33	+1.37	−149.13	+0.54	+2.19	+28.02	+4.16	+139.20
3: 3 yr/50 kgN	+0.33	+2.65	−148.07	+0.54	+2.19	+28.02	+4.17	+139.20
4: 3 yr/75 kgN	+0.33	+3.89	−147.09	+0.54	+2.19	+28.02	+4.16	+139.20
5: 3 yr/100 kgN	+0.32	+5.09	−145.43	+0.53	+2.19	+28.02	+4.12	+139.20
6: 3 y/150 kgN	+0.31	+7.37	−141.43	+0.53	+2.19	+28.02	+4.02	+139.20
7: 3 y/200 kgN	+0.31	+9.76	−138.30	+0.53	+2.19	+28.02	+4.02	+139.20
8: 7 yr/0 kgN	+0.30	+0.00	−167.27	+0.36	+2.19	+28.02	+5.55	+139.20
9: 7 yr/25 kgN	+0.30	+0.56	−166.23	+0.35	+2.19	+28.02	+5.51	+139.20
10: 7 yr/50 kgN	+0.30	+1.06	−165.81	+0.35	+2.19	+28.02	+5.51	+139.20
11: 7 yr/75 gN	+0.30	+1.55	−165.40	+0.35	+2.19	+28.02	+5.51	+139.20
12: 7 yr/100 kgN	+0.30	+2.04	−164.95	+0.35	+2.19	+28.02	+5.50	+139.20
13: 7 yr/150 kgN	+0.29	+2.99	−164.28	+0.35	+2.19	+28.02	+5.47	+139.20
14: 7 yr/200 kgN	+0.29	+3.95	−163.36	+0.35	+2.19	+28.02	+5.47	+139.20

(GJ) because an energy unit is needed to compare the results to various other combustion studies.

Statistical analysis

The relationships between *aboveground biomass* (*AGB*), *GWP*, *EP*, photosynthesis, total ecosystem respiration, N_2O emissions, and NO_3 leaching were explored by principal component analysis (PCA) (SIMCA-P v13, Umetrics, Umeå, Sweden). PCA was here employed for data mining and data description, where the resulting graphic plot (Fig. 4) summarized the largest variability in the data set and could be interpreted more easily than a matrix of data (Ghirardo *et al.*, 2005). The principles of PCA and its objectives can be found in detail elsewhere (Martens & Martens, 2001; Gottlieb *et al.*, 2004). Before computing the PCA, data were logarithmically transformed (log2), centered, and scaled with $1 \times SD^{-1}$. The resulting significant principal components were cross-validated using 7 validation rounds and 200 maximum iterations. Additionally, two-way ANOVA was carried out with a significance level of $\alpha = 0.05$ for all tests.

Results

Life cycle inventory

Aboveground biomass (*AGB*) under the 14 production chains ranged from 5.44 to 6.39 Mg_{dm} yr^{-1} ha^{-1} (Fig. 2). Plant productivity with a 7-year rotation cycle

was on average 10.4% higher than with a 3-year rotation cycle ($P = 0.016$). Highest biomass productivities were reached in the production chains with highest fertilization rates (chain 7: 3 yr/200 kgN and chain 14: 7 yr/200 kgN) (Fig. 2). Within the 3-year rotation cycles, the maximum production was reached in the second rotation of the plantations' lifetime, while in the 7-year rotation cycles, it was in the first rotation (Fig. S10). The application of fertilizer after each harvest had no significant influence on the total *AGB* of the poplar SRC; however, it lead to increased soil N_2O emissions and stimulated nitrate leaching, especially in the 3-year rotation cycles (Fig. 3).

Life cycle impact assessment

Effect of rotation cycle length. Our study shows that the *GWP* of the different production chains depended mostly on the length of the rotation cycles and successively on fertilization regimes, as indicated by the first and second principal components of the PCA, respectively (Fig. 4). The dependency of the *GWP* on rotation cycle length was found highly significant ($P < 0.001$). Cases with 7-year rotation cycles resulted in a lower, thus better, *GWP* (on average: 15.6 Mg CO_2-eq. ha^{-1}) than the 3-year cycles (on average: 39.4 Mg

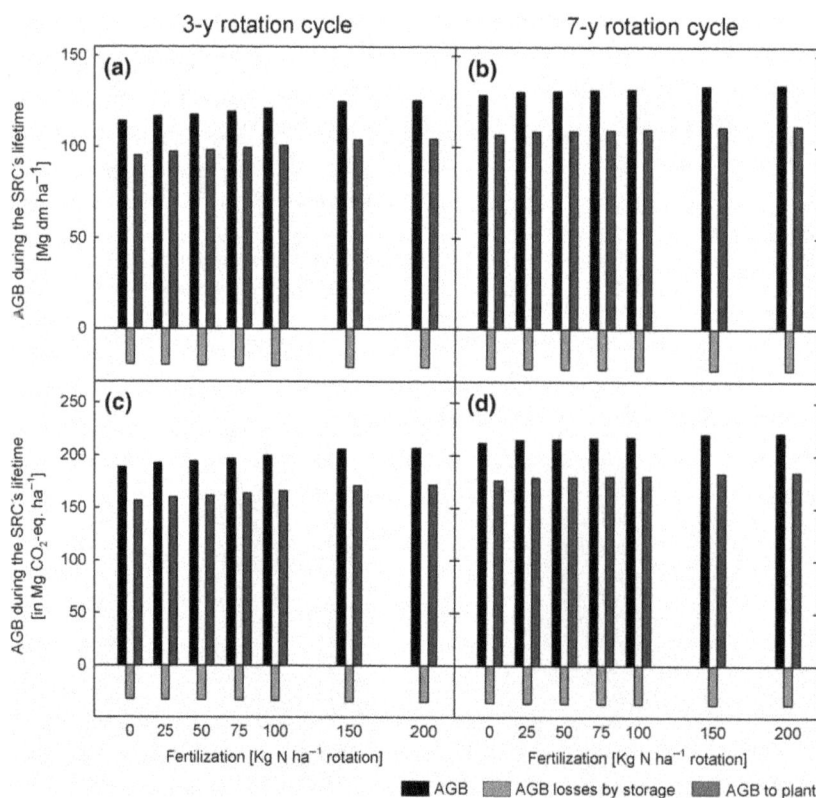

Fig. 2 Production of *aboveground biomass* (*AGB*) during the SRC's lifetime and losses during storage [in (a) and (b) Mg_{dm} ha^{-1} and (c) and (d) Mg CO_2-eq. ha^{-1}]. An overview of the 14 analyzed production chains can be found in Table 1.

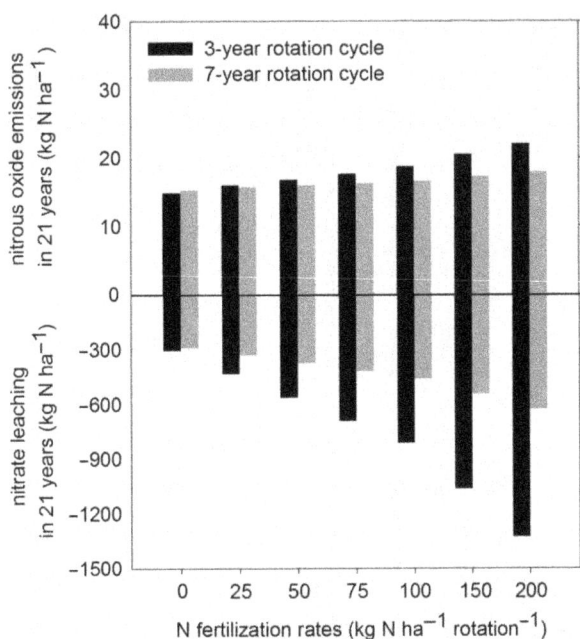

Fig. 3 Results of Life Cycle Inventory – soil N_2O emissions and NO_3 leaching per hectare during the plantations' lifetime, for all 14 production chains. An overview of the 14 analyzed production chains can be found in Table 1.

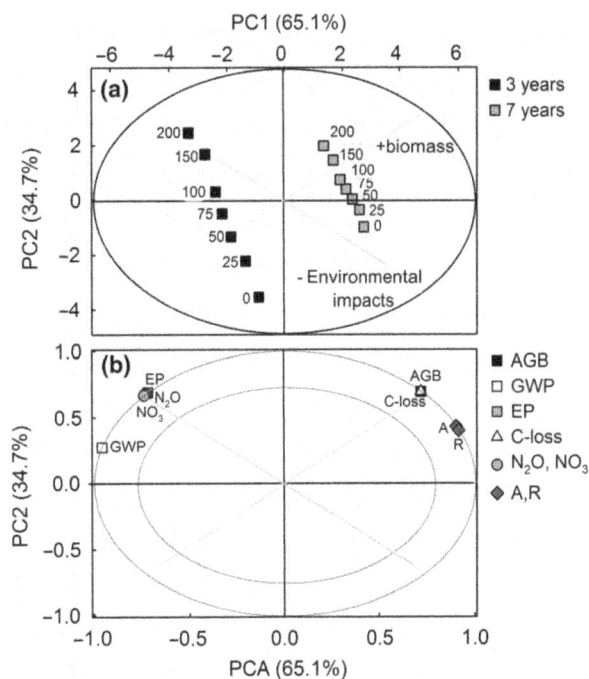

Fig. 4 Results of principal component analysis. Score (a) and correlation loading (b) plots of principal component analysis (PCA) of 7 different N fertilizer treatments (0, 25, 50, 75, 100, 150, 200 kg NH_4NO_3-N per hectare) and two alternative harvesting rotation cycles (no. harvest × years) of (7 × 3, in black) and (3 × 7, in gray). PCA was computed using *aboveground biomass* (*AGB*), *GWP*, *EP*, C loss during storage, net photosynthesis (A), ecosystem respiration (R), N_2O emissions, and NO_3 leaching data per unit ground area (hectare). In plot A, the Hotelling's T^2 ellipse denotes a significance level of $\alpha = 0.05$. In plot B, the loading values are normalized to 1 and the ellipses denote the 100% (outer) and 75% (inner) explained variance. Two gray arrows were added to the plots indicating the dimension related to (i) *AGB* and (ii) EP, NO_2, NO_3, respectively. Model fitness (referring to the first 2 principal components): cross-validated fraction of the total predicted variation $(Q^2) = 98.9\%$; explained total data variation $R^2 = 99.7\%$.

CO_2-eq. ha^{-1}) (Fig. 5). The lowest *GWP* was reached for the 7-year rotation cycle without fertilization (chain 8 (7 yr/0 kgN): 10.6 Mg CO_2-eq. ha^{-1}, Fig. 5), whereas the highest *GWP* corresponded to the 3-year rotation cycle with highest fertilization treatment (chain 7 (3 yr/200 kgN) with 56.5 Mg CO_2-eq. ha^{-1} (Fig. 5).

The *EP* was influenced by the length of the rotation cycles ($P = 0.0066$, Fig. 4). The lowest *EP* was reached with a 7-year rotation cycle and no fertilization treatment (chain 8 (7 yr/0 kgN): 195.6 kg PO_4-eq. ha^{-1}, Fig. 5). The *EP* ranged from 0.15 PO_4-eq. GJ^{-1} (chain 8: 7 yr/0 kgN) to 0.56 kg PO_4-eq. GJ^{-1} (chain 7: 3 yr/200 kg) (Fig. 5).

Effect of fertilization. The *GWP* was positively correlated with the fertilization rates within each rotation cycle length, meaning that the *GWP* increased with increasing fertilization rate. The *EP* showed the same behavior and tended to increase with increasing amount of fertilizer. There was a significant difference between the impacts in the lowest (chain 1: 3 yr/0 kgN & chain 8: 7 yr/0 kgN) and the highest (chain 7: 3 yr/200 kgN & chain 14:7 yr/200 kgN) fertilization treatments ($P = 0.007$).

Environmental impacts with respect to produced amount of aboveground biomass. When considering the amount of produced biomass, the increases in yield-scaled

emissions, that is, the ratios between *AGB* production and *GWP*, were much larger between the 3-year and the 7-year rotation cycles than those obtained by enhancing the fertilization rates from 0 to 200 kg N ha^{-1} rotation^{-1} (Fig. S13). The use of the 7-year rotation cycles decreased yield-scaled emissions by a factor of 2.2 ± 0.1 compared to the 3-year rotation cycles. Furthermore, fertilization increased significantly yield-scaled emissions (Fig. S13), that is, GHG emissions associated with fertilization increased faster as biomass production.

Environmental impacts per process step. Each process step of the production chain contributed differently to the *GWP* (Table 3). Most influencing was *Field-GHG* – as C sink. Therefore, we conducted a contribution analysis

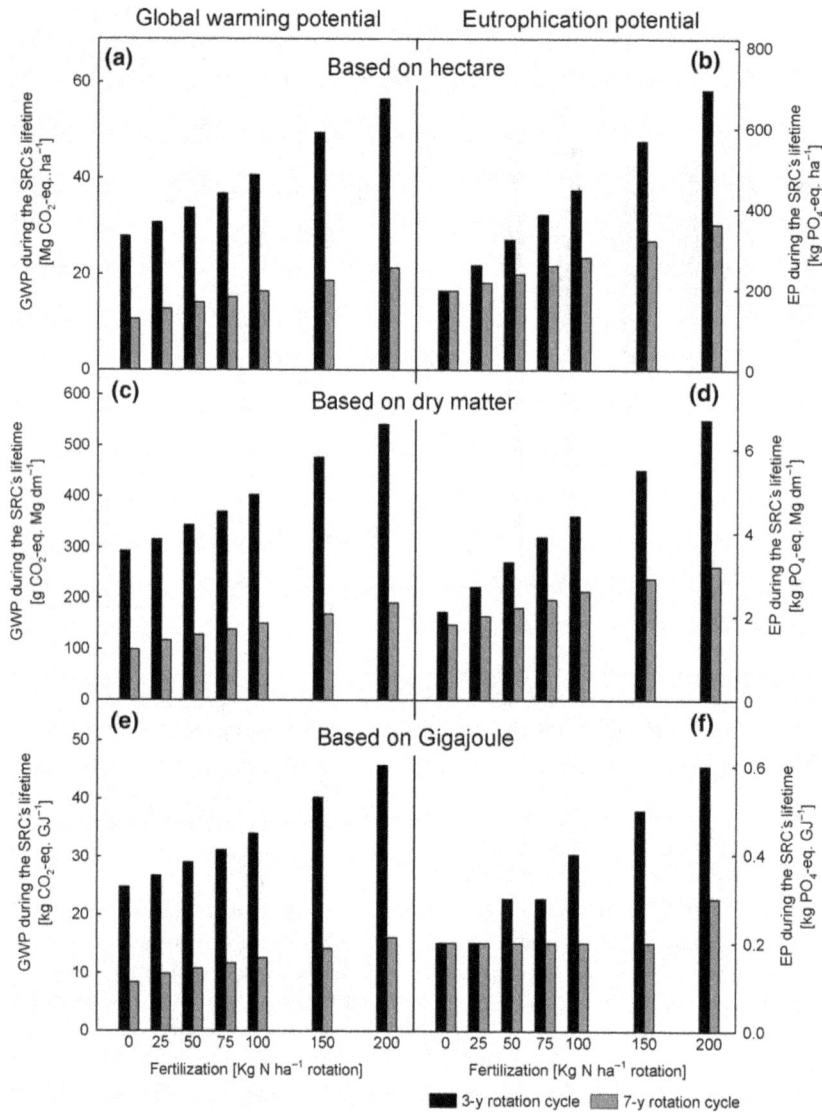

Fig. 5 *Global Warming Potential* and *Eutrophication Potential* for the production of poplar wood chips from SRC in 21 years, shown for all 14 production chains in different functional units. An overview of the 14 analyzed production chains can be found in Table 1.

and highlighted the CO_2 fluxes within *Field-GHG* for the most favorable production chain no. 8 (Fig. S1): Net ecosystem exchange was estimated to be −167.4 kg CO_2-eq. GJ^{-1}, which is derived from simulated ecosystem respiration of +399 kg CO_2-eq. GJ^{-1} and N_2O emissions of +8 kg CO_2-eq. GJ^{-1} (Fig. S1) on the one hand, and photosynthesis of −574 kg CO_2-eq. GJ^{-1} as well as CH_4 deposition of −0.4 kg CO_2-eq. GJ^{-1} on the other. Thus, in contrast to all other process steps, *Field-GHG* is acting as C sink (Fig. S1, Table 3). Table S2 presents more detailed emission data of all production chains for the process step *Field-GHG*.

On the other hand, *Combustion* is the major contributor for increasing the *GWP* ($P < 0.001$) (Table 3) by causing 75–79% of the total C emissions. Another significant

impact on *GWP* is caused by the process step *Storage*, as it is associated with significant C losses (+28 kg CO_2-eq. GJ^{-1}, Table 3). Emissions in *Removal* contributed with 6–33% to the *GWP* (+4.0–5.6 kg CO_2-eq. GJ^{-1}, Table 3). It has to be noted that 87–95% of this C release occurred after the elimination of plant roots from the soil (Table S8).

Among the technological processes, *Transport* caused the highest impact (+2.2 kg CO_2-eq. GJ^{-1}, Table 3). This aspect, however, strongly depended on the transport distance: The longer the way, the stronger the impact. Each additional kilometer (km) of transport with a lorry (20–28 t payload) emits +0.02 kg CO_2-eq. per GJ and km. Finally, the contribution of *Fertilization* to the *GWP* was very variable and depended on the management practice (Table 3). The more fertilizer was applied, the higher was the impact on GWP ha^{-1} – mainly due to upstream processes, in particular the production of fertilizer. Other processes (*Establishment and Maintenance, and Harvesting*) were of negligible magnitude (Table 3).

Due to the use of fuels, machineries, and fertilizer, all process steps contributed to EP (Table S5). In particular, *Field-GHG*, *Removal*, and *Fertilization* were the components causing 73–92% of the potential impacts (Table S5): *Field-GHG* and *Removal* due to nitrate leaching and *Fertilization* mainly due to upstream processes (i.e., fertilizer production). *Combustion* caused 7–25% of the burdens, mainly due to the disposal of rost ash in land farming (33 t yr^{-1}). All other process steps (*Establishment and Maintenance, Harvesting, Transport, and Storage*) were negligible (Table S5).

Carbon sources. The LCA showed that all process steps upstream and downstream of *Field-GHG* released CO_2 to the atmosphere (Fig. 6). By stepwise subtracting the impact of each process from the GWP savings gained in *Field-GHG*, the contribution of each process can be calculated, thereby allowing to assess the importance of each process to the overall GWP of poplar SRC *Field-GHG* reduction. We exemplified this calculation for four

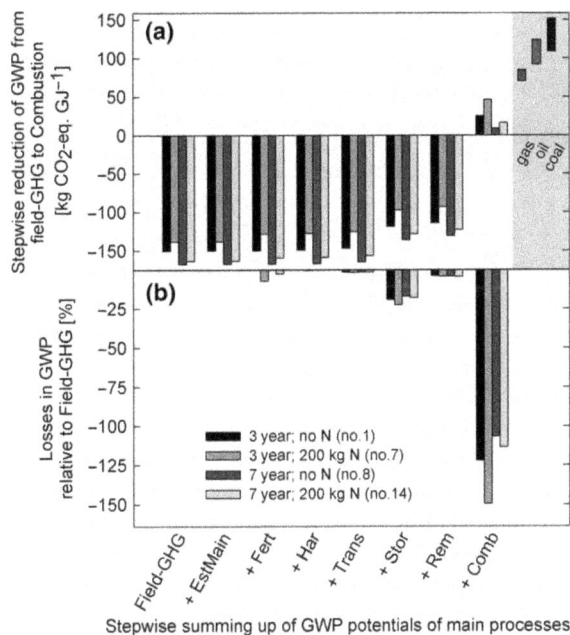

Fig. 6 Stepwise reduction of the beneficial *Global Warming Potential* (*GWP*) of the process biological production by other processes. (a) Summing up of the *GWP* starting with the process *Field-GHG*. On the right-hand side, the ranges of *GWP* from fossil sources (Cherubini *et al.*, 2009; Ecoinvent, 2010) are shown. (b) Relative contribution of each process to the decline in *GWP* saving potentials starting from *Field-GHG*. An overview of the here presented production chains can be found in Table 1.

selected production chains (chain 1: 3 yr/0 kgN, chain 7: 3 yr/200 kgN, chain 8: 7 yr/0 kgN, and chain 14: 7 yr/200 kgN; Fig. 6). In all cases, heat production from poplar SRC finally resulted in a moderate C release varying between 8 and 46 kg CO_2-eq. GJ^{-1} (which equals to 11–57 Mg CO_2-eq. ha^{-1}).

Discussion

Applied tools, data and assumptions

The combination of the *LCA-Umberto* with the process-based ecosystem model *LandscapeDNDC* demonstrated the analytical power of combining the two methodologies for embracing environmental and technological impacts of SRC production systems. In particular, the feature of *Umberto* to include 'own' data as well as data from the database *Ecoinvent* could be conveniently used for the integration of model outputs from *LandscapeDNDC*.

The quality of our comprehensive LCA depends very much on the reliability of the ecosystem simulations, which in the present study were evaluated with a large body of experimental data obtained from own field and laboratory experiments. It is, therefore, to a certain degree, specific for hybrid poplar SRC on marginal land under environmental conditions typical for southwest Germany.

However, our experimental investigations focused on the first 4-year period and included only one transition between rotation cycles. The extrapolation to multiple rotation cycles thus includes uncertainties regarding the long-term soil development and the impact of climatic events that may have not been observed within these four years. Particularly, the effort for removing the tree stumps in the end as well as the impact on soil emissions due to disturbance of the soil structure is prone to possible under- or overestimations. It should be noted that plant growth was well reproduced by the model during the first 2 years of the second rotation (Fig. S10). Likewise, the observed soil N_2O emissions, which are very difficult to be tracked by model predictions, were covered by *LandscapeDNDC* very well with a coefficient of determination of $r^2 = 0.41$ (Fig. S14). Other uncertain assumptions include the regeneration capacity of poplar plants after harvest and the combustion method. For example, the increase in productivity from the first- to the second-rotation cycle might have originated either from an initial lower investment of the plants into roots and soil microorganisms, and faster resprouting from already established root systems, or from unknown factors depending on the site-specific conditions (Hofmann-Schielle *et al.*, 1999; Verlinden *et al.*, 2015). On the other hand, it is not fully clear whether the growth capacity of hybrid poplars can be sustained during up to 7 rotation cycles. At an Italian site, growth of poplars persisted over 12 years and 3 cutting cycles (Nassi o di Nasso et al., 2010), while specific hybrid poplars performed poorly after the fourth rotation on marginal soils in Belgium (Dillen et al., 2013). However, the chosen time period of 21 years seems reasonable. The resprouting ability of poplars is indeed declining with age, but reports indicate that the mortality rate is small after 16 years (at least for some clones) (Dillen et al., 2013), and reports of long-term studies indicate that growth vigor can even increase after 15 years of repeated harvesting. However, poplar SRCs are more profitable when harvested several times without replanting and thus praxis oriented. Additionally, similar studies (e.g., Deckmyn et al., 2004) have chosen comparable time periods (25 years) for growing poplar coppice in a 3-year rotation system, which is in line with the present investigation.

Potential impacts on GWP and EP

The *Global Warming Potentials* (*GWPs*) and *Eutrophication Potentials* (*EPs*) associated with the heat production from poplar SRC on marginal land ranged between 8–46 kg CO_2-eq. GJ^{-1} and 0.15–0.56 kg PO_4-eq. GJ^{-1}, respectively. This span is very large and can be explained by the 14 simulated management scenarios covering fertilization rates varying between 0 and 1.4 t NH_4NO_3 ha^{-1} in 21 years. These values are considerably higher than the results of previous studies (Rödl, 2010; Bacenetti et al., 2012; Fiala & Bacenetti, 2012; González-García et al., 2012a,b; Gabrielle et al., 2013; Miguel et al., 2015). As noticed, studies simultaneously addressing technological, agronomic as well as environmental aspects of SRC production have not been performed so far. Also, some studies use literature data only (e.g., Rugani et al., 2015). For example, González-García et al. (2012a) and Bacenetti et al. (2016) focused only on technological processes when analyzing environmental impacts of woody biofuel production in the Po Valley, Italy. In the case of Bacenetti et al. (2016), the estimated *GWP* was 24.7–49.6 kg CO_2-eq. Mg_{dm}^{-1} compared to 98.9–541.4 kg CO_2-eq. Mg_{dm}^{-1} in our study. Keeping in mind that main C sources as storage for up to several weeks, combustion and long-distance transport processes were not considered by Bacenetti et al. (2016), and the higher *GWP* herein can be explained. Also, inputs varied between the studies, for example, González-García et al. (2012a) assumed a diesel consumption of 92 l ha^{-1} for soil cultivation while it was up to 423 l ha^{-1} in our case (Burger, 2010).

The same is true for *EP*: The resulting *EP* for two management regimes for willow SRC in Sweden (González-García et al., 2012b) was much lower (5.9–159.5 kg PO_4-eq. ha^{-1}) than our results (195.6–694.4 kg PO_4-eq. ha^{-1}). In our case, 92–95% of the emissions occurred in the process step *Field-GHG* due to NO_3 leaching, and another 1–4% resulted from the removal of ashes in the process step *Combustion*. The latter was not considered by González-García et al. (2012a). González-García and colleagues included the leaching of nutrients, using modeled data following the literature recommendations. From their analysis, they concluded that NO_3 leaching is an important component and that environmental assessments would profit from the field measurement and modeling data (e.g., Díaz-Pines et al., 2016). The study by Murphy et al. (2014) evaluated the environmental impacts associated with cultivation, fertilization (max. 800 kg N ha^{-1}), harvest, and transport of willow biomass on *Field-GHG*. They considered the transport process (50 km), however, not the impact of the combustion process. The omission of the combustion process resulted therefore in lower *GWP* values (5.8–11.7 kg CO_2-eq. GJ^{-1}) compared to our study (8.4–45.7 kg CO_2-eq. GJ^{-1}).

In conclusion, the somehow higher *GWP* and *EP* values found herein result mainly by our holistic approach that aimed to address technological, agronomic as well as environmental aspects and, thus, by having different system boundaries compared to other studies and by higher level of details concerning the data input.

Effect of rotation cycle length

The combination of LCA and PCA clearly showed that the main factor controlling the biomass production and the environmental impact was the rotation cycle length. The biomass production from SRC was higher in 7-year rotation cycles compared to the 3-year cycles, conversely to the impacts on *GWP*, which decrease by increasing the rotation cycle length. Also in other studies, longer rotation cycles were related to higher biomass yields (Guidi et al., 2009; Nassi o di Nasso et al., 2010; Bacenetti et al., 2012; Rugani et al., 2015) which corroborate our modeling study. It has to be noted, however, that the initial planting density was equal in all studies although shorter rotation cycles might be associated with higher densities than the longer cycles. The growth potential would probably be reached faster, but the outcome of the simulations also depends on other factors (e.g., N availability). Thus, different plant densities were not considered (c.f. Nassi o di Nasso et al., 2010), as it would lead to decreasing comparativeness and increasing uncertainties (e.g., representation of competition, speed of crown expansion).

The benefit of longer rotation cycles mainly originates from the fact that leaf area index tends to be smaller in the first year of regrowth than in the later stages and that these years are less frequent in the 7-year rotation cycles (DeBell *et al.*, 1996; Fang *et al.*, 1999). Such a development has been reproduced with *LandscapeDNDC* also at the experimental site (R. Grote & K. Block, unpublished data). Coppicing poplars in longer periods are visibly positive not only because of the higher biomass accumulation. Further benefits concern the N cycle: In a 7-year rotation period, N cycling within the system is enhanced due to a larger (average) litter fall and intensified N uptake (due to in average larger requirements) decreasing the N loss. In addition, less N inputs are required due to only 3 fertilization events (instead of 7). Furthermore, fewer harvests lead to less organizational effort for the farmer, and thus, SRC is easier to be adopted. A more extensive management also leads to lower environmental impacts (Fig 5) due to lower fuel consumption in field and transport operations (Tables S2 and S3) and due to a reduced requirement for N input. From our results, we recommend to establish hybrid poplar SRC with longer rotation cycles to minimize the environmental impacts and to maximize the biomass production.

Effect of fertilization treatments

Although less important compared to the rotation cycle length, the present study indicates that the studied fertilization regimes affect the SRC biomass production while negatively impacting the environment. Fertilizers are commonly applied in SRC to improve the plant biomass growth (Rewald *et al.*, 2016). However, generally, the effect of fertilization of hybrid poplars is largely variable reaching from extremely relevant (Luo & Polle, 2009) to minor importance or not detectable at all (e.g., Scholz & Ellerbrock, 2002; Balasus *et al.*, 2012). In the present study, biomass yields responded to the fertilizer N rates very modest, indicating that other parameters were limiting. The biomass growth in the *LandscapeDNDC* simulations is limited by three factors: (i) photosynthesis, (ii) soil water, and (iii) nutrient availability, while the two latter are coupled. As the response to different N fertilization rates is weak, we assume that our system was not nitrogen limited, and therefore, additional N inputs will not pronounce plant growth. This assumption is supported by leaf (around 2.5% N), bark (around 0.5% N), and wood (0.12–0.16% N) total N contents (data not shown), indicating no clear fertilization effects. Only leaves of *cv.* Monviso showed a small increase in leaf total N contents from $2.31 \pm 0.42\%$ (controls) to $2.83 \pm 0.52\%$ (fertilized trees). Additional nitrogen sources are dry deposition, the high soil nutrient pools from the land-use management change, and the mobilization from litter decomposition.

Also, the fertilization effects on growth depend next to the initial N availability on the time course of N

depletion, indicating that the fertilization effect is often only visible in later rotation cycles (Hofmann-Schielle *et al.*, 1999; Jug *et al.*, 1999). Short rotations profit particularly if initial N is low, while otherwise, much of the fertilization gets lost (Balasus *et al.*, 2012), and the effect of additional N input is only visible in later rotation cycles when the soil is already more depleted. Another important reason why the response to N was weak is because we applied the fertilization once per rotation cycle. A yearly application was not considered because farmers aim to minimize the labor input and costs by cultivating extensive SRC. The supply of fertilizer had a strong influence on environmental impacts. In particular, the *EP* increased with increasing application of fertilizer resulting from stimulated nitrate leaching. This has been reflected by the LCA and is well in accordance with other field investigations (e.g., Balasus *et al.*, 2012). In the present study, *EP* ranged from 0.15 to 0.56 kg PO_4-eq. GJ^{-1} (chain 8: 7 yr/0 kgN & chain 7: 3 yr/200 kgN, respectively). An input of 50 kg N ha^{-1} rotation^{-1} led to an increase in *EP* by a factor of 1.2–1.6, and an input of 100 kg N ha^{-1} rotation^{-1} increased *EP* by a factor 1.4–2.3. Also, N_2O emission increased significantly with fertilization, adding another environmental trade-off to the relative small gain in biomass production. The difference between C sequestration and release was highest when the rotation cycle was longer (7 years) and fertilization was omitted (chain 8: 7 yr/0 kgN). According to our results, fertilization cannot be recommended during the first-rotation period of hybrid poplar cultivation and should be considered only in small amounts in later cycles of the plantation's lifetimes.

Environmental impacts per process step

The two most relevant process steps along the production chains are plant growth as such (*Field-GHG*, acting as C sink) and combustion procedures (*Combustion*, acting as C source), the latter because fixed C is released. In this respect, it should be noted that the process step *Combustion* can considerably contribute to the *EP* due to the disposal of rost ash in land farming. As its main component is calcium, it has an eutrophication effect, which, however, could be mitigated when used as limestone.

When excluding *Field-GHG* and *Combustion* from the LCA, it turned out that the *Storage* of wood chips is the main emission source causing 62–78% of the total burden. Nevertheless, considering storage with accompanied drying of wood chips is necessary because small- to medium-sized heating plants usually require wood chips with low water content to increase heat efficiency. Unfortunately, this process also implies a substantial

loss of C to the atmosphere (approx. 17%) and, consequently, a loss in terms of energy efficiency. The measured C loss rate is well in line with previous findings
(e.g., Lenz *et al.*, 2015 (17–22%) or Manzone & Balsari,
2016 (10%)). If the wood chips would not be dried, considerably less energy would be produced, compensating
the gain in C to feed the power plants. However, the
optimum balance between losses and gains is an ongo-
ing discussion. Possible options to decrease losses
include outdoor drying (Lenz *et al.*, 2015), different chip
sizes or pile heights (Jirjis, 2005; Scholz *et al.*, 2005; Pari
et al., 2015), and the application of technological assistant systems such as ventilation.

Among the technological processes, the transport
operation caused the highest environmental impacts. Of
course, this result strongly depends on the transport
distance (here 50 km). However, it is well known that a
regional use of wood chips can be favored and that
either a reduction of WC (Schweier *et al.*, 2016) or a densification process (Adams *et al.*, 2015) before the transport operation would highly reduce the environmental
impacts.

Effect of substitution

To conclude, LCA results show that in all cases, heat
production from hybrid poplar SRC finally resulted in a
moderate C release (8–46 kg CO_2-eq. GJ^{-1}). However,
the use of poplar wood chips for bioenergy production
is still much more favorable compared to heat production from fossil fuels (Fig. 6, Hansen *et al.*, 2013). The
impacts of the most frequently used fossil energy on
GWP (Fig. 6a right bars) vary between 70–85 kg CO_2-
eq. GJ^{-1}_{heat} (natural gas), 90–120 kg CO_2-eq. GJ^{-1}_{heat}
(oil), and 110–150 kg CO_2-eq. GJ^{-1}_{heat} (coal) (Cherubini
et al., 2009; Ecoinvent, 2010). Generation of heat from
the most favorable production chain 8 (7y/0kgN) (*GWP*
of 8.4 kg CO_2-eq. GJ^{-1}_{heat}) substituting the same
amount produced by fossil oil (*GWP* of 90–120 kg CO_2-
eq.GJ^{-1}_{heat}, Fig. 6a) will result in a CO_2-saving potential
of ~97 kg (82–112) CO_2-eq. GJ^{-1}_{heat} (which equals
123 Mg CO_2-eq. ha^{-1}).

In addition, it should be noted that environmental
impacts from poplar SRC cultivation could be easily
offset to assure a carbon-neutral system, for example,
by incorporating 4–8 t C rotation $cycle^{-1}$. Another
option may be the use of belowground biomass for
energy production. So far, we assumed that it was
taken out at the end of the plantations' lifetime, but
simply remained in the field. The additional biomass
(5.3–6.3 Mg_{dm} ha^{-1}) could be either used for heat production in the plant or upgraded to biochar and then
put on the site, the last one favoring the increase in soil
organic C stocks.

Acknowledgment

This work was carried out in the framework of the PROBIOPA
project ('Sustainable production of biomass from poplar short-
rotation coppice on marginal land') funded by the Ministry of
Education and Research (BMBF) in the frame of the program
Bioenergie2021 (Förderzeichen 0315412). Funding from the
'Sustainable Bioeconomy' portfolio program of the Karlsruhe
Institute of Technology is acknowledged. The authors thank
Sylvestre Njakou Djomo (Aarhus University, Denmark) for
advice in LCA-modeling, Alexander Ac (Czech Globe, Czech
Republic) for supporting the argumentation in the introduction
part of an earlier version of the manuscript and Martin Brun-
smeier (University Freiburg, Germany) for supporting the data
collection for combustion in a heating plant.

References

Adams PWR, Shirley JEJ, McManus MC (2015) Comparative cradle-to-gate life cycle
assessment of wood pellet production with torrefaction. *Applied Energy*, **138**, 67–380.

Anderson-Teixeira KJ, Masters MD, Black CK, Zeri M, Hussain MZ, Bernacchi CJ,
De Lucia EH (2013) Altered belowground carbon cycling following land-use
change to perennial bioenergy crops. *Ecosystems*, **16**, 508–520.

Aust C, Schweier J, Brodbeck F, Sauter UH, Becker G, Schnitzler JP (2014) Woody
biomass production potential of short rotation coppices on agricultural land in
Germany. *Global Change Biology Bioenergy*, **6**, 521–533.

Aylott MJ, Casella E, Tubby I, Street NR, Smith P, Taylor G (2008) Yield and spatial
supply of bioenergy poplar and willow short-rotation coppice in the UK. *New
Phytologist*, **178**, 358–370.

Bacenetti J, González-García S, Mena A, Fiala M (2012) Life Cycle Assessment. An
application to poplar for energy cultivated in Italy. *Journal of Agricultural Engineering*, **11**, 72–78.

Bacenetti J, Bergante S, Facciotto G, Fiala M (2016) Woody biofuel production from
short rotation coppice in Italy: environmental-impact assessment of different species and crop Management. *Biomass and Bioenergy*, **94**, 209–219.

Balasus A, Bischoff WA, Schwarz A, Scholz V, Kern J (2012) Nitrogen fluxes during
the initial stage of willows and poplars in short rotation coppices. *Journal of Plant
Nutrition and Soil Science*, **175**, 729–738.

Behnke K, Grote R, Brüggemann N *et al.* (2012) Isoprene emission-free poplars – a
chance to reduce the impact from poplar plantations on the atmosphere. *New
Phytologist*, **194**, 70–82.

Brilli F, Gioli B, Fares S *et al.* (2016) Rapid leaf development drives the seasonal pattern of volatile organic compound (VOC) fluxes in a "coppiced" bioenergy poplar
plantation. *Plant, Cell and Environment*, **39**, 539–555.

Burger F (2010) *Bewirtschaftung und Ökobilanzierung von Kurzumtriebsplantagen* [dis-
sertation]. TU München (DE), [German].

Butterbach-Bahl K, Kiese R (2013) Biofuel production on the margins. *Nature*, **493**,
483–485.

Cherubini F, Bird ND, Cowie A, Jungmeier G, Schlamadinger B, Woess-Gallasch S
(2009) Energy- and greenhouse gas-based LCA of biofuel and bioenergy systems:
Key issues, ranges and recommendations. *Resources, Conservation and Recycling*,
53, 434–447.

Cherubini F, Strømman AH (2011) Life cycle assessment of bioenergy systems: state
of the art and future challenges. *Bioresource Technology*, **102**, 437–451.

DeBell DS, Clendenen GW, Harrington CA, Zasada JC (1996) Tree growth and stand
development in short-rotation *Populus* plantings: 7-year results for two clones at
three spacings. *Biomass and Bioenergy*, **11**, 253–269.

Deckmyn G, Laureysens I, Garcia J, Muys B, Ceulemans R (2004) Poplar growth and
yield in short rotation coppice: model simulations using the process model
SECRETS. *Biomass and Bioenergy*, **26**, 221–227.

Denman K, Brasseur G, Chidthaisong A *et al.* (2007) Couplings between changes in
the climate system and biogeochemistry. In: *Climate Change 2007: The Physical
Science Basis* (eds Solomon S, Qin D, Manning M, Chen Z, Marquis M, Averyt K,
Tignor M, Miller H), pp. 499–587. Cambridge University Press, Cambridge, UK
and New York, NY, USA.

Díaz-Pines E, Molina-Herrera S, Dannenmann M *et al.* (2016) Nitrate leaching and
nitrous oxide emissions diminish with time in a hybrid poplar short-rotation coppice in southern Germany. *Global Change Biology Bioenergy*. in press, doi: 10.1111/
gcbb.12367.

Dillen SY, Djomo SN, Al Afas N, Vanbeveren S, Ceulemans R (2013) Biomass yield and energy balance of a short-rotation poplar coppice with multiple clones on degraded land during 16 years. *Biomass and Bioenergy*, 56, 157–165.

Dirnböck T, Kobler J, Kraus D, Grote R, Kiese R (2016) Impacts of management and climate change on nitrate leaching in a forested karst area. *Journal of Environmental Management*, 165, 243–252.

Djomo SN, El Kasmioui O, Ceulemans R (2010) Energy and greenhouse gas balance of bioenergy production from poplar and willow: a review. *Global Change Biology Bioenergy*, 3, 181–197.

van den Driessche R, Thomas B, Kamelchuk D (2008) Effects of N, NP, and NPKS fertilizers applied to four-year old hybrid poplar plantations. *New Forests*, 35, 221–233.

Ecoinvent (2010) *Swiss Centre for Life Cycle Inventories*. Ecoinvent Centre, Empa, St.Gallen, Switzerland. Available at: http://www.ecoinvent.org/ (accessed 1 October 2014).

Fang S, Xu X, Lu S, Tang L (1999) Growth dynamics and biomass production in short-rotation poplar plantations: 6-year results for three clones at four spacings. *Biomass and Bioenergy*, 17, 415–425.

Fang S, Xue J, Tang L (2007) Biomass production and carbon sequestration potential in poplar plantations with different management patterns. *Journal of Environmental Management*, 85, 672–679.

Fiala M, Bacenetti J (2012) Economic, energetic and environmental impact in short rotation coppice harvesting operations. *Biomass and Bioenergy*, 42, 107–113.

Fortier J, Truax B, Gagnon D, Lambert F (2015) Plastic allometry in coarse root biomass of mature hybrid poplar plantations. *BioEnergy Research*, 8, 1691–1704.

Gabrielle B, Nguyen The N, Maupu P, Vial E (2013) Life cycle assessment of eucalyptus short rotation coppices for bioenergy production in southern France. *Global Change Biology Bioenergy*, 5, 30–42.

Gasol CM, Gabarrell X, Anton A, Rigola M, Carrasco J, Ciria P, Rieradevall J (2009) LCA of poplar bioenergy system compared with *Brassica carinata* energy crop and natural gas in regional scenario. *Biomass and Bioenergy*, 33, 119–129.

Ghirardo A, Sørensen HA, Petersen M, Jacobsen S, Søndergaard I (2005) Early prediction of wheat quality: analysis during grain development using mass spectrometry and multivariate data analysis. *Rapid Communications in Mass Spectrometry*, 19, 525–532.

Ghirardo A, Koch K, Taipale R, Zimmer I, Schnitzler J-P, Rinne J (2010) Determination of *de novo* and pool emissions of terpenes from four common boreal/alpine trees by $^{13}CO_2$ labelling and PTR-MS analysis. *Plant, Cell and Environment*, 33, 781–792.

Ghirardo A, Gutknecht J, Zimmer I, Brüggemann N, Schnitzler J-P (2011) Biogenic volatile organic compound and respiratory CO_2 emissions after ^{13}C-labeling: online tracing of C translocation dynamics in poplar plants. *PLoS ONE*, 6, e17393.

Ghirardo A, Wright LP, Bi Z et al. (2014) Metabolic flux analysis of plastidic isoprenoid biosynthesis in poplar leaves emitting and nonemitting isoprene. *Plant Physiology*, 165, 37–51.

González-García S, Bacenetti J, Murphy R, Fiala M (2012a) Present and future environmental impact of poplar cultivation in Po valley (Italy) under different crop management systems. *Journal of Cleaner Production*, 26, 56–66.

González-García S, Mola-Yudego B, Dimitrou I, Aronsson P, Murphy R (2012b) Environmental assessment of energy production based on long term commercial willow plantations in Sweden. *Science of the Total Environment*, 421–422, 201–219.

Gottlieb DM, Schultz J, Bruun SW, Jacobsen S, Søndergaard I (2004) Multivariate approaches in plant science. *Phytochemistry*, 65, 1531–1548.

Grote R, Lehmann E, Bruemmer C, Brueggemann N, Szarzynski J, Kunstmann H (2009) Modelling and observation of biosphere-atmosphere interactions in natural savannah in Burkina Faso, West Africa. *Physics and Chemistry of the Earth*, 34, 251–260.

Grote R, Kiese R, Gruenwald T, Ourcival J-M, Granier A (2011a) Modelling forest carbon balances considering tree mortality and removal. *Agricultural and Forest Meteorology*, 151, 179–190.

Grote R, Korhonen J, Mammarella I (2011b) Challenges for evaluating process-based models of gas exchange at forest sites with fetches of various species. *Forest Systems*, 20, 389–406.

Guidi W, Tozzini C, Bonari E (2009) Estimation of chemical traits in poplar short rotation coppice at stand level. *Biomass and Bioenergy*, 33, 1703–1709.

Haas E, Klatt S, Fröhlich A et al. (2013) LandscapeDNDC: a process model for simulation of biosphere–atmosphere–hydrosphere exchange processes at site and regional scale. *Landscape Ecology*, 28, 615–636.

Haberl H, Erb KH, Krausmann F et al. (2007) Quantifying and mapping the human appropriation of net primary production in Earth's terrestrial ecosystems. *Proceedings of the National Academy of Sciences of the United States of America*, 104, 12942–12947.

Hansen A, Meyer-Aurich A, Prochnow A (2013) Greenhouse gas mitigation potential of a second generation energy production system from short rotation poplar in Eastern Germany and its accompanied uncertainties. *Biomass and Bioenergy*, 56, 104–115.

Hartmann H (2009) Grundlagen der thermo-chemischen Umwandlung biogener Festbrennstoffe. In: *Energie aus Biomasse - Grundlagen, Techniken und Verfahren* (ed. Kaltschmitt M et al.) (Hrsg.), pp. 333–374. Springer Verlag, Berlin.

Hellebrand HJ, Scholz V, Kern J (2008) Fertiliser induced nitrous oxide emissions during energy crop cultivation on loamy sand soils. *Atmospheric Environment*, 42, 8403–8411.

Heller M, Keoleian G, Volk T (2003) Life cycle assessment of a willow bioenergy cropping system. *Biomass and Bioenergy*, 25, 147–165.

Hofmann-Schielle C, Jug A, Makeschin F, Rehfuess KE (1999) Short-rotation plantations of balsam poplars, aspen and willows on former arable land in the Federal Republic of Germany. I. Site-growth relationships. *Forest Ecology and Management*, 121, 41–55.

Holst J, Grote R, Offermann C, Ferrio JP, Gessler A, Mayer H, Rennenberg H (2010) Water fluxes within beech stands in complex terrain. *International Journal of Biometeorology*, 54, 23–36.

IFU (2011) *Umberto 5.6*. Institut für Umweltinformatik, Hamburg, Germany.

Jirjis R (2005) Effects of particle size and pile height on storage and fuel quality of comminuted *Salix viminalis*. *Biomass and Bioenergy*, 28, 193–201.

Jug A, Hofmann-Schielle C, Makeschin F, Rehfuess KE (1999) Short-rotation plantations of balsam poplars, aspen and willows on former arable land in the Federal Republic of Germany. II. Nutritional status and bioelement export by harvested shoot axes. *Forest Ecology and Management*, 121, 67–83.

Kavdir Y, Hellebrand HJ, Kern J (2008) Seasonal variations of nitrous oxide emission in relation to nitrogen fertilization and energy crop types in sandy soil. *Soil and Tillage Research*, 98, 175–186.

Kern J, Hellebrand HJ, Scholz V, Linke B (2010) Assessment of nitrogen fertilization for the CO_2 balance during the production of poplar and rye. *Renewable and Sustainable Energy Reviews*, 14, 1453–1460.

Kim Y, Berger S, Kettering J, Tenhunen J, Haas E, Kiese R (2014) Simulation of N_2O emissions and nitrate leaching from plastic mulch radish cultivation with LandscapeDNDC. *Ecological Research*, 29, 441–454.

Kim Y, Seo Y, Kraus D, Klatt S, Haas E, Tenhunen J, Kiese R (2015) Estimation and mitigation of N_2O emission and nitrate leaching from intensive crop cultivation in the Haean catchment, South Korea. *Science of the Total Environment*, 529, 40–53.

Kraus D, Weller S, Klatt S, Haas E, Wassmann R, Kiese R, Butterbach-Bahl K (2015) A new LandscapeDNDC biogeochemical module to predict CH_4 and N_2O emissions from lowland rice and upland cropping systems. *Plant and Soil*, 386, 125–149.

Lenz H, Idler C, Hartung E, Pecenka R (2015) Open-air storage of fine and coarse wood chips of poplar from short rotation coppice in covered piles. *Biomass and Bioenergy*, 83, 269–277.

Li C, Frolking S, Frolking TA (1992) A model of nitrous oxide evolution from soil driven by rainfall events: 1. Model structure and Sensitivity. *Journal of Geophysical Research*, 97, 9759–9776.

Li C, Aber J, Stange F, Butterbach-Bahl K, Papen H (2000) A process-oriented model of N_2O and NO emissions from forest soils: 1. Model development. *Journal of Geophysical Research: Atmospheres*, 105, 4369–4384.

Luo Z, Polle A (2009) Wood composition and energy content in a poplar short rotation plantation on fertilized agricultural land in a future CO_2 atmosphere. *Global Change Biology*, 15, 38–47.

Manzone M, Balsari P (2016) Poplar woodchip storage in small and medium piles with different forms, densities and volumes. *Biomass and Bioenergy*, 87, 162–168.

Manzone M, Bergante S, Facciotto G (2014) Energy and economic evaluation of a poplar plantation for woodchips production in Italy. *Biomass and Bioenergy*, 60, 164–170.

Martens H, Martens M (2001) *Multivariate Analysis of Quality – An Introduction*. John Wiley & Sons Ltd, Chichester.

McBride AC, Dale VH, Baskaran LM et al. (2011) Indicators to support environmental sustainability of bioenergy systems. *Ecological Indicators*, 11, 1277–1289.

Miguel GS, Corona B, Ruiz D et al. (2015) Environmental, energy and economic analysis of a biomass supply chain based on a poplar short rotation coppice in Spain. *Journal of Cleaner Production*, 94, 93–101.

Molina-Herrera S, Grote R, Santabárbara-Ruiz I et al. (2015) Simulation of CO_2 fluxes in European forest ecosystems with the coupled soil-vegetation process model "LandscapeDNDC". *Forests*, 6, 1779–1809.

Molina-Herrera S, Haas E, Klatt S et al. (2016) A modeling study on mitigation of N_2O emissions and NO_3 leaching at different agricultural sites across Europe using LandscapeDNDC. *Science of the Total Environment*, 553, 128–140.

Murphy F, Devlin G, McDonnel K (2014) Energy requirements and environmental impacts associated with the production of short rotation willow (*Salix* sp.) chip in Ireland. *Global Change Biology Bioenergy*, 6, 727–739.

Nassi o di Nasso N, Guidi W, Ragaglini G, Tozzini C, Bonari E (2010) Biomass production and energy balance of a 12-year-old short-rotation coppice poplar stand under different cutting cycles. *Global Change Biology Bioenergy*, **2**, 89–97.

Nemecek T, Kägi T (2007) *Life Cycle Inventories of Agricultural Production Systems*. Final report Ecoinvent v2.0 No.15. Swiss Centre for Life Cycle Inventories, Dübendorf, CH.

Pari L, Brambilla M, Bisaglia C, Del Giudice A, Croce S, Salerno M, Gallucci F (2015) Poplar wood chip storage: effect of particle size and breathable covering on drying dynamics and biofuel quality. *Biomass and Bioenergy*, **81**, 282–287.

Pecenka R, Balasus A, Scholz V, Kern J, Lenz H (2013) Long term yields and gas emissions from poplar and willows grown on agricultural land in dependence to nitrogen. *Agricultural Engineering*, **45**, 27–37.

Quartucci F, Schweier J, Jaeger D (2015) Environmental analysis of Eucalyptus timber production from short rotation forestry in Brazil. *International Journal of Forest Engineering*, **26**, 225–239.

Ramanathan V, Feng Y (2008) On avoiding dangerous anthropogenic interference with the climate system: Formidable challenges ahead. *Proceedings of the National Academy of Sciences of the United States of America*, **105**, 14245–14250.

Rewald B, Kunze ME, Godbold DL (2016) NH_4:NO_3 nutrition influence on biomass productivity and root respiration of poplar and willow clones. *Global Change Biology Bioenergy*, **8**, 51–58.

Rödl A (2010) Production and energetic utilization of wood from short rotation coppice - a life cycle assessment. *International Journal of Life Cycle Assessment*, **15**, 567–578.

Rösch C, Aust C, Jorissen J (2013) Envisioning the sustainability of the production of short rotation coppice on grassland. *Energy, Sustainability and Society*, **3**, 7.

Rugani B, Golkowska K, Vázquez-Rowe I, Koster D, Benetto E, Verdonckt P (2015) Simulation of environmental impact scores within the life cycle of mixed wood chips from alternative short rotation coppice systems in Flanders (Belgium). *Applied Energy*, **156**, 449–464.

Sabbatini S, Arriga N, Bertolini T *et al.* (2016) Greenhouse gas balance of cropland conversion to bioenergy poplar short-rotation coppice. *Biogeosciences*, **13**, 95–113.

Schaap M, Wichink Kruit R, Kranenburg R, Segers A, Builtjes P, Banzhaf S, Scheuschner T (2015) *Atmospheric Deposition to German Natural and Seminatural Ecosystems During 2009*. In: Report to PINETI II Project (Project N° 371263240-1). Umweltbundesamt, Dessau-Roßlau, Germany.

Schnitzler JP, Becker G, Butterbach-Bahl K, Brodbeck F, Palme K, Rennenberg H (2014) Verbundprojekt: BioEnergie 2021: Nachhaltige PROduktion von BIOmasse mit Kurzumtriebsplantagen der PAppel auf Marginalstandorten (PRO-BIOPA) [in German]. Project report for the Federal Ministry of Education and Research (BMBF), grant Number (0315412), 75 pp.

Scholz V, Ellerbrock R (2002) The growth productivity, and environmental impact of the cultivation of energy crops on sandy soil in Germany. *Biomass and Bioenergy*, **23**, 81–92.

Scholz V, Idler C, Daries W, Egert J (2005) Schimmelpilzentwicklung und Verluste bei der Lagerung von Holzhackschnitzeln (Development of mould and losses during storage of wood chips). *Holz als Roh- und Werkstoff*, **63**, 449–455. [in German].

Schweier J (2013) *Production From Energy Wood From Short Rotation Coppice on Agricultural Marginal Land in South-West Germany - Environmental and Economic Assessment of Alternative Supply Concepts With Particular Regard to Different Harvesting Systems*. Publisher Dr. Hut, München. [in German].

Schweier J, Becker G (2012) New Holland forage harvester's productivity in short rotation coppice - Evaluation of field studies from a German perspective. *International Journal of Forest Engineering*, **23**, 82–88.

Schweier J, Becker G, Schnitzler JP (2016) Life Cycle Analysis of the technological production of wood chips from poplar short rotation coppice plantations on marginal land in Germany. *Biomass and Bioenergy*, **85**, 235–242.

Stange F, Butterbach-Bahl K, Papen H, Zechmeister-Boltenstern S, Li C, Aber J (2000) A process-oriented model of N_2O and NO emissions from forest soils: 2. Sensitivity analysis and validation. *Journal of Geophysical Research: Atmospheres*, **105**, 4385–4398.

Strömgren M, Mjöfors K, Holmström B, Grelle A (2012) Soil CO_2 flux during the first years after stump harvesting in two Swedish forests. *Silva Fennica*, **46**, 67–79.

Styles D, Jones M (2007) Energy crops in Ireland: quantifying the potential life-cycle greenhouse gas reductions of energy-crop electricity. *Biomass and Bioenergy*, **31**, 759–772.

UNFCCC (2015) *Adoption of the Paris Agreement*. Proposal by the President. United Nations Framework Convention on Climate Change, Conference of the Parties. FCCC/CP/2015/L.9/Rev.1. United Nations Office, Geneva, Switzerland.

Vanzo E, Jud W, Li Z *et al.* (2015) Facing the future: effects of short-term climate extremes on Isoprene-Emitting and Non-emitting Poplar. *Plant Physiology*, **169**, 560–575.

Verlinden MS, Broeckx LS, Ceulemans R (2015) First vs. second rotation of a poplar short rotation coppice: Above-ground biomass productivity and shoot dynamics. *Biomass and Bioenergy*, **73**, 174–185.

Walter K, Don A, Flessa H (2015) Net N_2O and CH_4 soil fluxes of annual and perennial bioenergy crops in two central German regions. *Biomass and Bioenergy*, **81**, 556–567.

Werner C, Haas E, Grote R, Gauder M, Graeff-Hönninger S, Claupein W, Butterbach-Bahl K (2012) Biomass production potential from *Populus* short rotation systems in Romania. *Global Change Biology Bioenergy*, **4**, 642–653.

Zhang W, Liu C, Zheng X *et al.* (2015) Comparison of the DNDC, LandscapeDNDC and IAP-N-GAS models for simulating nitrous oxide and nitric oxide emissions from the winter wheat–summer maize rotation system. *Agricultural Systems*, **140**, 1–10.

Zona D, Janssens IA, Aubinet M, Gioli B, Vicca S, Fichot R, Ceulemans R (2013a) Fluxes of the greenhouse gases (CO_2, CH_4 and N_2O) above a short-rotation poplar plantation after conversion from agricultural land. *Agricultural and Forest Meteorology*, **169**, 100–110.

Zona D, Janssens IA, Gioli B, Jungkunst HF, Serrano MC, Ceulemans R (2013b) N_2O fluxes of a bio-energy poplar plantation during a two years rotation period. *GCB Bioenergy*, **5**, 536–547.

3

Estimating product and energy substitution benefits in national-scale mitigation analyses for Canada

CAROLYN SMYTH[1], GREG RAMPLEY[2], TONY C. LEMPRIÈRE[2], OLAF SCHWAB[2] and WERNER A. KURZ[1]

[1]*Natural Resources Canada, Canadian Forest Service, 506 Burnside Road West, Victoria, BC V8Z 1M5, Canada,* [2]*Natural Resources Canada, Canadian Forest Service, 580 Booth Street, Ottawa, ON K1A 0E4, Canada*

Abstract

The potential of forests and the forest sector to mitigate greenhouse gas (GHG) emissions is widely recognized, but challenging to quantify at a national scale. Mitigation benefits through the use of forest products are affected by product life cycles, which determine the duration of carbon storage in wood products and substitution benefits where emissions are avoided using wood products instead of other emissions-intensive building products and energy fuels. Here we determined displacement factors for wood substitution in the built environment and bioenergy at the national level in Canada. For solid wood products, we compiled a basket of end-use products and determined the reduction in emissions for two functionally equivalent products: a more wood-intensive product vs. a less wood-intensive one. Avoided emissions for end-use products basket were weighted by Canadian consumption statistics to reflect national wood uses, and avoided emissions were further partitioned into displacement factors for sawnwood and panels. We also examined two bioenergy feedstock scenarios (*constant supply* and *constrained supply*) to estimate displacement factors for bioenergy using an optimized selection of bioenergy facilities which maximized avoided emissions from fossil fuels. Results demonstrated that the average displacement factors were found to be similar: product displacement factors were 0.54 tC displaced per tC of used for sawnwood and 0.45 tC tC^{-1} for panels; energy displacement factors for the two feedstock scenarios were 0.47 tC tC^{-1} for the *constant supply* and 0.89 tC tC^{-1} for the *constrained supply*. However, there was a wide range of substitution impacts. The greatest avoided emissions occurred when wood was substituted for steel and concrete in buildings, and when bioenergy from heat facilities and/or combined heat and power facilities was substituted for energy from high-emissions fossil fuels. We conclude that (1) national-level substitution benefits need to be considered within a systems perspective on climate change mitigation to avoid the development of policies that deliver no net benefits to the atmosphere, (2) the use of long-lived wood products in buildings to displace steel and concrete reduces GHG emissions, (3) the greatest bioenergy substitution benefits are achieved using a mix of facility types and capacities to displace emissions-intensive fossil fuels.

Keywords: Canada's managed forest, displacement factor, forest products, GHG emissions, local bioenergy, substitution impacts

Introduction

Forest-related carbon (C) mitigation strategies offer important and viable pathways towards climate stabilization through increased use of harvested wood products (HWPs) that store C and avoid the consumption of emissions-intensive materials such as concrete and steel, and avoid emissions from burning fossil fuels for electricity or heat production (Pacala & Socolow, 2004; Böttcher *et al.*, 2008; Sathre & O'Connor, 2010; Werner *et al.*, 2010; Lundmark *et al.*, 2014; Smyth *et al.*, 2014). The potential greenhouse gas (GHG) emission reductions that can be achieved through substitution of wood products for other products and fossil fuels need to be quantified because substitution impacts are part of a larger systems approach which includes changes in forest C, HWPs tracking and substitution benefits (Smyth *et al.*, 2014). Use of a systems perspective highlights trade-offs between activities aimed at increasing carbon storage in the ecosystem, increasing carbon storage in HWPs or increasing the substitution benefits of using wood in place of fossil fuels or more emissions-intensive products (Lemprière *et al.*, 2013).

Displacement factors are used to describe the substitution benefit in mitigation strategies when wood is used instead of some other material (Schlamadinger & Marland, 1996). For wood products in the built

Correspondence: Carolyn Smyth
e-mail: Carolyn.Smyth@canada.ca

environment, displacement factors are calculated from an end-use product, which typically includes building components or a complete building (e.g. Lippke *et al.*, 2004). However national-level mitigation strategies require a broader scope and need to derive a displacement factor for primary wood products (e.g. sawnwood and panels) based on a range of end-use products (e.g. homes, manufacturing, furniture), but displacement factors using this broader scope have not been estimated for Canada.

Our first objective in this study was to develop a methodology for estimating displacement factors for primary wood products. Many different products can provide the same service, and the competition of wood products with other types of products creates a number of potential substitution effects involving the forest products industry. For example, consider a single-family home, and comparison of emissions for two functionally equivalent buildings: a more wood-intensive building would use wood-framing, and a less wood-intensive building would use concrete-framing (Gustavsson *et al.*, 2006). In recent decades, the substitution impacts for housing construction has been especially well studied, but with varying results depending on the assumptions on forest C, and end-of-life options (Upton *et al.*, 2008). Gustavsson *et al.* (2006) found that the production of wood-framed buildings in Scandinavian countries emits less than the production of functionally equal concrete-framed buildings. A meta-analysis of 21 studies by Sathre & O'Connor (2010) calculated displacement factors ranged from a low of −2.3 to a high of 15.0 (tonnes of carbon of emission reduction per tonne of carbon used in wood product) with an average value of 2.1. The variability in displacement factors occurred because they considered a variety of end-use products and a variety of system definitions. Nonetheless, their average value of 2.1 has been used in other studies (e.g. Malmsheimer *et al.*, 2011; Macintosh *et al.*, 2015), or 1.1 if the biogenic emissions were removed (Keith *et al.*, 2015), or a range of displacement factors has been assumed (Hennigar *et al.*, 2008; Soimakallio *et al.*, 2016). The product displacement factor was estimated to be 1.5 by Knauf *et al.* (2015) for Germany, and their study included 16 estimates of displacement factors, which were volume weighted based on a material flow analysis, and a single substitution factor was obtained. Here we use a similar method to estimate displacement factors for sawnwood and panels for Canada by comparing emissions from functionally equivalent products that are weighted based on the national statistics of the broad uses of wood in Canada.

For substitution benefits from bioenergy products, a number of studies have examined GHG reductions and have found that impacts depend on the feedstock source, conversion efficiency and displaced fossil fuel characteristics (e.g. Lemprière *et al.*, 2013). Bioenergy substitution impacts have been assessed for specific fuel types separately (e.g. Schlamadinger & Marland, 1996; Guest *et al.*, 2012; Zanchi *et al.*, 2012) and for specific regions (Ralevic *et al.*, 2010; McKechnie *et al.*, 2011; Ter-Mikaelian *et al.*, 2011) but there are few national-level studies of fossil fuel displacement (Werner *et al.*, 2010; Whittaker *et al.*, 2011; Lundmark *et al.*, 2014; Smyth *et al.*, 2014), and, to date, no national studies have considered regional fossil fuel and feedstock availability together with optimized choices about bioenergy facilities to maximize substitution benefits. In Canada, substantial regional heterogeneity exists in the feedstock supply and energy demand, and the emissions intensity of future electricity production varies markedly between jurisdictions. Given the substantial variation in energy demand, fibre availability and fossil fuel use, we anticipate large variations in the regional displacement factor.

The second objective of this study was to estimate regional bioenergy displacement factors for local heat and electricity production. Strategic and operational-level decision-making requires consideration of relevant factors that can vary substantially across the country, meaning that decision-making about the objectives and feasibility of forest-based bioenergy must take into account local or regional conditions. In this study, we employed an optimization technique to maximize avoided emissions by selecting (from nine candidate bioenergy facilities) the type, size and number of bioenergy facilities that would displace the highest-emitting fossil fuels with a given supply of harvest residues as feedstock.

In the following section, we introduce the methodologies that were applied consistently across the country to identify and examine the differences generated by local circumstances. These results will be useful for policy-makers considering mitigation portfolios, both at the national level and at the regional level.

Materials and methods

Analytical framework

In this study, product displacement factor calculations included emissions associated with extraction, transportation of raw materials and manufacturing. We assumed that the emissions associated with transporting the finished products to the consumer were the same for wood and nonwood products and did not therefore include these estimates.

For energy displacement factors, we constrained bioenergy production to be local, within a forest management unit (FMU), of which we include 634 in Canada's managed forest that are similar to the spatial analysis units used by Stinson *et al.* (2011, Fig. 1). Conversion efficiencies for nine selected

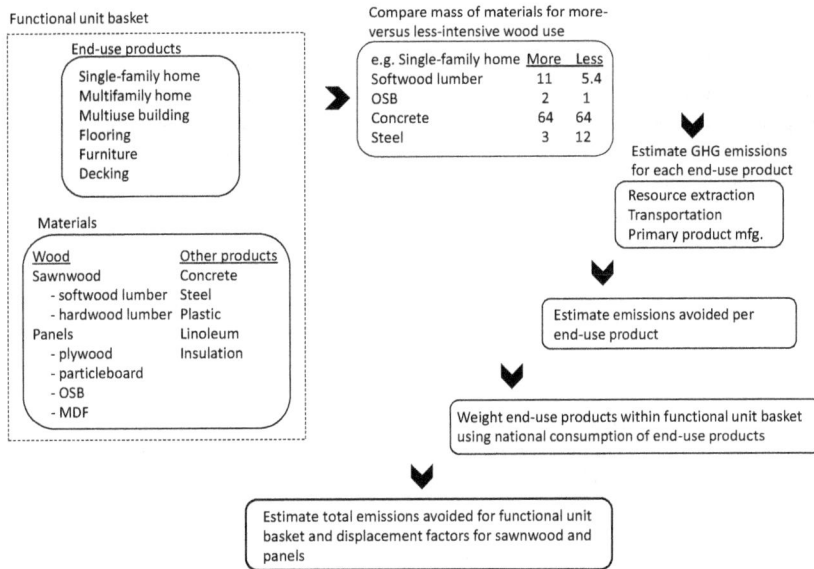

Fig. 1 Overview of the method for calculating wood product displacement factors.

bioenergy facilities were used to estimate the amount of heat and electricity that could be produced, with assumed complete combustion. We did not include processing emissions associated with grinding and loading of the extracted harvest residues or transportation emissions because we assumed these emissions to be minimal relative to the combustion emissions (Jones *et al.*, 2010). Avoided fossil fuel emissions were based on published emissions intensities that included extraction, transportation of raw materials and conversion to heat or electricity.

Greenhouse gas emissions and removals associated with forest ecosystem C dynamics, emissions from instant oxidation of bioenergy and release of C from the processing of HWPs and from postconsumer emissions were not included in this analysis because they were addressed in other system components (Smyth *et al.*, 2016). Displacement factors estimated here are used to estimate the avoided emissions per unit of wood used. They can only be used within the larger system framework and must not be used in isolation as such use would fail to account the impacts of harvesting on ecosystem C balances and the emissions from HWPs.

Product substitution

Sawnwood and panels are traditionally used to manufacture a variety of end-use products, each with a different GHG implication. Displacement factors for sawnwood and panels were determined by considering a basket of end-use products. We have defined each end-use product as a functional unit, which describes the service delivered by the product (e.g. single-family home). For each functional unit, a comparison of GHG emissions based on the construction materials for two functionally equivalent products was estimated. Each end-use product was then weighted based on national consumption statistics to ensure that the basket reflected national usage. Finally, displacement factors were estimated for

sawnwood and panels, the main wood commodities contained in the end-use products. The overall process is described in the schematic in Fig. 1.

The basket of end-use products included buildings (single-family home, multifamily home, six-storey multiuse building), residential flooring, furniture and decking. The amount of wood and other materials needed for each end-use product was estimated for a more wood-intensive product, relative to a less wood-intensive product. Operational emissions for buildings can account for the majority of GHG emissions (Sharma *et al.*, 2011), but estimating these emissions was beyond the scope of this study, and we assumed that the end-use products for both scenarios would have the same operational functional life and operational emissions and that differences in emissions between the scenarios were solely the result of material selection and construction. We further assumed that all solid wood products had the same specific gravity.

Materials' emissions factors were taken from published values for each end-use product (Schmidt *et al.*, 2004, Marceau *et al.*, 2007; Athena Sustainable Materials Institute, 2008a,b, 2009a,b,c; National Renewable Energy Laboratory, 2008, Bala *et al.*, 2010). We preferentially selected material lists and emissions factors for end-use products manufactured within North America, where available, to ensure consistency between end-use products and national consumption statistics.

For each end-use product, f, the net emissions avoided, N_f, for a more wood-intensive product relative to a less wood-intensive product was estimated as:

$$N_f = \sum_{i=1}^{n} \Delta m_i (x + t + s)_i, \quad (1)$$

where n is the total number of materials in each end-use product; Δm is the difference in mass of a material for (more wood-intensive minus less wood-intensive) for the two comparative products; and x, t and s are emissions for resource extraction, transportation and primary product manufacturing, respectively.

A weighting factor, W_f, was applied to weight each end-use product's avoided emissions within the functional unit basket. It was estimated from the proportion of wood consumed at the national level for each broad wood use (buildings, residential improvement, furniture and manufacturing) divided by the proportion of wood within the functional unit basket. The weighting factor for each end-use product was defined as:

$$W_f = \frac{A_f}{m_{Mf} / \sum_{f=1}^{6} m_{Mf}},\qquad(2)$$

where A_f is the percentage of the 2005–2010 average annual consumption of primary solid wood products in each end-use product as reported by the Forest Economic Advisors' statistics for Canada (FEA, 2011), and m_{Mf} is the mass (m) of wood material in the more (M) wood-intensive end-use products.

The weighted avoided emissions were then estimated as:

$$N_{fD} = N_f W_f,\qquad(3)$$

which represented the emissions avoided for each weighted end-use product within the functional unit basket weighted by national consumption levels.

Displacement factors were estimated for each primary solid wood product, p (representing sawnwood and panels) based on the percentage of primary products reported by FEA for Canada within each of the six end-use products (K_{fp}) in the functional unit basket:

$$DF_p = \frac{\sum_{f=1}^{6} N_{fD} K_{fp}}{D_p},\qquad(4)$$

$$D_p = \sum_{f=1}^{6} \Delta M_{fp} W_f,\qquad(5)$$

where the total avoided emissions per basket for each functional unit component were converted to total avoided emissions for sawnwood and panels by weighting N_{fD} using the shares of sawnwood and panels in each functional unit component, respectively. Then, the weighted avoided emissions for sawnwood and panels were divided by the incremental increase in wood mass, D_p, for those two products, respectively, to calculate the displacement factors. The final values were converted from tCO_2 t^{-1} wood to tC tC^{-1}. We did not partition the displacement factor for panels into nonstructural and structural components because of the HWP commodity tracking framework uses an aggregated half-life for structural and nonstructural panels (Smyth et al., 2014).

Energy substitution

Substitution emissions from using bioenergy (bioenergy scenario) in place of fossil fuels (business-as-usual scenario) were estimated by comparing fossil fuel emissions to bioenergy emissions for combinations of nine bioenergy facilities, based on an assumption that bioenergy would substitute for the most emission-intensive fuel source first, and then proceed to successively less emission-intensive fuels. An overview of the process

is shown in the schematic in Fig. 2. A linear programming (LP) model was created to determine the optimal configuration (type, size and number) of regional bioenergy facilities that maximized avoided emissions. Regions were defined based on 502 FMUs in which harvesting and silvicultural activities are undertaken (Stinson et al., 2011) and for which mitigation estimates are projected (Smyth et al., 2016).

Energy demand for heat and electricity combined within each region was estimated from each jurisdiction's per capita energy (heat and electricity) usage (National Energy Board, 2013) multiplied by the region's population. We assumed that one-third of the energy usage was for electricity and two-thirds for heat (National Energy Board, 2010). Population within a region was estimated from census data (Statistics Canada, 2011) by overlaying the FMU boundaries with population dissemination blocks. These blocks are the smallest geographic area for which population and dwelling counts are disseminated.

Fossil fuel sources for electricity production were based on the projected fuel mix for each jurisdiction (National Energy Board, 2013) averaged over the period 2017–2035. Fossil fuel sources for heat production were based on contemporary estimates of heat fuel sources were used (Office of Energy Efficiency, 2015) because projections were unavailable. Each jurisdiction's energy use was compiled by fuel type for space heating in the residential and commercial sector, and process heating for the industrial sector. There were limited data available on process heat, and the fuel mix without electricity was used as a proxy of the energy mix for heat. When electricity was excluded, over 90% of industrial energy was used for process heating (boilers or heaters) (National Energy Board, 2013).

Each FMU was assumed to have the average fuel mix for electricity and heat of the province or territory. However, some FMUs contain remote (off-grid) communities with a different fuel mix (Natural Resources Canada, 2015). In such instances, the FMU-level fuel mix was adjusted to reflect remote communities' fuel usage, based on the relative population sizes of the remote communities and the FMU.

Fossil fuel emissions intensities were taken from published values (Office of Energy Efficiency, 2015). Heat generated from electricity was assumed to use the average grid emissions intensity. We assumed that only fossil fuels would be displaced, and therefore, it was not necessary to quantify emissions for nuclear, hydro-electricity, wind, tide and existing biomass energy sources. Emissions intensities for electricity were highest for coal at 1 tCO_2e MWh^{-1}, followed by fuel oil and diesel at 0.8 tCO_2e MWh^{-1} and natural gas as 0.45 tCO_2e MWh^{-1}. Heating fuel emissions ranged from 0.438 tCO_2e MWh^{-1} for coal and petcoke, to 0.361 tCO_2e MWh^{-1} for fuel oil, and 0.255 tCO_2e MWh^{-1} for natural gas. Heat emissions from electricity were based on the fuel mix for electricity within a given jurisdiction and ranged from 0.0012 to 0.585 tCO_2e MWh^{-1} (Table S2).

For the bioenergy facilities, nine facilities were selected from the literature as representative of potential installations applicable to variety of Canadian regions and for which information was available on emissions and costs (Pröll et al., 2011, Wood & Rowley, 2011; Biopathways, 2012, RETScreen International, 2015). The nine selected bioenergy facilities had a range of

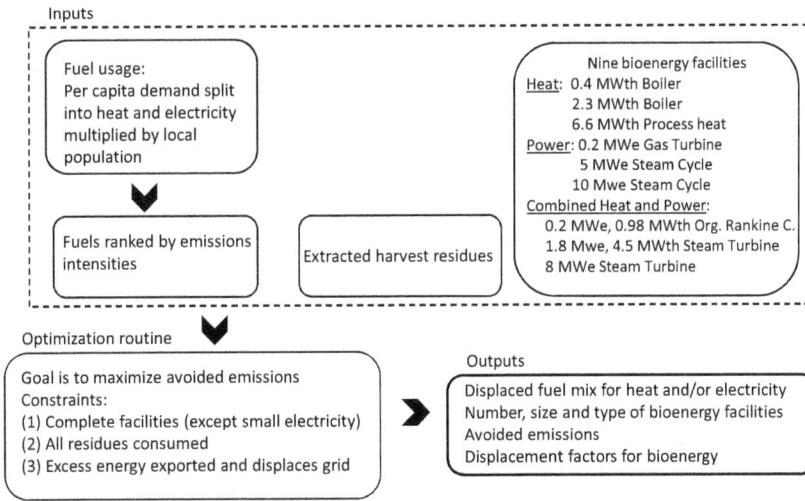

Fig. 2 Overview of the method for calculating bioenergy displacement factors. Transportation emissions associated with transport of residues were not included.

200–400 kW for small facilities, and 7–10 MW for large facilities (Table 5). Biomass demand for the facilities ranged from 0.8 to 2 kodt yr^{-1} for small facilities, and for large facilities biomass demand ranged from 47 to 64 kodt yr^{-1}. Energy facilities included boilers for district and process heat production, and steam and gas turbines for CHP and power production. The functional units for energy were 1 MWh of electricity and 1 MWh of heat.

The LP model maximized avoided emissions, E_a,

$$E_a = \max\left(\sum H_j I_j + \sum V_k I_k\right), \qquad (6)$$

where H and V are the amounts of energy (in MWh) produced by each fuel that would be displaced by harvest residues used in heat and electricity production, respectively; I is the emissions intensity for j = 4 heat fuels: (1) coal and petcoke, (2) fuel oil, (3) natural gas and (4) electricity; and k = 5 electricity fuels: (1) coal, (2) fuel oil, (3) diesel, (4) natural gas and (5) grid. LP model inputs were the extracted harvest residues, regional energy demand, and existing fuel sources and emissions intensities for heat and electricity production. We assumed that all extracted harvest residues would be used to generate energy locally, within the FMU. We did not include raw material transportation emissions, and these would likely vary across the FMUs given the differences in FMU sizes. The total amount of energy that could be substituted was constrained by the amount of extracted harvest residues within each FMU, and only complete facilities were permitted except in the case of the small electricity facility where partial facilities were permitted to ensure all residues were consumed. Heat production was constrained by the local heat demand. If the amount of bioenergy produced exceeded local demand, then excess biomass was converted to electricity and exported to the electricity grid.

A bioenergy displacement factor, DF$_e$ (tC avoided per tC used), was estimated for each region as the total maximum avoided emissions divided by the C in extracted harvest residues:

$$\mathrm{DF_e} = E_a\left(\frac{12}{44}\right)R^{-1}, \qquad (7)$$

where E_a is the avoided emissions in tCO$_2$e, R is the C in extracted harvest residues in tC, and the factor of 12/44 converts from CO$_2$ to C.

Two examples were selected to estimate the range in displacement factors within the optimization results. In the first example, *constant supply*, displacement factors for each FMU were estimated for a fixed biomass feedstock of 64 thousand oven-dried tonnes (kodt), which matched the fibre demand of the largest electricity facility or six medium CHP facilities. In the second example, *constrained supply*, extracted harvest residues were limited to meet each FMU's actual demand for heat from fossil fuels to (1) assess the change in displacement factors relative to a *constant supply* example and to (2) determine the amount of extracted harvest residues needed to displace fossil fuel based heat production. We included heat produced from electricity as fossil fuel based heat if the grid emissions exceeded 400 kg CO$_2$e MWh^{-1}. If the residues available in the *constrained supply* case could not support a small heat facility locally, we did not estimate a displacement factor in that FMU.

Results

Wood product substitution

Table 1 shows the six end-use products compiled from the literature that were selected for the functional unit basket, and their associated material uses for a more wood-intensive scenarios as compared to a less wood-intensive scenario. The comparative material lists showed that sawnwood predominantly substituted for steel in single-family homes, sawnwood substituted for concrete in multifamily homes, and sawnwood and

Table 1 The composition and associated material mass (in tonnes) of the six comparative end-use products for the functional unit basket

End-use product	Material	Mass more wood-intensive (t)	Mass less wood-intensive (t)
Single-family home*	Sawnwood (softwood lumber)	10.8	5.4
	Panel: Oriented Strand Board (OSB)	2.2	1.1
	Concrete	63.5	64.0
	Steel Beams	3.0	11.6
Multifamily home†	Sawnwood (softwood lumber)	59.0	33.0
	Panel: Particleboard	18.0	17.0
	Panel: Plywood	21.0	20.0
	Concrete	223	1352
	Steel Beams	16.0	25.0
	Insulation	21.0	25.0
Multiuse building‡	Sawnwood (softwood lumber)	75.0	0.0
	Panel: Particleboard	17.0	3.0
	Panel: Plywood	21.0	0.0
	Concrete	236	1430
	Steel Beams	550	703
Flooring§ (Residential upkeep)	Sawnwood	0.7	0.077
	Linoleum	0	0.18
Furniture¶	Panel: Medium Density Fibreboard	0.011	0
	Plastic [High-density polyethylene (HDPE)]	0	0.0034
Decking‖ (Manufacturing)	Sawnwood (softwood lumber)	0.36	0
	Plastic (HDPE)	0	0.21
	Wood flour (saw dust)	0	0.05

*A typical two-storey house in Minneapolis with a basement and total floor area of 192 m^2. Design consisted of solid wood-framing members except for composite floor I-joists, OSB sheathing for roof, walls and floor, and pre-engineered roof trusses (Lippke et al., 2004).

†Four-storey building with 16 apartments with a floor area of 1190 m^2 (Gustavsson et al., 2006).

‡A 7300 m^2, six-storey university building. The bottom three floors and basement are used as classrooms and open-plan offices, the top three floors are used as hotel rooms (Scheuer et al., 2003).

§100 m^2 of flooring (Nedermark, 1998).

¶TV chassis (Beovision Avant) (Jönsson et al., 1997).

‖Deck surface (29.7 m^2) assuming a 10-year service life.

panels substituted for steel and concrete in multiuse buildings. For the other three end-use products (flooring, furniture and decking), sawnwood and panels replaced plastic in the comparison studies selected.

Estimates of GHG material emissions (Table 2) were compiled from published studies, with preference given to North American products and uses. For solid wood products, lumber (sawnwood) and plywood have lower unit emissions than particleboard, Oriented Strand Board and Medium Density Fibreboard (MDF), due to the higher manufacturing and transportation emissions.

Based on the published comparative studies for the six end-use products, multiuse buildings had the highest avoided emissions (Table 3) mainly due to the substitution of wood for large amounts of steel and concrete. Multifamily home and single-family home components also had steel and concrete substitution,

but avoided emissions were lower than in multiuse buildings because of the lower material demands. The avoided emissions for manufacturing (decking in our analysis) were small, as were the avoided emissions for residential upkeep. Furniture avoided emissions, represented as MDF vs. HDPE plastic, were found to have small negative avoided emissions, indicating the use of fibreboard increased emissions relative to plastic for the selected comparative study due to its higher emission intensity (Table 2).

National consumption statistics were used to partition sawnwood and panels within the six end-use products. Sawnwood was primarily used in residential upkeep (represented by flooring in this analysis) at 35% of the total, manufacturing (represented by decking) at 27% and single-family homes at 16% (Table 3). Panels were used in most end-use products, but had modest volume percentages that ranged from 1% to 4%.

Table 2 Greenhouse gas emissions for materials used in the functional unit basket

Primary product	Total (kg CO_2e t^{-1})	x Extraction (kg CO_2e t^{-1})	t Transportation (kg CO_2e t^{-1})	s Manufacturing (kg CO_2e t^{-1})
Lumber*	111.8	32.7	17.9	61.3
MDF†	2646	62.2	159.8	2424
OSB‡	586	82.0	51.5	452
Particleboard§	1191	41.6	98.8	1050.4
Plywood¶	240.3	53.5	28.0	158.9
Concrete‖	88.0	0.8	2.3	84.9
Steel (from ore)**	2797	1180	30.0	1587
Insulation††	214.0	214.0	0	0
Linoleum flooring‡‡	608.8	0.0	0.0	608.8
High-density polyethylene§§	1800	0.0	0.0	1800

MDF, Medium Density Fibreboard; OSB, Oriented Strand Board.
*Based on 2.3597 m³ (thousand board feet) (Athena Sustainable Materials Institute 2009c).
†Based on 92.903 m³, 19 mm basis (thousand square feet, 3/4 inch basis) (Athena Sustainable Materials Institute 2009a).
‡Based on 92.903 m³, 9.5 mm basis (thousand square feet, 3/8 inch basis) (Athena Sustainable Materials Institute 2008a).
§Based on 92.903 m³, 19 mm basis (thousand square feet, 3/4 inch basis) (Athena Sustainable Materials Institute 2009b).
¶based on 92.903 m³, 9.5 mm basis (thousand square feet, 3/8 inch basis) (Athena Sustainable Materials Institute 2008b).
‖Marceau et al. (2007).
**National Renewable Energy Laboratory (2008) and Iosif et al. (2010).
††Schmidt et al. (2004).
‡‡Based on 100 m² (Jönsson et al., 1997).
§§Bala et al. (2010).

Table 3 End-use product avoided emissions, Canadian consumption statistics and weighting factors

End-use product	N_f Avoided emissions (tCO$_2$ per end-use product)	A_f Canadian wood consumption (% volume, sawnwood, structural panels, nonstructural panels)	m_{MF} Mass of wood material in the more wood-intensive end-use product (t)	W_f Weighting factor
Single-family home	22.8	19.0 (15.8, 3.3, 0.1)	13	3.3
Multifamily home	121.1	5.3 (3.6, 1.6, 0.1)	98	0.12
Multiuse building	503.0	4.0 (3.0, 0.8, 0.1)	113	0.08
Flooring (residential upkeep)	0.03	39.7 (35.1, 3.4, 0.8)	0.74	121
Furniture	−0.02	3.2 (0, 0, 3.2)	0.011	662
Decking (manufacturing)	0.3	28.8 (27.2, 1.9, 0)	0.36	180

Weighting factors estimated for each end-use product within the functional unit basket ranged from 0.1 to 662 (Table 3). These weighting factors were applied to the mass of wood in each end-use product, to estimate the total avoided emissions for each end-use product in the functional unit basket (Table 4). The highest avoided emissions for end-use products weighted by national consumption were for single-family homes (75 tCO$_2$ per home), manufacturing (62 tCO$_2$ per deck) and multiuse buildings (40 tCO$_2$ per building).

Partitioning the avoided emissions into sawnwood and panels found that overall displacement factor for sawnwood was 0.99 tCO$_2$ avoided per tonne of sawnwood or 0.54 tC tC^{-1} and for panels the displacement factor was 0.83 tCO$_2$ avoided per tonne of panels or 0.45 tC tC^{-1}.

Bioenergy substitution

Nine bioenergy facilities (Table 5) included three facilities for heat production, three for electricity production and three for combined heat and power production. Energy demand and fuel mix (projected electricity and contemporary heat) varied substantially across the

Table 4 Weighted avoided emissions by end-use product and proportion of sawnwood and panels in the incremental change in mass for more-intensive vs. less-intensive end-use products

End-use product	N_{fD} Weighted Avoided Emissions in the basket (tCO$_2$ per end -use product)	Per cent of incremental wood change (K_{fp})		
		Sawnwood (%)	Structural panels (%)	Nonstructural panels (%)
Single-family home	75.1	83	17	0
Multifamily home	14.8	93	4	4
Multiuse building	39.8	68	19	13
Flooring (residential upkeep)	4.2	100	0	0
Furniture	−15.1	0	0	100
Decking (manufacturing)	62.3	100	0	0

Table 5 Description of the three sizes (small, medium, large) of three types (district heat, power and combined heat and power) for the nine selected bioenergy facilities. Assuming 340 operating days, 24 h per day operating hours and a wood energy content of 20 GJ odt^{-1}

Facility type	Facility description	Biomass demand (kodt yr^{-1})	Electrical conversion rate (MWh odt^{-1})	Thermal conversion rate (GJ odt^{-1})	Assumed electrical efficiency (%)	Assumed thermal efficiency (%)	Implied overall efficiency (%)
Heat	0.4 MWth boiler for district heating*	0.783	–	15.0	–	75	75
	2.3 MWth boiler for district heating†	3.97	–	17.0	–	85	85
	6.62 MWth process heat via syngas‡	11.58	–	16.8	–	84	84
Power	0.2 MWe gas turbine§	1.60	1.02	–	18	–	18
	5 MWe steam cycle‡	34.97	1.17	–	21	–	21
	10 MWe steam cycle‡	63.86	1.28	–	23	–	23
CHP	0.2 MWe, 0.98 MWth Organic Rankine Cycle¶	2.09	0.78	14.0	14	70	84
	1.8 MWe and 4.5 MWth steam turbine‖	10.58	1.39	10.8	25	54	79
	8 MWe CHP steam turbine‡	46.87	1.39	5.88	25	29	54

*Retscreen International (2015).
†Retscreen International (2015).
‡Biopathways (2012).
§Arena et al. (2010).
¶Wood & Rowley (2011).
‖Pröll et al. (2011).

country (Table 6) and reflected many different influences including available resources, population distribution, climate and resource development.

In the first example, *constant supply*, we chose a fixed wood residue supply of 64 kodt yr^{-1}, which could supply a large electricity facility, or six medium CHP facilities to demonstrate that regions with the same amount of harvest residue consumed could have very different avoided emissions. Estimated displacement factors for 502 FMUs (total of 32 Modt biomass feedstock) had an average displacement factor of 0.47 tC tC^{-1} (range of 0.001–1.85 tC tC^{-1}), with a clear S-shaped trend of greater displacement factors with greater populations (Fig. 3a). Produced heat had to be consumed within the FMU and could not exceed demand, while excess electricity could be exported to the grid. Thus, heat consumption increased with population size (Fig. 3b) while energy exports decreased (Fig. 3d). There were two offset S-shape trends, with the second S-shape having higher displacement factors at all population levels.

Table 6 Average provincial or territorial per capita energy consumption and energy fuel mix for projected electricity production (*E*) and heat production (*H*). Percentages have been rounded and may not add to 100%. Additional details can be found in Table S1

Province or Territory	*E*: consumed (MWh per person)	*H*: consumed (MWh per person)	*E*: coal (%)	*E*: fuel oil (%)	*E*: NG* (%)	*E*: rest† (%)	*H*: coal and petcoke (%)	*H*: fuel oil (%)	*H*: NG (%)	*H*: electricity (%)	*H*: biomass (%)
Alberta	33.5	67.1	32		59	9	21	5	68	1	5
British Columbia	27.7	55.5			13	88	5	9	42	8	35
Manitoba	16.2	32.4				100	2	8	64	15	11
New Brunswick	14.8	29.6	16		11	74	3	37	9	32	19
Newfoundland and Labrador	16.3	32.6		1	2	98	1	30	0	44	25
Northwest Territories	30.8	61.6		46	13	41	11	53	11	5	16
Nova Scotia	11.8	23.7	30		34	35	2	69	0	2	28
Ontario	13.0	25.9			19	80	26	5	57	4	8
Prince Edward Island	12.4	24.8				100	4	50	5	16	25
Quebec	15.1	30.2			2	99	10	13	34	19	24
Saskatchewan	33.4	66.9	33		43	23	14	7	70	3	6
Yukon	10.8	21.6		19		81	11	53	11	5	16

*Natural Gas (NG).

†'Rest' includes power generation from hydro-electricity, wind and tide, biomass and uranium.

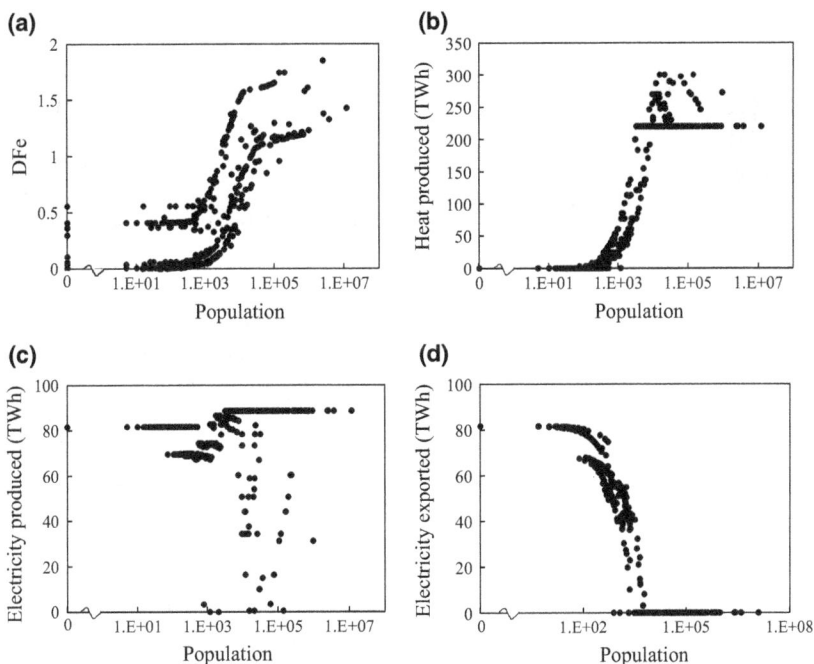

Fig. 3 Displacement factors, annual heat and electricity production, and exported electricity as a function of population within each of the 502 forest management units, assuming a constant annual biomass feedstock of 64 kodt yr^{-1}.

These higher displacement factors were associated with high grid emissions intensities for some regions. Figure 3c shows the electricity produced had three distinct levels of electricity production related to (1) a large electricity facility selected within low population regions, (2) a mix of heat, electricity and CHP facilities in midrange population regions and (3) the highest electricity production associated with six medium CHP facilities in high population regions. For the same amount of feedstock, the six medium CHP facilities produced 220 TWh yr^{-1} of heat as well as more electricity than a large electricity facility due to higher conversion efficiency.

In the second example, *constrained supply*, the amount of bioenergy feedstock was reduced by limiting extracted harvest residues to match the demand of fossil fuel-based heat. The total extracted harvest residues fell to 11.0 Modt and residues were collected in 327 of the 502 FMUs. Heat generated from bioenergy was used to displace heat generated from fuel oil, natural gas and coke and petcoke (Fig. 4a). Electricity was predominantly generated by CHP facilities and displaced electricity generated from natural gas and coal. There was a much smaller proportion of electricity grid than in the *constant supply* example. A minor proportion of wood biomass used for heat production was displaced in the constant supply because all fossil fuel sources had been consumed, but substitution of wood biomass did not contribute to avoided emissions. The average displacement factor was 0.89 in the *constrained supply* example, almost twice as high as in the *constant supply* example (Fig. 4b). Avoided emissions in the *constrained supply*

example were equivalent to 80% of the avoided emissions in the *constant supply* example while consuming only 34% of the biomass residues.

Discussion

The carbon neutrality assumption for bioenergy that forest bioenergy emits no C to the atmosphere as long as the postharvest forest regrows to its preharvest C level fails to properly assess the GHG emissions of bioenergy (Johnson *et al.*, 2009; Ter-Mikaelian *et al.*, 2015). The potential GHG emission reductions that can be achieved through wood use need to be quantified in a systems approach which includes changes in forest C, HWPs tracking and substitution benefits to prevent the development of policies that develop no net benefit to the atmosphere. For product substitution, climate benefits occur in a different sector because emission reductions would occur in manufacturing and transportation sectors due to the reduction in steel or concrete, or plastic production. Policies generally do not include substitution impacts directly, but instead include reductions in overall GHG emissions or promote changes in construction, such as the Wood First Act (BC JTST, 2015).

Product substitution

Wood product displacement factors in this study are lower than the mean displacement factor estimated by Sathre & O'Connor (2010), hereafter S&O. Their meta-analysis estimated an average displacement factor of 2.1 tC tC^{-1} based on 21 studies with considerable range in estimates (-2.3 to 15 tC tC^{-1}), but most factors were between 1 and 3 tC tC^{-1}. There are differences in methodology between this study and S&O which prevent direct comparison. In S&O, the system boundaries were not the same for all studies, with some studies including forest ecosystem emissions, some including operational emissions, as well as postconsumer emissions from landfills and fossil fuels substitution. In this study, we examined product substitution for harvesting, transportation of raw materials and manufacturing, and tracked changes in forest ecosystem emissions and HWP commodity lifetimes and postconsumer treatment in other system components. We estimated displacement factors from four of the studies included by S&O that had the same end-use products as our study (Petersen & Solberg, 2004; Gustavsson *et al.*, 2006; John *et al.*, 2009; Salazar & Meil, 2009). To ensure the system boundaries were the same, we removed the forest ecosystem, operational and postconsumer emissions from the four selected studies and weighted the avoided emissions per end-use product. We found that S&O's displacement factors were 0.51 tC tC^{-1} for sawnwood

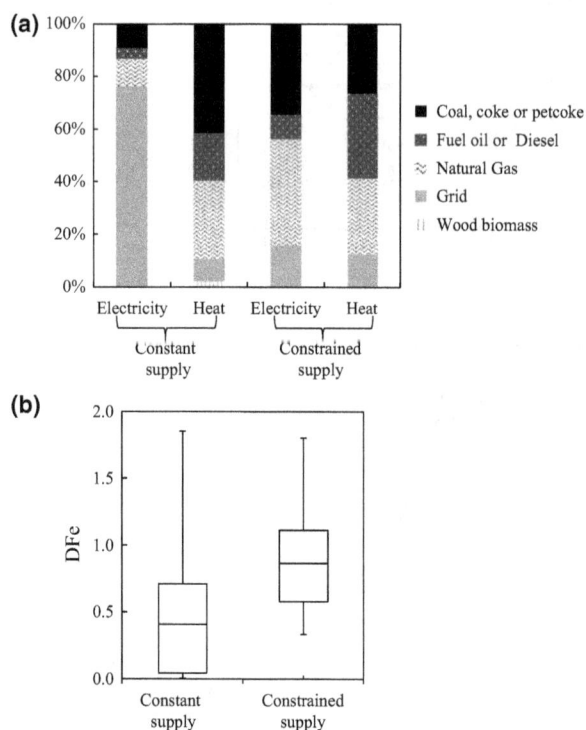

Fig. 4 Comparison of (a) displaced fuel sources percentages for heat and electricity for the *constant supply* example and *constrained supply* example. (b) Boxplots of displacement factors for the two examples with first, third and median values indicated by the boxes, and minima and maxima represented by error bars.

and 0.14 tC tC^{-1} for panels. These compared reasonably well to our estimates of 0.54 tC tC^{-1} sawnwood and 0.45 tC tC^{-1} for panels.

Additional comparative studies and information on end-uses are needed to reduce the uncertainty in the displacement factor estimates. We are constrained by the availability of comparative studies of product end-uses that include complete material information. One of the key areas of uncertainty is the use of solid wood products in residential upkeep and manufacturing, and corresponding comparative studies that identify emissions in a wood-intensive product vs. plastic products. Also, displacement factors for the primary solid wood products are strongly influenced by the building material lists, and additional data estimating 'typical' building (house, apartment building, office building, etc.) composition for all regions in Canada would be an asset. Our end-use categories are quite broad, and as a result, the selection of comparative end-uses is a source of uncertainty.

Energy substitution

In the first example, a constant biomass supply for all regions revealed that the energy demand for the FMUs was highly variable and that displacement factors generally increased with population due to an increased heat production and decreased export of electricity outside of the FMU. Displacement factors for the same biomass supply and population varied by up to ~0.5, which reflected the different emissions intensities of the jurisdictions' projected electricity fuel sources (Table S2).

In the second example, biomass consumption was reduced so as to only displace heat generated from fossil fuels. Displacement factors increased in this example, but the number of participating regions dropped by almost 35%, and the extracted harvest residues decreased by over 50% relative to other estimates (Smyth *et al.*, 2016). The *constrained supply* example identified regions where local bioenergy production from extracted harvest residues could generate the highest substitution benefits from heat and/or CHP facilities, and it would be of interest to use these optimized feedstock levels in national-level mitigation analyses to assess uncertainty in mitigation estimates.

Our analysis focused on smaller-scale facilities that could support smaller communities, or many facilities could be combined to support cities. We had initially included a 17MWe electricity facility, but this facility could only be supported in a few FMUs with sufficient harvest residues and was never selected by the LP optimization model, so it was replaced by a 10 MWe facility. Adding facilities capable of also utilizing

agricultural residues could change the potential facility scale. As noted by Cleary & Caspersen (2015), large electricity facilities may not be as useful as CHP facilities because large electricity facilities are generally not located near population centres, so the heat that is produced cannot be used, and although potentially more efficient, large facilities tend to require additional pelletization processing and larger feedstock transportation distances which offsets increased efficiency.

Our results are consistent with McKechnie *et al.* (2011) who found displaced emissions were higher when harvest residues displaced a high GHG intensity fuel (coal in cofiring) rather than a lower GHG intensity fuel (ethanol in their case, natural gas in ours). Other studies (Guest *et al.*, 2012; Zanchi *et al.*, 2012; Cintas *et al.*, 2015; Cleary & Caspersen, 2015) have also found the magnitude of the displaced emissions depends critically on which fuel source is displaced. Most of the above-mentioned studies focus on electrical systems, but the type of energy produced is also a critical factor and explains why heat is often a better substitution than electricity. This analysis is unique because we considered local use of bioenergy within FMUs for the managed forest of Canada and included remote community fuel usage in the projections of electricity fuel mix.

Two assumptions, (1) the omission of transportation-related emissions for bioenergy and (2) the calculation of avoided emissions from fossil fuels based on current emissions intensities and fixed fuel cost effects, result in an overestimate of the mitigation potential. In the first case, emissions associated with transporting harvest residues to the bioenergy facilities within an FMU were not included. In this study, the transportation emissions are assumed to be small, an assumption that has been in other studies where fixed haul distances or circular transportation distances are assumed (Thakur *et al.*, 2014; Laganière *et al.*, 2015). When short haul distances (<150 km) are used, the emissions associated with transportation are a small percentage of the total emissions from harvesting, transportation and combustion. For example, Jones *et al.* (2010) that emissions from the entire biomass delivery process of collecting, grinding and hauling biomass comprised only 3.2–3.9% of the total emissions, and Domke *et al.* (2012) found that transportation emissions were less than 2% of the total emissions. Transportation costs are often the limiting factor for residue capture, and thus, costs associated with hauling biomass are included in economic modelling (e.g. review by Shabani *et al.*, 2013). It is beyond the scope of the present analysis to estimate transportation-related emissions, but their omission overestimates the mitigation potential if transportation emissions of wood exceed those of fossil fuels. In the second case, we assumed bioenergy would displace the most emissions-intensive fossil fuels without consideration of policies that would target particular fuel types (e.g. closing coal facilities or reducing diesel use). If regions with high-intensity fossil fuels switched to lower emissions fossil fuels, the bioenergy mitigation potential would be overestimated. Further, we did not consider nonfossil fuels in the substitution impacts, but it is possible that bioenergy could be used to substitute for photovoltaics or hydro-electricity, in which case the substitution impacts would be overestimated.

Estimated average displacement factors ranged from 0.47 to 0.89 tC tC^{-1}, with a maximum value of 1.85 tC tC^{-1}. These values are similar to the value of 0.97 tC tC^{-1} by Gan & Smith (2006) for electricity generation from harvest residues and snags displacing coal. More recently, Rajagopal & Plevin (2013) using an economic model suggested that globally the displacement factor varies from 0.4 to 1.0 tC tC^{-1} for ethanol and oil-based transportation fuels.

We did not consider changes in per capita fuel usage over time, nor did we consider changes in population. A potential source of uncertainty is that the energy demand (including industrial energy demand) was assumed to be correlated to population, but industrial facilities can be large consumers not necessarily linked to population centres. Additionally, only local use of bioenergy was considered, and pelletization of residues and export from the FMU was not analysed. Assessing pellet export would require in-depth data on transportation emissions and was beyond the scope of this analysis.

Conclusions

The results of this study highlight the importance of wood use options in mitigating climate change. Forests and forest products can contribute to mitigating climate change, and this analysis presented methods and analyses to estimate substitution benefits from using wood products in the built environment and energy production to displace higher emissions products and fuels at the national level. Wood product substitution benefits should be included when using a systems approach to evaluate mitigation strategy potential, both for domestic and global reductions in GHG emissions. This study examined wood product substitution, and displacement factors from this analysis must be combined with ecosystem and HWP C dynamics to quantify the overall mitigation benefits (e.g. Smyth et al., 2014, 2016).

Displacement factors for solid wood substitution were estimated from a functional unit basket that included three buildings, and three other end-use products for which avoided emissions were weighted by national consumption levels. Displacement factors were found to be 0.54 tC displaced per tC of sawnwood used, and 0.45 tC tC^{-1} for panels.

Displacement factors for bioenergy products were estimated for two feedstock supply levels based on 502 FMUs across Canada, and an optimization routine which displaced highest emissions-intensity fuels in heat, electricity or CHP bioenergy facilities.

Nine bioenergy facilities were selected to represent a variety of energy products (heat, power, CHP), conversion technologies, efficiencies (boilers and turbines) and facility sizes (0.2–10 MW). Displacement factors for a *constant supply* for each region had an average of 0.47 tC tC^{-1} and displaced mostly electricity grid and coal for heat production with 27 MtCO$_2$e yr^{-1} displaced from 32 Modt yr^{-1} consumed. Displacement factors for feedstock supply that was constrained to meet fossil fuel heat demand had a higher average value of 0.89 tC tC^{-1} and avoided fossil fuel emissions of 22 MtCO$_2$e from 11 Modt yr^{-1} harvest residues consumed.

Acknowledgements

This study would not have been possible without strong cooperation between provincial, territorial and federal government agencies. We thank all members (past and present) of the National Forest Sinks Committee and their colleagues. Funding for this study was provided by the Government of Canada's Clean Air Agenda, the Program of Energy Research and Development, and in-kind contributions from provincial and territorial governments. We thank our colleague Élizabeth Walsh for her help in compiling energy data, and we thank the anonymous reviewers for their insightful comments.

References

Arena U, Di Gregorio F, Santonastasi M (2010) A techno-economic comparison between two design configurations for a small scale, biomass-to-energy gasification based system. *Chemical Engineering Journal*, **162**, 580–590.

Athena Sustainable Materials Institute (2008a) *A Cradle-to-gate Life Cycle Assessment of Canadian Oriented Strand Board*, pp. 87. Athena Sustainable Materials Institute, Ottawa, ON.

Athena Sustainable Materials Institute (2008b) *A Cradle-to-gate Life Cycle Assessment of Canadian Softwood Plywood Sheathing*, pp. 86. Athena Sustainable Materials Institute, Ottawa, ON.

Athena Sustainable Materials Institute (2009a) *A Cradle-to-gate Life Cycle Assessment of Canadian Medium Density Fiberboard (MDF)*, pp. 95. Athena Sustainable Materials Institute, Ottawa, ON.

Athena Sustainable Materials Institute (2009b) *A Cradle-to-gate Life Cycle Assessment of Canadian Particleboard*, pp. 93. Athena Sustainable Materials Institute, Ottawa, ON.

Athena Sustainable Materials Institute (2009c) *A Cradle-to-gate Life Cycle Assessment of Canadian Softwood Lumber*, pp. 143. Athena Sustainable Materials Institute, Ottawa, ON.

Bala A, Raugei M, Benveniste G, Gazulla C, Fullana-I-Palmer P (2010) Simplified tools for global warming potential evaluation: when 'good enough' is best. *The International Journal of Life Cycle Assessment*, **15**, 489–498.

BC JTST (2015) Wood first initiative: wood first act, ministry of jobs tourism and skills training and responsible for labour. Available at: http://www.bclaws.ca/Recon/document/ID/freeside/00_09018_01 accessed (May, 2016)

Biopathways (2012) Biopathways model. Available at: http://www.nr-can.gc.ca/forests/industry/bioproducts/13321 (accessed May 2012).

Böttcher H, Kurz WA, Freibauer A (2008) Accounting of forest carbon sinks and sources under a future climate protocol – factoring out past disturbance and management effects on age–class structure. *Environmental Science & Policy*, **11**, 669–686.

Cintas O, Berndes G, Cowie AL, Egnell G, Holmström H, Ågren GI (2015) The climate effect of increased forest bioenergy use in Sweden: evaluation at different spatial and temporal scales. *Wiley Interdisciplinary Reviews: Energy and Environment*, **5**, 351–369.

Cleary J, Caspersen JP (2015) Comparing the life cycle impacts of using harvest residue as feedstock for small- and large-scale bioenergy systems (part I). *Energy*, **88**, 917–926.

Domke GM, Becker DR, D'amato AW, Ek AR, Woodall CW (2012) Carbon emissions associated with the procurement and utilization of forest harvest residues for energy, northern Minnesota, USA. *Biomass and Bioenergy*, **36**, 141–150.

FEA (2011) Forest economic advisors LLC, Quarterly Lumber Forecast Service. Availabe at: https://www.getfea.com/lumber/quarterly-forecast-service/forecast-summary (accessed February 2011).

Gan J, Smith CT (2006) Availability of logging residues and potential for electricity production and carbon displacement in the USA. *Biomass and Bioenergy*, **30**, 1011–1020.

Guest G, Cherubini F, Strømman AH (2012) Climate impact potential of utilizing forest residues for bioenergy in Norway. *Mitigation and Adaptation Strategies for Global Change*, **11**, 667–691.

Gustavsson L, Pingoud K, Sathre R (2006) Carbon dioxide balance of wood substitution: comparing concrete- and wood-framed buildings. *Mitigation and Adaptation Strategies for Global Change*, **11**, 667–691.

Hennigar CR, Maclean DA, Amos-Binks LJ (2008) A novel approach to optimize management strategies for carbon stored in both forests and wood products. *Forest Ecology and Management*, **256**, 786–797.

Iosif A-M, Hanrot F, Birat J-P, Ablitzer D (2010) Physicochemical modelling of the classical steelmaking route for life cycle inventory analysis. *The International Journal of Life Cycle Assessment*, **15**, 304–310.

John S, Nebel B, Perez N, Buchanan A (2009) *Environmental Impacts of Multi-Storey Buildings Using Different Construction Materials*. Department of Civil and Natural Resources Engineering, University of Canterbury, Christchurch, New Zealand.

Johnson R, Ramseur JL, Gorte RW (2009) *Estimates of Carbon Mitigation Potential from Agricultural and Forestry Activities*. Congressional Research Service.

Jones G, Loeffler D, Calkin D, Chung W (2010) Forest treatment residues for thermal energy compared with disposal by onsite burning: emissions and energy return. *Biomass and Bioenergy*, **34**, 737–746.

Jönsson Å, Tillman AM, Svensson T (1997) Life cycle assessment of flooring materials: case study. *Building and Environment*, **32**, 245–255.

Keith H, Lindenmayer D, Macintosh A, Mackey B (2015) Under what circumstances do wood products from native forests benefit climate change mitigation? *PLoS One*, **10**, e0139640.

Knauf M, Kohl M, Mues V, Olschofsky K, Fruhwald A (2015) Modeling the CO-effects of forest management and wood usage on a regional basis. *Carbon Balance and Managment*, **10**, 13.

Laganière J, Paré D, Thiffault E, Bernier PY (2015) Range and uncertainties in estimating delays in greenhouse gas mitigation potential of forest bioenergy sourced from Canadian forests. *GCB Bioenergy*, **9**, 358–369.

Lemprière TC, Kurz WA, Hogg EH *et al.* (2013) Canadian boreal forests and climate change mitigation. *Environmental Reviews*, **21**, 293–321.

Lippke B, Wilson J, Perez-Garcia J, Bowyer J, Meil J (2004) CORRIM: lifecycle environmental performance of renewable building materials. *Forest Products Journal*, **54**, 8–19.

Lundmark T, Bergh J, Hofer P *et al.* (2014) Potential roles of Swedish forestry in the context of climate change mitigation. *Forests*, **5**, 557–578.

Macintosh A, Keith H, Lindenmayer D (2015) Rethinking forest carbon assessments to account for policy institutions. *Nature Climate Change*, **5**, 946–949.

Malmsheimer RW, Bowyer JL, Fried JS *et al.* (2011) Managing forests because carbon matters: integrating energy, products, and land management policy. *Journal of Forestry*, **109**, S7–S51.

Marceau ML, Nisbet MA, Vangeem MG (2007) *Life Cycle Inventory of Portland Cement Concrete*. Portland Cement Association, Skokie, IL.

McKechnie J, Colombo S, Chen J, Mabee W, Maclean HL (2011) Forest bioenergy or forest Carbon? Assessing trade-offs in greenhouse gas mitigation with wood-based fuels. *Environmental Science & Technology*, **45**, 789–795.

National Energy Board (2010) Industrial energy use in Canada emerging trends. *Energy Briefing Note*, pp. 21. Calgary, AB. Available at: https://www.neb-one.gc.ca/nrg/ntgrtd/ftr/2013/ppndcs/ppndcs-eng.html.

National Energy Board (2013) Canada's energy future 2013 – energy supply and demand projections to 2035 – appendices. Available at: https://www.neb-one.gc.ca/nrg/ntgrtd/ftr/2013/ppndcs/ppndcs-eng.html (accessed May 2013).

National Renewable Energy Laboratory (2008) US Life Cycle Inventory Database: hot rolled sheet, steel, at plant. Available at: https://www.lcacommons.gov/nrel/process/show/592324a2-5765-42b5-a5af-f3e6ae87b4ac

Natural Resources Canada (2015) Remote Communities Database. Available at: http://www2.nrcan.gc.ca/eneene/sources/rcd-bce/index.cfm?fuseaction=admin.home1 (accessed May 2014).

Nedermark R (1998) Ecodesign at Bang and Olufsen. In: *Product Innovation and Eco-Efficiency. Twenty-Three Industry Efforts to Reach the Factor 4* (ed. Klostermann J, Tukker A), pp. 233–240. Springer.

Office of Energy Efficiency (2015) Comprehensive Energy Use Database. Available at: http://oee.nrcan.gc.ca/corporate/statistics/neud/dpa/menus/trends/comprehensive_tables/list.cfm (accessed May 2015).

Pacala S, Socolow R (2004) Stabilization wedges: solving the climate problem for the next 50 years with current technologies. *Science*, **305**, 968–972.

Petersen A, Solberg B (2004) Greenhouse gas emissions and costs over the life cycle of wood and alternative flooring materials. *Climatic Change*, **64**, 143–167.

Pröll T, Rauch R, Aichernig C, Hofbauer H (2011) Fluidized bed steam gasification of solid biomass – performance characteristics of an 8 MWth combined heat and power plant. *International Journal of Chemical Reactor Engineering*, **5**, A54.

Rajagopal D, Plevin RJ (2013) Implications of market-mediated emissions and uncertainty for biofuel policies. *Energy Policy*, **56**, 75–82.

Ralevic P, Ryans M, Cormier D (2010) Assessing forest biomass for bioenergy: operational challenges and cost considerations. *Forestry Chronicle*, **86**, 43–50.

Retscreen International (2015) RETScreen Project Database. Available at: http://www.retscreen.net/ang/software_and_data.php (accessed May 2015).

Salazar J, Meil J (2009) Prospects for carbon-neutral housing: the influence of greater wood use on the carbon footprint of a single-family residence. *Journal of Cleaner Production*, **17**, 1563–1571.

Sathre R, O'Connor J (2010) Meta-analysis of greenhouse gas displacement factors of wood product substitution. *Environmental Science & Policy*, **13**, 104–114.

Scheuer C, Keoleian GA, Reppe P (2003) Life cycle energy and environmental performance of a new university building: modeling challenges and design implications. *Energy and Buildings*, **35**, 1049–1064.

Schlamadinger B, Marland G (1996) The role of forest and bioenergy strategies in the global carbon cycle. *Biomass and Bioenergy*, **10**, 275–300.

Schmidt A, Jensen A, Clausen A, Kamstrup O, Postlethwaite D (2004) A comparative life cycle assessment of building insulation products made of stone wool, paper wool and flax. *The International Journal of Life Cycle Assessment*, **9**, 122–129.

Shabani N, Akhtari S, Sowlati T (2013) Value chain optimization of forest biomass for bioenergy production: a review. *Renewable and Sustainable Energy Reviews*, **23**, 299–311.

Sharma A, Saxena A, Sethi M, Shree V (2011) Life cycle assessment of buildings: a review. *Renewable and Sustainable Energy Reviews*, **15**, 871–875.

Smyth CE, Stinson G, Neilson E, Lemprière TC, Hafer M, Rampley GJ, Kurz WA (2014) Quantifying the biophysical climate change mitigation potential of Canada's forest sector. *Biogeosciences*, **11**, 3515–3529.

Smyth C, Kurz WA, Rampley GJ, Lemprière TC, Schwab O (2016) Climate change mitigation potential of local use of harvest residues for bioenergy in Canada. *GCB Bioenergy*, **9**, 817–832.

Soimakallio S, Saikku L, Valsta L, Pingoud K (2016) Climate change mitigation challenge for wood utilization – the case of Finland. *Environmental Science & Technology*, **50**, 5127–5134.

Statistics Canada (2011) Dissemination block boundary file. Available at: http://www12.statcan.ca/census-recensement/2011/geo/bound-limit/bound-limit-2011-eng.cfm (accessed May 2014).

Stinson G, Kurz WA, Smyth CE *et al.* (2011) An inventory-based analysis of Canada's managed forest carbon dynamics, 1990 to 2008. *Global Change Biology*, **17**, 2227–2244.

Ter-Mikaelian M, Mckechnie J, Colombo S, Chen J, Maclean H (2011) The carbon neutrality assumption for forest bioenergy: a case study for northwestern Ontario. *The Forestry Chronicle*, **87**, 644–652.

Ter-Mikaelian MT, Colombo SJ, Chen J (2015) The burning question: does forest bioenergy reduce carbon emissions? A review of common misconceptions about forest carbon accounting. *Journal of Forestry*, **113**, 57–68.

Thakur A, Canter CE, Kumar A (2014) Life-cycle energy and emission analysis of power generation from forest biomass. *Applied Energy*, **128**, 246–253.

Upton B, Miner R, Spinney M, Heath LS (2008) The greenhouse gas and energy impacts of using wood instead of alternatives in residential construction in the United States. *Biomass and Bioenergy*, **32**, 1–10.

Werner F, Taverna R, Hofer P, Thürig E, Kaufmann E (2010) National and global greenhouse gas dynamics of different forest management and wood use scenarios: a model-based assessment. *Environmental Science & Policy*, **13**, 72–85.

Whittaker C, Mortimer N, Murphy R, Matthews R (2011) Energy and greenhouse gas balance of the use of forest residues for bioenergy production in the UK. *Biomass and Bioenergy*, **35**, 4581–4594.

Wood SR, Rowley PN (2011) A techno-economic analysis of small-scale, biomass-fuelled combined heat and power for community housing. *Biomass and Bioenergy*, **35**, 3849–3858.

Zanchi G, Pena N, Bird N (2012) Is woody bioenergy carbon neutral? A comparative assessment of emissions from consumption of woody bioenergy and fossil fuel. *GCB Bioenergy*, **4**, 761–772.

4

Climate change mitigation potential of local use of harvest residues for bioenergy in Canada

CAROLYN SMYTH[1], WERNER A. KURZ[1], GREG RAMPLEY[2], TONY C. LEMPRIÈRE[2] and OLAF SCHWAB[2]

[1]Natural Resources Canada, Canadian Forest Service, 506 Burnside Road West, Victoria, BC, V8Z 1M5 Canada, [2]Natural Resources Canada, Canadian Forest Service, 580 Booth Street, Ottawa, ON, K1A 0E4 Canada

Abstract

We estimate the mitigation potential of local use of bioenergy from harvest residues for the 2.3×10^6 km^2 (232 Mha) of Canada's managed forests from 2017 to 2050 using three models: Carbon Budget Model of the Canadian Forest Sector (CBM-CFS3), a harvested wood products (HWP) model that estimates bioenergy emissions, and a model of emission substitution benefits from the use of bioenergy. We compare the use of harvest residues for local heat and electricity production relative to a base case scenario and estimate the climate change mitigation potential at the forest management unit level. Results demonstrate large differences between and within provinces and territories across Canada. We identify regions with increasing benefits to the atmosphere for many decades into the future and regions where no net benefit would occur over the 33-year study horizon. The cumulative mitigation potential for regions with positive mitigation was predicted to be 429 Tg CO$_2$e in 2050, with 7.1 TgC yr^{-1} of harvest residues producing bioenergy that met 3.1% of the heat demand and 2.9% of the electricity demand for 32.1 million people living within these regions. Our results show that regions with positive mitigation produced bioenergy, mainly from combined heat and power facilities, with emissions intensities that ranged from roughly 90 to 500 kg CO$_2$e MWh^{-1}. Roughly 40% of the total captured harvest residue was associated with regions that were predicted to have a negative cumulative mitigation potential in 2050 of −152 Tg CO$_2$e. We conclude that the capture of harvest residues to produce local bioenergy can reduce GHG emissions in populated regions where bioenergy, mainly from combined heat and power facilities, offsets fossil fuel sources (fuel oil, coal and petcoke, and natural gas).

Keywords: bioenergy, Canada's managed forest, Carbon Budget Model of the Canadian Forest Sector, climate change mitigation, GHG emissions, harvest residues

Introduction

Global efforts to reduce the rate of increase in atmospheric greenhouse gas (GHG) concentrations require both a reduction in GHG emissions and an increase in removals of carbon dioxide (CO$_2$) from the atmosphere. It is anticipated that global bioenergy production will increase (IPCC, 2011) and there is significant interest in promoting bioenergy use to meet national and international GHG emission reduction goals (e.g. DECC, 2012, O'Neill, 2012). In Canada, there are national and provincial/territorial emission reduction targets for 2020 and 2030 (Government of Canada 2013, 2015), to which bioenergy production could contribute.

In this analysis, we estimate the climate change mitigation potential of local (i.e. forest management unit) use of bioenergy in Canada from harvest residues.

Earlier bioenergy studies have used a variety of methodologies to examine the biophysical potential, generally related to specific activities at smaller scales (Valente *et al.*, 2011; Domke *et al.*, 2012; Repo *et al.*, 2012; Röder *et al.*, 2015), but few studies have attempted to determine national mitigation potential (Werner *et al.*, 2010; Whittaker *et al.*, 2011; Lundmark *et al.*, 2014). Determination of the mitigation potential of forest-derived products is complex because the forest sector interacts with energy and industrial products sectors, and a systems approach to analysis is required (Naburs *et al.*, 2007; Obersteiner *et al.*, 2010; White, 2010; Lemprière *et al.*, 2013). The assumption that bioenergy is 'carbon neutral' (i.e. that it has no net GHG emissions) must be avoided, and bioenergy emissions must be estimated quantitatively. We previously examined harvest of live trees for bioenergy (Smyth *et al.*, 2014), but did not find this strategy to be effective for mitigation, which is consistent with other studies (Colombo *et al.*, 2005; Ralevic *et al.*, 2010; McKechnie *et al.*, 2011;

Correspondence: Carolyn Smyth
e-mail: Carolyn.Smyth@canada.ca

Ter-Mikaelian *et al.*, 2015), although it could have potential in remote communities where fossil fuels are transported over long distances. Harvest residues for bioenergy have also been examined previously, and this feedstock source had a higher mitigation potential than harvest of live trees because there is no forgone sequestration to consider. If residues are not used for bioenergy, they would progressively decay over time, or be burned for fuel hazard management.

In this analysis, we consider bioenergy from harvest residues using a systems approach to determine mitigation potential by considering the forest ecosystem, harvested wood products (HWP), bioenergy emissions, and displaced emissions when bioenergy is substituted for other energy sources. Displaced emissions considered the local jurisdictional-average heat and electricity fuel mix including, where available, the fuel mix of remote (off-grid) communities. The electricity infrastructures of Canadian off-grid remote communities are diverse and vary depending on access to energy resources, remoteness of location, and impact of climate. However, with the exception of a few local hydro grid-tied communities, the vast majority of remote communities across Canada rely on diesel generators for the production of electricity (Royer, 2011). We compared the emissions associated with capturing and burning harvest residues locally for bioenergy (converted to heat and/or

electricity) to the alternate scenario of *in situ* decay and/or slashburning (Fig. 1) to determine whether the use of captured harvest residues resulted in a net reduction of GHG emissions from the forest and energy sector over the 2017-to-2050 period.

Recognizing that the use of harvest residues may not result in a decrease in GHG emissions in all locations, our first objective was to identify regions where local use of forest-derived bioenergy from harvest residues results in positive mitigation. The second objective was to determine the time it takes to reach the break-even point (*i.e.* the point at which the fossil and bioenergy sources have the same impact on the atmosphere) for regions where there is a reduction in GHG emissions. This is an important indicator because bioenergy does not have to be C neutral (*i.e.* C uptake in the forest offsets the emissions from burning the biomass for energy) to achieve a mitigation benefit; it just has to generate fewer emissions than the base case energy source. For example, emissions from bioenergy include the emissions from bioenergy burning [112 kg CO_2e GJ^{-1} based on the IPCC default value (Gómez *et al.*, 2006)] and the forgone release of C from decay and/or slashburning. These are compared to fossil fuel emissions which range from 56.1 to 94.6 kg CO_2e GJ^{-1} depending on the fossil fuel type (IPCC default values). The time at which the break-even point is reached is important, because early

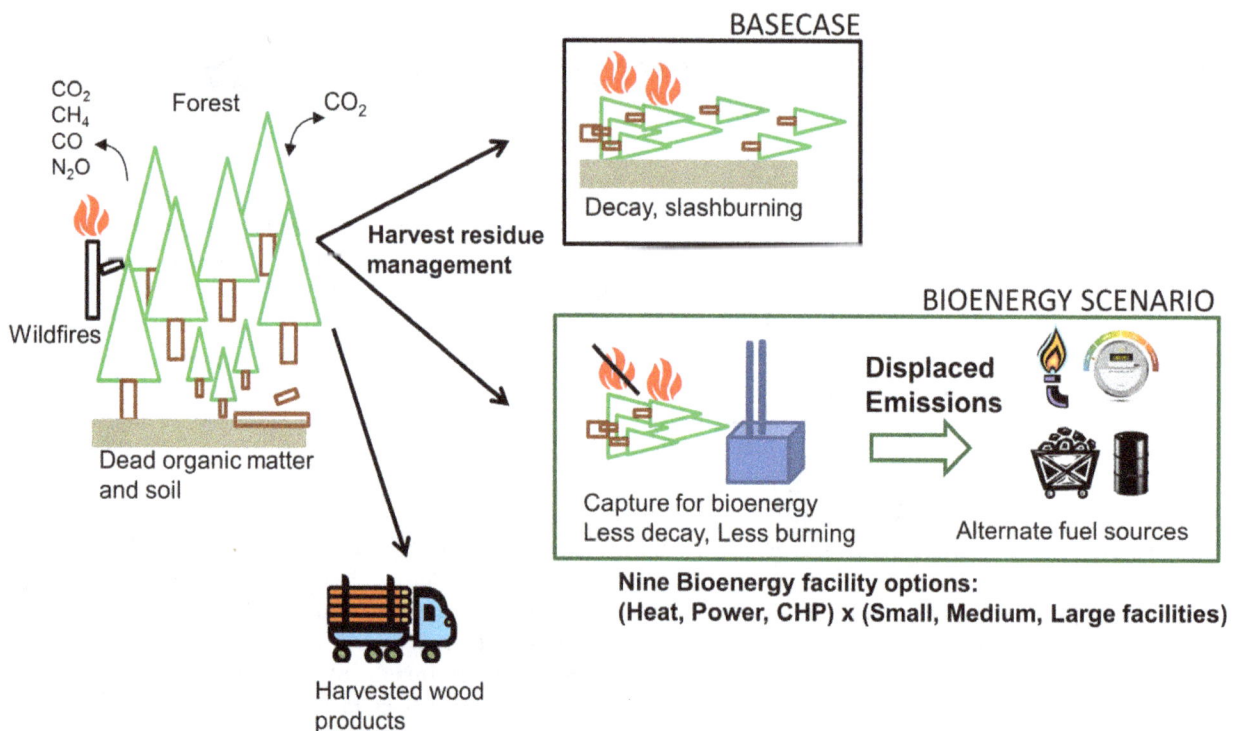

Fig. 1 Schematic of C flows in the *Base Case* and *Bioenergy Scenario*. The *Bioenergy Scenario* differs from the *Base Case* forest management assumptions by reducing slashburning (where applicable) and utilizing harvest residues for bioenergy.

emission reductions contribute to achieving emission reduction targets and limit future climate change. Our third objective was to determine what proportion of the national heat and electricity demand could be produced from bioenergy derived from harvest residue for regions with positive mitigation.

This study is the first comprehensive integrated analysis of the climate change mitigation potential for bioenergy from captured harvest residues for Canada's managed forest. It builds on the integrated analysis of the climate change mitigation potential for Canada's managed forest for seven forest management strategies and two harvested wood products strategies examined in Smyth *et al.* (2014). Comparisons are made to the results from the earlier analyses to assess the mitigation potential for bioenergy from harvest residues relative to three strategies: increasing the longevity of HWPs, better utilization that alters forest management practices, and harvesting less. Results from this national study can be used to inform bioenergy policy decisions. There are currently no federal regulations on minimum contributions from forest biomass and only two regulations for biofuels: fuel producers and importers must have an average renewable content of at least 5% based on the volume of gasoline and at least 2% of average renewable content based on the volume of diesel fuel and heating distillate oil (Environmental Protection Act Bill C-33, 2008).

Materials and methods

Greenhouse gas emissions from the forest ecosystem and HWP are based on the same models and data sets used to produce estimates for Canada's GHG National Inventory Report (NIR2014). We used the same forest management units (FMUs), Fig. S1, and the same historical (1990–2012) assumptions for harvests, wildfire, insects, deforestation and afforestation (Environment Canada, 2014). We chose these frameworks and data sets because they are well documented, the models and data sets have been peer-reviewed, and methods for estimating forest GHG emissions and removals have been reviewed, refined and revised as part of the annual National Inventory Reporting process since 2004.

Base case and bioenergy scenarios

Our analysis examined how changes in Canada's forest sector activities could reduce GHG emissions relative to a base case. The *Base Case* was defined as forest management (FM) activity levels and energy use that would occur in the absence of mitigation activity. In the *Bioenergy Scenario*, a proportion of the harvest residues (which varied by province and territory, Table 1) was recovered and used for bioenergy instead of decaying or being burned, starting in 2017.

In the rest of the time period, 2013–2050, both scenarios included the same wildfire and harvest projections for each

Table 1 Indicators for the *Base Case* and *Bioenergy Scenario*. Some parameters have ranges, indicating that implementation varied by jurisdiction

Scenario	Residue recovered* (%)	Residue recovered (Tg C yr^{-1})	Slashburning (% of harvested area)
Base Case	0–10	1.0	0–50
Bioenergy Scenario	25–60	11.9	0–30

*Per cent of harvest residues recovered for bioenergy feedstock.

FMU. Projected annual burned-areas for wildfire were estimated for each FMU from the historical burned area averaged from 1990 to 2012. Future harvest volumes were based on information provided by provincial and territorial government experts in response to detailed questionnaires (personal communications, February 2014). Future spatial allocation of projected harvesting within a jurisdiction was estimated from historical (2000–2011) disturbance change information in 250-m resolution remotely-sensed products (Guindon *et al.*, 2014) overlain with FMU boundaries. These estimates of harvest area were weighted by C harvest density to convert area allocations into the volume allocations required for modelling. The density factor took into account C density in the FMU and was estimated from average harvested merchantable C per hectare over the historical period (1990–2012), normalized to one for each jurisdiction. For the province of British Columbia, a different method was used because the impact of mountain pine beetle has significantly influenced the historical harvest allocation: future spatial harvest allocation was based on the forecast of future Allowable Annual Cut (AAC) levels (http://www.for.gov.bc.ca/hts/aactsa.htm, Accessed May 2014).

Clearcut harvesting assumed utilization rates of 85–97% of the merchantable stem biomass present at the time of harvest, with the remainder left on site as logging residue along with trees below merchantable size. In the *Base Case*, most of harvest residues progressively decayed over time, but in some regions harvest residues are piled and burned for fuel hazard management or a small amount is captured for bioenergy. In the *Bioenergy Scenario*, up to 60% of harvest residues are captured for bioenergy (depending on the province/territory) and slashburning activities are reduced or stopped. Additional details on model assumptions are found in Table S2.

We assumed that emissions associated with local (within FMU) transportation of harvest residues were roughly the same as those associated with transportation of base case fuels. We did not consider global climate change impacts on forest growth, decomposition, or disturbance regimes.

Analytical frameworks

The system boundary of the analysis included FM, HWPs, bioenergy, and corresponding fossil emissions displaced in the energy sector. We assessed 634 FMUs and identified 502 where harvest activity could support bioenergy production.

Forest ecosystem C dynamics

Forest ecosystem C dynamics were analysed using the National Forest Carbon Monitoring, Accounting and Reporting System (NFCMARS) data sets and its core modelling engine, the Carbon Budget Model of the Canadian Forest Sector (CBM-CFS3). See Stinson *et al.* (2011) for a description of NFCMARS data sets and Kurz *et al.* (2009) for a description of CBM-CFS3. Model simulations were conducted for Canada's managed forest, which included lands managed for sustainable harvest, lands under protection from natural disturbances, and areas managed to conserve forest ecological values. Forest inventory data included stand attributes (age, species types) and merchantable volume yield tables for each of the hardwood (broadleaf) and softwood (coniferous) components. CBM-CFS3 tracks C stocks in ten biomass pools (hardwood and softwood versions of merchantable stemwood, foliage, coarse roots, fine roots, and 'other' which includes branches and non-merchantable-sized trees), 11 dead organic matter pools (which include woody litter, the soil organic horizon and mineral soil), and emissions of carbon dioxide (CO_2), methane (CH_4), carbon monoxide (CO) from slashburning and wildfires. Nitrous oxide (N_2O) is also included, using an emissions factor that is applied to the CO_2 emissions resulting from burning. Additional background information on forest ecosystem modelling is in the Supplementary Information.

Harvested wood-product emissions

Carbon Budget Model of the Canadian Forest Sector outputs describing the quantities of C transferred to HWPs and bioenergy were inputs to the Carbon Budget Modelling Framework for Harvested Wood Products (CBMF-HWP), an analytical tool that tracks the fate of harvested C through manufacturing, use, and end-of-life treatment. All emissions associated with forest C harvested in Canada were tracked in the analysis using the production approach, irrespective of whether the HWPs were exported, in keeping with internationally agreed upon approaches for HWP C accounting (IPCC, 2014), and are consistent with the 2014 National Inventory Report (Environment Canada, 2014). HWP commodity and postconsumer parameters (Table S2) were the same for both scenarios and therefore HWPs other than bioenergy had no impact on mitigation potential. Bioenergy production from captured harvest residues was also tracked in the CBMF-HWP, and these emissions were higher in the *Bioenergy Scenario*.

Displaced emissions

Displaced emissions are defined as emissions that would have occurred if *Base Case* energy sources had been used. Displaced emissions were estimated by multiplying the captured harvest residues by a regionally determined displacement factor for each FMU. The displacement factor was estimated for each FMU by selecting the type, size and number of bioenergy facilities that maximized displaced emissions based on the (1) available harvest residues, (2) regional energy consumption, and (3) *Base Case* fuel mix. See Smyth *et al.* (2016) for a complete description of the displacement factor estimates.

Base Case energy sources for electricity were from projected energy sources (Table S3) for each province/territory (National Energy Board, 2013) and sources of heat were from contemporary (2012) energy sources (Office of Energy Efficiency, 2015) because projections were not available. For FMUs that contained a remote community, the regional fuel mix was estimated from the provincial or territorial average and then adjusted to include the contemporary remote community fuel mix using to a weighted-proportion of the population of the remote community and regional population (Natural Resources Canada, 2014). Remote communities are defined as those that are not connected to an electricity grid and that therefore have a different fuel mix than the jurisdictional-average fuel mix.

Energy consumption for heat and electricity was estimated from each jurisdiction's per capita energy use and contemporary population estimates from census data (Statistics Canada, 2011).

We assumed that all captured harvest residues were first used within their FMU to produce heat and electricity to meet local demand, which was estimated from per capita use multiplied by the population within the FMU. Heat production was constrained to local demand, and any excess harvest residues (beyond that needed for local demand) were consumed to generate grid electricity which was assumed to displace the average electricity fuel mix.

The nine bioenergy facilities included three different types of facilities (heat, power, and combined heat and power) and three different sizes of facilities, ranging from 200 kW turbines to a 10 MWh steam cycle power facility, Table 2. These facilities were selected for analysis because they cover a wide range of heat and power production scales, and because information was available on annual fibre demand and operating and capital costs for these facilities.

Mitigation indicators

Mitigation was defined as the difference between *Base Case* emissions and *Bioenergy Scenario* emissions:

$$M = E_{\text{Base}} - E_{\text{BioE}}, \qquad (1)$$

where M is the mitigation, E_{Base} is the *Base Case* emissions, and E_{BioE} is the *Bioenergy Scenario* emissions. Evaluating the mitigation scenario relative to the *Base Case* in this way and applying *Base Case* and the mitigation scenario to the same forest inventory data factors out age-class legacy effects (Böttcher *et al.*, 2008) on contemporary C dynamics. Simulating the same base level of natural disturbance in the *Base Case* and the *Bioenergy Scenario* also causes the impacts of natural disturbances assumed to occur from 2013 onward to be almost completely factored out, with slight differences caused by interactions between forest management and natural disturbance activities.

Total emissions in both the *Base Case* (E_{Base}) and the *Bioenergy Scenario* (E_{BioE}) were estimated as the sum of emissions from three components:

$$E = E_{\text{Forest}} + E_{\text{HWP}} - E_{\text{Displaced}}, \qquad (2)$$

where E_{Forest} is GHG removals from the forest due to C uptake and the emissions from the forest due to heterotrophic decay

Table 2 Bioenergy facilities description adapted from Smyth *et al.* (2016)

Facility type	Facility description	Biomass delivered (odt)	Electrical conversion rate (kWh odt^{-1})	Thermal conversion rate (kWh odt^{-1})	Assumed electrical efficiency (%)	Assumed thermal efficiency (%)	Böttcher overall efficiency (%)
Heat	0.4 MWth boiler for district heating*	783	–	15.00	–	75	75
	2.3 MWth boiler for district heating*	3974	–	17.00	–	85	85
	6.62 MWth process heat via syngas†	11 576	–	16.80	–	84	84
Power	0.2 MWe gas turbine‡	1600	1020	–	18	–	18
	5 MWe steam cycle†	34 971	1167	–	21	–	21
	10 MWe steam cycle†	63 861	1278	–	23	–	23
CHP	0.2 MWe, 0.98 MWth organic rankine cycle§	2098	778	14.00	14	70	84
	1.8 MWe, 4.5MWth steam turbine¶	10 575	1389	10.80	25	54	79
	8 MWe CHP steam turbine†	46 870	1393	5.88	25	29	54

*RETScreen International (2015).
†Biopathways (2015).
‡Arena *et al.* (2010).
§Wood & Rowley (2011).
¶Pröll *et al.* (2011).

and disturbances; E_{HWP} is emissions from bioenergy produced from harvest residues, bioenergy from harvested roundwood, and postconsumer emissions; and $E_{Displaced}$ is the emissions displaced by substituting bioenergy for alternate fuel sources.

Indicators included projections of future harvest residue availability for bioenergy from 2017 to 2050 for each FMU, identification of FMUs where there is a positive mitigation benefit of using residues for bioenergy, and identification of the point in time at which this positive benefit begins to occur (i.e. the 'break-even' point). After the break-even point, continuing to displace the *Base Case* energy sources results in a net GHG emission reduction.

Break-even points were also estimated based on a cumulative radiative forcing indicator which considers the temporal dynamics of atmospheric CO_2. We use the method by Sathre & Gustavsson (2011), where the atmospheric decay of each annual pulse emissions is estimated as

$$(CO_2)_t = (CO_2)_0 \left[0.217 + 0.259e^{\frac{-t}{172.9}} + 0.338e^{\frac{-t}{18.51}} + 0.186e^{\frac{-t}{1.186}} \right], \quad (3)$$

where t is the number of years since the pulse emissions, $(CO_2)_0$ is the mass of CO_2 emitted at year 0 for the *Base Case* minus the *Bioenergy* scenario, and $(CO_2)_t$ is the mass of CO_2 remaining in the atmosphere at year t. The total change in atmospheric mass of CO_2 for each year of the study period is then determined by summing the emissions occurring during that year plus the emissions of all previous years minus their decay during the intervening years.

The change in atmospheric mass of CO_2 is then converted to change in atmospheric concentration, based on the molecular mass of CO_2 (44.0095 g mol^{-1}), the molecular mass of dry air

(28.95 g mol^{-1}), and the total mass of the atmosphere (5.148 × 10^{21} g). Annual changes in instantaneous radiative forcing due to the CO_2 concentration changes are then estimated using

$$F_{CO_2} = \frac{3.7}{\ln 2} \ln \left[1 + \frac{\Delta CO_2}{CO_{2R}} \right], \quad (4)$$

where F_{CO_2} is instantaneous radiative forcing in W m^{-2}, ΔCO_2 is the change in atmospheric concentration of CO_2 in units of ppmv, and the reference concentration, CO_{2R}, is 383 ppmv. Positive radiative forcing tends to warm the earth's surface, while negative radiative forcing tends to cool it. We then estimate the cumulative radiative forcing (CRF) occurring each year in units of W-s m^{-2}, by multiplying the instantaneous radiative forcing of each annual period by the number of seconds in a year and determine the break-even point, after which there is a net reduction in the cumulative radiative forcing.

Sensitivity analysis

A sensitivity analysis was performed to assess the change in cumulative mitigation potential by omitting the remote community fuel mix and assuming higher efficiencies in bioenergy conversion resulting in different displacement factors. These sensitivity runs were modelled using two different sets of displacement factors. In the first case, the remote community fuel mix was omitted, and the jurisdiction-average fuel mix was assumed in all FMUs. In the second case, a set of nine generic bioenergy facilities with higher energy conversion efficiencies and different facility sizes (Table S4) was used to determine displacement factors for the sensitivity analysis.

Results

Results are presented include (1) national-level stocks and flows that are spatially and temporally averaged, (2) national-level emissions time series that are spatially averaged, and (3) FMU-level mitigation indicators that are temporally averaged.

Averaged stocks and flows

National-scale estimates of average stocks and flows are shown in Fig. 2 for the *Base Case* and *Bioenergy Scenarios.*

Most of the C is stored in dead organic matter, with 71.8% of the total C stocks in litter, dead wood, and soil. Live biomass storage accounts for 27.1% of the total C stock, and 1.2% is contained within HWPs that have been harvested since 1990 and are in use.

The largest C flows in the system are part of the annual cycle of uptake of C from the atmosphere, turnover of biomass stocks, and release of C through heterotrophic respiration (consistent with Stinson *et al.*, 2011). Smaller disturbance emissions are found for wildfires and slashburning. HWP flows release C to the atmosphere from instant oxidation of postconsumer products

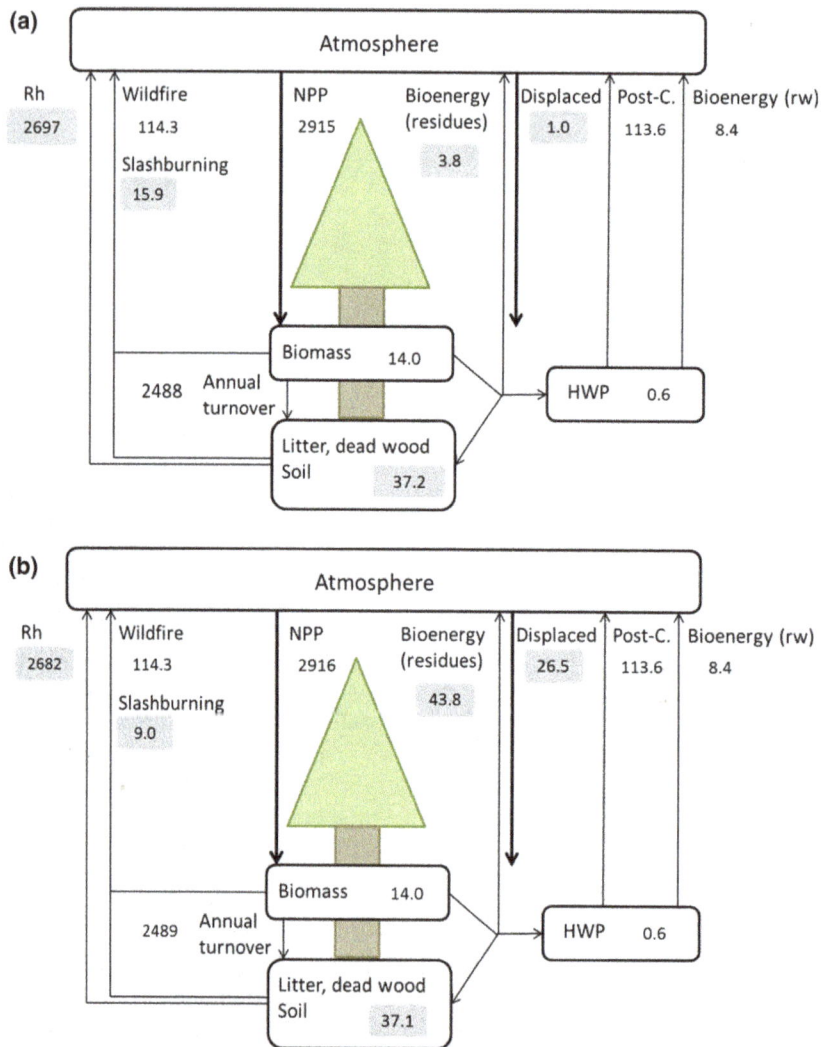

Fig. 2 Stocks and flows temporally averaged over 2017–2050 for Canada's 2.3 × 10^6 km^2 managed forest in the (a) *Base Case* and (b) *Bioenergy Scenario.* Changes in forest management activities affect certain stocks and flows (grey boxes). Stocks for biomass, dead organic matter (litter, dead wood, and soil), and in-use harvested wood products (HWP) are in PgC. Annual flows, in TgCO$_2$e yr^{-1}, show uptake of C through net primary productivity (NPP), transfer of C through annual turnover, and release of C through heterotrophic respiration (Rh). Release of C also occurs as a result of disturbances and through release of C from postconsumer HWP (Post C.) and bioenergy from roundwood (rw). Emissions from wood harvested before 1990 are not included. Harvest transfers an average of 41.5 TgC yr^{-1} (152.2 TgCO$_2$ yr^{-1}) from live biomass and dead organic matter to harvested wood products.

and mill residues, and bioenergy production from harvest residues and roundwood.

The *Base Case* and *Bioenergy Scenarios* differ in the treatment of harvest residues which affects emissions related to bioenergy and slashburning, heterotrophic respiration, and dead organic matter stocks. In the *Base Case*, harvest residues are either (1) piled and burned, and most of the C is instantly released to the atmosphere, or (2) left in the forest to progressively decompose, and most of the C is released to the atmosphere more slowly over time or incorporated into the soil as more stable C compounds. In the *Bioenergy Scenario*, harvest residues are used to produce bioenergy, and carbon that would have been gradually released from the dead wood and litter pools or stored in the soil is released immediately to the atmosphere. The *Bioenergy Scenario* has lower heterotrophic respiration, lower dead organic matter stocks (dead wood, litter and soil), and lower slashburning emissions relative to the *Base Case*, for regions where slashburning is present (Table S2).

Harvest transfers an average of 41.5 TgC yr^{-1} (152.2 TgCO$_2$ yr^{-1}) from live biomass and dead organic matter from 2017 to 2050. In the *Bioenergy Scenario*, an average of 11.9 TgC yr^{-1} of harvest residues are used for bioenergy, and the displaced emissions from using bioenergy in place of another energy source are 28.6 TgCO$_2$e yr^{-1}.

The *Base Case* includes a small amount of harvest residues used for bioenergy and small associated displaced emissions, but these emissions are also contained in the *Bioenergy Scenario* and therefore cancel out of the mitigation estimate.

Mitigation timeseries

Timeseries of emissions/removals associated with forest, HWP, and displaced emissions components are shown in Fig. 3a for both scenarios. Emissions/removals associated with growth, transfer, decay, and disturbances (including wildfires and slashburning) are contained within the forest component.

During the historical period, the forest ecosystem shows large fluctuations in emissions and removals due to the impacts of natural disturbances (large and intermittent wildfire activity and mountain pine beetle epidemic from 2000 to 2010 (Kurz *et al.*, 2008; Stinson *et al.*, 2011)).

For the future period, forest ecosystem removals are enhanced in the *Bioenergy Scenario* relative to the *Base Case* due to lower slashburning emissions and lower heterotrophic respiration. Bioenergy emissions, product emissions, and postconsumer product emissions are included in the HWP component. HWP emissions are

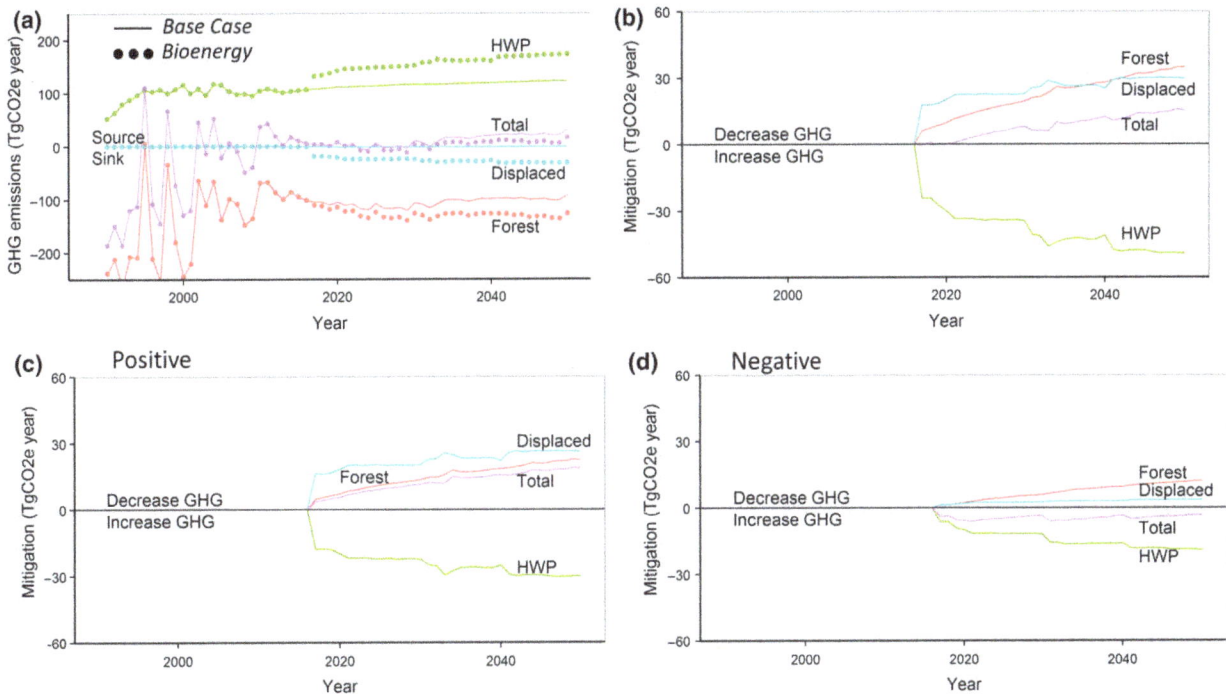

Fig. 3 Time series of (a) net GHG emissions/removals from the forest ecosystem, HWP emissions including bioenergy, displaced emissions and the total emissions for the *Base Case* and *Bioenergy Scenarios*, and (b) total climate change mitigation potential for all FMUs and the contribution from each component. Time series of the total and components for (c) FMUs with positive mitigation in 2050, and (d) FMUs with negative mitigation in 2050.

higher in the *Bioenergy Scenario* relative to the *Base Case* because of the capture and burning of harvest residues for bioenergy. The timeseries show a stepwise increase in HWP emissions due to a stepwise increase in proportion of harvest residues captured. Displaced emissions also show a stepwise increase over time and represent a reduction in emissions to the atmosphere. Total emissions are a small source of GHG emissions to the atmosphere of 11.5 TgCO$_2$e yr^{-1} for the *Base Case* and a smaller source of 3.4 TgCO$_2$e yr^{-1} for the *Bioenergy Scenario*, on average from 2017 to 2050.

The potential mitigation time series was obtained by subtracting the *Bioenergy* emissions/removals from the *Base Case*, Fig. 3b. The increase in emissions associated with production of bioenergy from harvest residues was offset by enhanced removals in the forest ecosystem and displaced emissions, resulting in an overall positive mitigation potential which increased over time. Of the 502 FMUs contained in the overall mitigation estimate, 278 had a positive cumulative mitigation (reduction in emissions relative to the *Base Case*) in 2050 and 224 FMUs had an increase in emissions (negative mitigation), resulting from increased use of harvest residues for bioenergy, Fig. 3c, d and Table 3.

Mitigation indicators for FMUs

We examined patterns of captured harvest residues, energy demand, displacement factors, and total cumulative mitigation potential to understand the spatial distribution of FMUs with positive and negative mitigation (Fig. 4). Captured harvest residues depend on assumptions about the proportion of harvest residues that can be extracted for bioenergy for each province or territory (See Supplementary Information), the allocation of future harvest within each jurisdiction, and the characteristics of the forest. For most jurisdictions, captured harvest residues are found in the northern and central regions of the managed forest (Fig. 4a), whereas energy demand, using population density as a proxy, is concentrated in the southern part of the jurisdictions (Fig. 4b). Results show that captured harvest residues and energy demand are not always co-located and the potential for bioenergy production exceeds local heat and electricity demand in many FMUs with small populations. Displacement factors (Fig. 4c) estimated from the displaced emissions resulting from the substitution of the suite of bioenergy facilities varied widely between 0 and 1.8 tonnes of C displaced per tonne of biogenic C utilized (tC/tC), with an average value of 0.5. Displacement factors

Table 3 Summary information for forest management units (FMUs) with positive and negative cumulative mitigation in 2050

Description	Positive mitigation		Negative mitigation	
Number of included FMUs	278		224	
Residues captured for bioenergy (Tg C yr^{-1})	7.1		4.8	
Total population (millions)	32.1		0.2	
Remote community population (thousands)	103.9		0	
Forest area ($\times 10^6$ km^2)	1.4		0.8	
Cumulative mitigation (TgCO$_2$e)	2020: 21.2			
			2020: −20.5	
	2030: 115.7			
			2030: −67.1	
	2040: 255.1			
			2040: −114.2	
	2050: 429.1			
			2050: −152.5	
Average break-even time (years)	6.2		−	
Heat demand met by bioenergy (% of local demand)	3.1		77.8	
Power demand met by bioenergy (% of local demand)	2.9		308.6	
Displaced energy fuel mix (total 100%)	Heat		Heat	
	Fuel Oil	23	Fuel Oil	3.6
	Coal and Petcoke	18.1	Coal and Petcoke	4.8
	Natural Gas	17.1	Natural Gas	17.1
	Electricity	9.4	Electricity	3.9
	Electricity		Electricity	
	Grid	12.6	Grid	63.8
	Natural Gas	8.7	Natural Gas	1.8
	Coal	7.7	Coal	0.7
	Diesel	1.7	Diesel	0.3

(a)

Average captured residues
(m³ yr⁻¹)

Legend
- ○ Remote communities
- ▦ Nonmanaged forest
- ▨ Not included
- 1 – 1,000
- 1,001 – 20,000
- 20,001 – 50,000
- 50,001 – 300,000
- 300,001 – 4,500,000

(b)

Population density as
proxy for energy demand

Population/Harvest area

Legend
- ○ Remote communities
- ▦ Nonmanaged forest
- ▨ Not included
- 0
- 1 – 5
- 6 – 20
- 21 – 50
- 51 – 100
- 101 – 500
- 501 – 71000

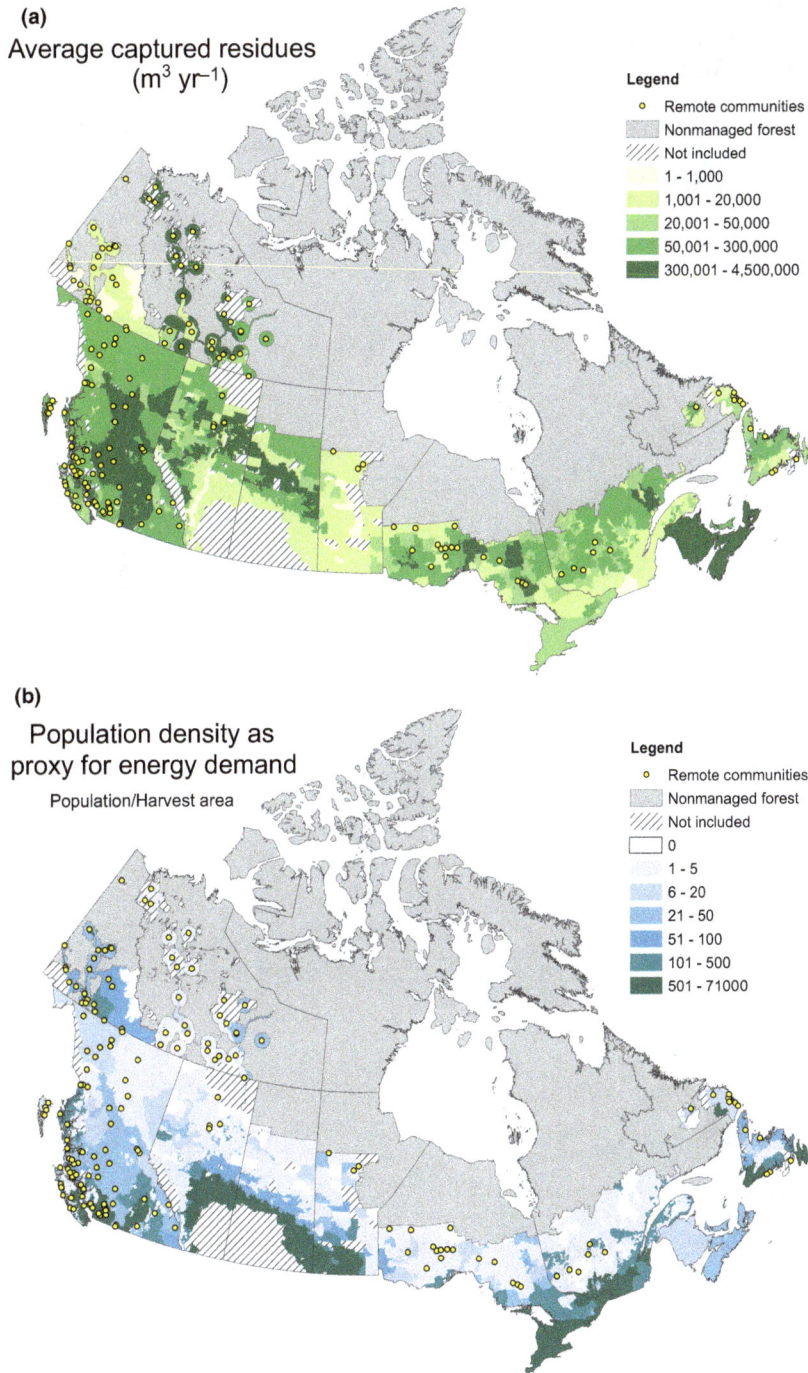

Fig. 4 Maps of average captured residues between 2017 and 2050, population density as a proxy for energy demand, displacement factor, and national cumulative mitigation from 2017 to 2050. Only local use of bioenergy was considered, with heat production constrained by local demand, and excess bioenergy exported to the electricity grid.

varied between and within jurisdictions, and depended upon the *Base case* fuel mix, the energy conversion efficiency, and the local energy demand. Low displacement factors were found in regions where bioenergy exceeded local demand, and excess residues were converted to electricity and subsequently used to displace grid electricity produced by low-emission sources (predominantly hydro-electricity). High-displacement factors were found in regions (such as the Atlantic maritime and Boreal plains ecozones) where bioenergy displaced high-emissions fossil fuel sources (coal and fuel oil).

(c)
Displacement factor
(tC avoided/tC used)

(d)
Cumulative mitigation 2050
MtCO2e

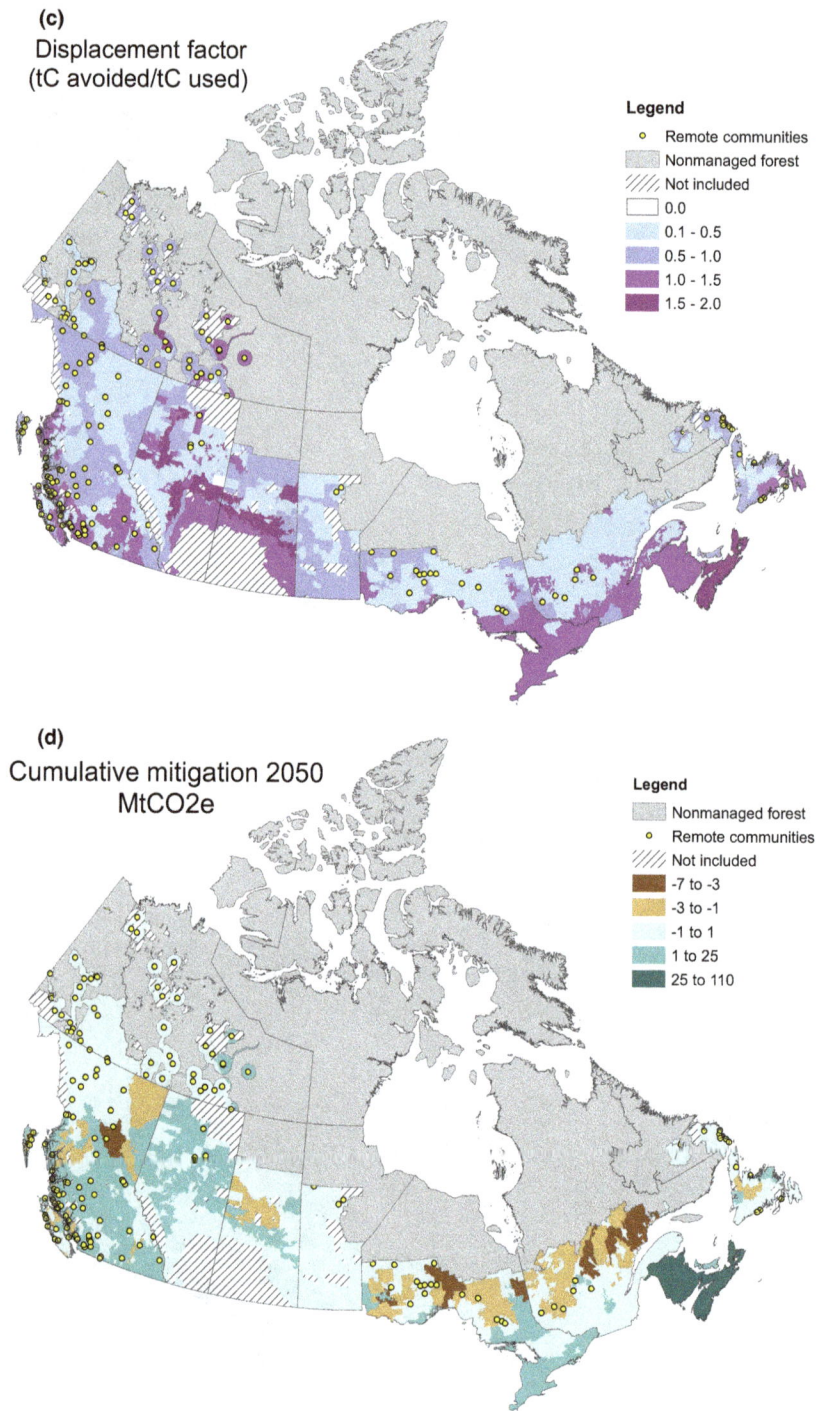

Fig. 4 Continued.

The cumulative mitigation from 2017 to 2050 is shown in Fig. 4d. Regions with positive cumulative mitigation (*i.e.* net reduction in GHG emissions to the atmosphere) generally produced less bioenergy than local energy demand and displaced high-emissions fossil fuels. There are many regions where the cumulative mitigation was negative, indicating the use of harvest residues for bioenergy increased GHG emissions to the atmosphere. These regions corresponded to regions where bioenergy production exceeded local demand and displaced emissions were low because excess bioenergy displaced a low-emission electricity-grid mix. The total national cumulative mitigation from 2017 to 2050 was 429 TgCO$_2$e for the 278 regions with positive mitigation potential (Table 3). This contribution came from 1.4 × 10^6 km^2 of managed forest in which 7.1 TgC yr^{-1} of harvest residues (27 Mm^3 yr^{-1}) produced bioenergy that met 3.1% of the heat demand and 2.9% of the electricity demand for 32.1 million people living within these regions. The average break-even time for regions with positive mitigation was 6.2 years, with a standard deviation of 8.6 years. Many regions had break-even times of zero (43% of all regions), and the majority of regions (63%) had break-even times less than 10 years. However, 25% of the regions with positive mitigation had break-even times between 10 and 20 years.

When the timing-adjusted emissions were considered, the number of regions with positive mitigation potential dropped to 210, and the break-even times increased to 9.6 years (SD 9.5 year). However, the cumulative mitigation potential from the 210 regions was very similar to the previous estimate at 428.5 TgCO$_2$e. For many regions, the break-even time was zero, and for these regions the timing-adjusted emissions had no impact.

Mitigation sign and magnitude were significantly affected by the magnitude of the displaced emissions and was predominantly affected by the jurisdiction-average fuel mix. The emissions intensity range of the displaced energy sources ranged from 0 for hydro-, solar-, and wind-generated electricity to 1000 kg CO$_2$e MWh^{-1} for coal-generated electricity. The

emissions intensity for bioenergy from residues (estimated as the bioenergy and forest mitigation potential divided by the energy produced) is compared to the displaced emissions intensity in Fig. 5. There was a wide range of emissions intensities for the harvest residues, reflecting differences in energy conversion efficiencies, number and types of facilities selected for substitution, and the alternate fate of residues in the forest (decay and/or slashburning in the Base Case). Regions with positive mitigation were associated with Base Case emissions intensities that ranged from ~200 kg CO$_2$e MWh^{-1} to 1000 kg CO$_2$e MWh^{-1}; these were displaced with Bioenergy Scenario emissions intensities (taking into account bioenergy and forest emissions) ranging from ~90 to 500 kg CO$_2$e MWh^{-1}. Regions with negative mitigation were associated with displaced emissions generally less than 220 kg CO$_2$e MWh^{-1}. Most of the projected electricity-grid emissions intensities (Table S3) are below this level, thus explaining why many regions have a negative mitigation when excess fibre was converted to electricity and used to displace the average grid fuel mix.

Sensitivity: impact of remote communities

The total population in remote communities was estimated at 109 366 people. Of these, the fuel mix was different from their jurisdiction average for 72 394 people in remote communities in 54 FMUs (Table S3). The impact of the remote community fuel mix was assessed

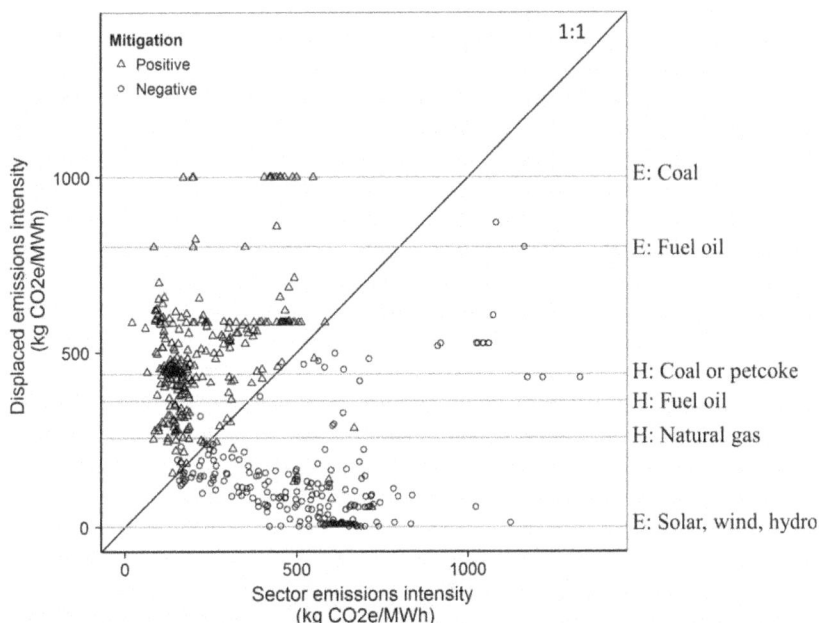

Fig. 5 Displaced emissions intensity versus the sector (Forest + HWP mitigation) emissions intensity for regions with positive mitigation and negative mitigation. Emissions intensities for alternate fuels are shown for reference.

by re-estimating the mitigation potential without the remote community fuel mix and comparing it to the previous results. Removing the fuel mix associated with remote communities decreased the total cumulative mitigation from 429 TgCO₂e to 397 TgCO₂e for the 278 FMUs originally estimated to have positive mitigation in 2050. Including only FMUs with positive mitigation when the remote community fuel mix was removed reduced the number of FMUs from 278 to 259 and decreased the total cumulative mitigation to 413 TgCO₂e.

Overall, taking remote community fuel mix into consideration increased the national mitigation potential because the fuel mix for a significant number of remote communities contained a larger proportion of higher emission fuels than the fuel mix for grid-connected communities. These results suggest that using residues for bioenergy in remote communities would result in a cumulative mitigation of 16 TgCO₂e, which is high considering that these communities only represent 0.3% of Canada's population.

Sensitivity: impact of bioenergy facility selection

The magnitude of the cumulative mitigation is expected to increase if energy conversion efficiency increases due to technological advances (Gustavsson *et al.*, 2015). The nine bioenergy facilities used in this analysis were selected because associated information on capital and operating costs was available that allowed us to estimate mitigation cost efficiency (Rampley *et al.*, 2016). Switching to a different set of nine types of bioenergy facilities (Table S4) with a higher conversion efficiency decreased the number of FMUs with positive mitigation from 278 to 268, which seems contradictory, but the bioenergy facilities have different fibre demands, and different sizes or types of bioenergy facilities with different conversion efficiencies were selected in some cases. A comparison of the displacement factors in the two cases revealed much lower displacement factors in regions with small residue availability when the alternate facilities were used. The alternate facilities had a higher annual fibre demands for heat and small CHP facilities than the generic facilities, and in some cases only small electricity facilities were selected. Total cumulative mitigation in 2050 increased from 429 to 552 TgCO₂e using the nine generic facilities. The largest difference between the nine selected facilities and the generic facilities was for the medium-sized CHP (1.8 MWe/4.5 MWth selected, 1.0 MWe/5 MWth generic) which produced most of the heat (67% and 85% for the selected and generic facilities, respectively) and much of the electricity (40% and 34%). The total heat produced from the nine generic facilities was 35% larger

than that produced by the selected facilities, while the total electricity production was roughly the same (i.e. within 0.5% of each other).

Mitigation potential comparisons

Results of this study cannot be compared directly to strategies examined in an earlier study (Smyth *et al.*, 2014) because of slight differences in the modelling assumptions made regarding projected harvest levels, start dates for the mitigation activities, and differences in displacement factors for bioenergy. However, these studies are similar enough that general comparisons of the magnitudes of the cumulative mitigation potential by 2050 can be made at the ecozone level. Figure 6 shows the present estimates of the cumulative mitigation potential in comparison with four of the nine strategies previously examined (Smyth *et al.*, 2014). The present results have much higher mitigation potential than a green harvest for bioenergy strategy because we found in the previous study that green harvest resulted in a negative mitigation potential. Compared to a strategy of using more of the stem harvest for longer-lived wood products, the national-level mitigation estimates were similar to results from this study (435 TgCO₂e vs. 429 TgCO₂e) but the allocation by ecozone was different: roughly half of the cumulative mitigation potential in this study was concentrated in one ecozone (Atlantic Maritimes), whereas the contribution from longer-lived wood products was

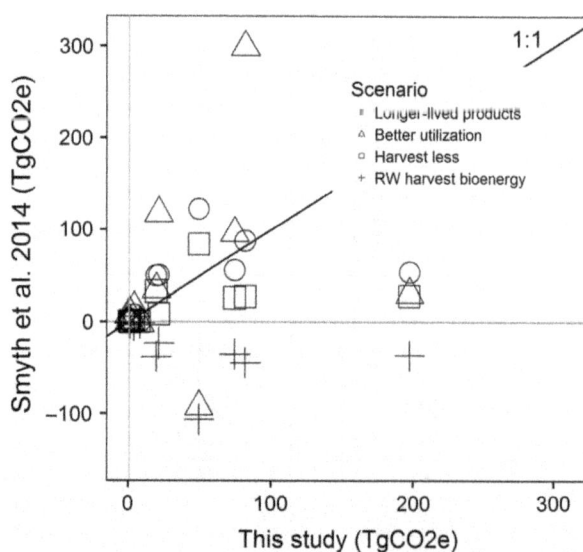

Fig. 6 Comparison of mitigation strategies from Smyth *et al.* (2014) to results from this study. Ecozones with higher mitigation potential from production of bioenergy from harvest residues (present results) are below the 1 : 1 line.

distributed over many ecozones. Overall, the longer-lived wood products strategy had a higher mitigation potential than capturing residues for bioenergy in all but two ecozones. Similarly, the better utilization strategy (involving increasing the harvest utilization rate of merchantable-sized trees, increasing the proportion of salvage harvest, capturing residues for bioenergy and omitting or reducing slashburning) had a higher mitigation potential than this study in all but two ecozones. As in the previous analysis, it would be possible to combine mitigation activities by creating a portfolio and selecting the best combination of activities to maximize the mitigation potential in each region.

Improvements to the earlier analyses (Smyth et al., 2014) include higher spatial resolution in the harvest allocation and improvements to the displaced emissions estimates. The bioenergy facilities and displaced fuel sources in the earlier analysis were selected based on expert judgement, but in this study we considerably refined displaced emissions estimates based on an optimized selection of the type, number and size of bioenergy facilities (see Smyth et al., 2016) that would maximize displaced emissions.

Discussion

Our results demonstrate a significant potential for climate change mitigation from Canada's forest sector through use of harvest residues for bioenergy. This type of quantitative analysis has never been carried out at the national level for Canada, in particular for displaced emissions based on multiple fuel sources and optimized facility selection.

These results should be interpreted as an upper limit in some regions because we have not included economic considerations, or regulatory or market barriers (Roach & Berch, 2014). Forests provide a range of services and co-benefits, and forest managers are required to manage for multiple objectives, some of which could come into conflict with mitigation objectives and may limit the level of mitigation strategy implementation (Golden et al., 2011). We assumed that harvest occurred primarily for the production of wood commodities, but 6% of harvest in 2050 was for the production of bioenergy where it may be less likely to capture harvest residues.

Results of the study found that 278 of the 502 regions investigated had a positive mitigation, in which 7.1 TgC yr^{-1} of harvest residues produced bioenergy that displaced a higher-emitting fossil fuel. For the other regions, there were significant volumes of residues (roughly 40%) that would not reduce GHG emissions if captured for bioenergy. In many of these regions,

captured residues exceeded local demand for heat and power, and excess residues were used to generate electricity. This is not to say that these regions could not produce positive mitigation benefits because each region has an optimal amount of captured residues that would maximize displaced emissions. The optimal amount of captured residues would depend on local heat demand and regional fuel mix, and we anticipate amount of residues necessary to meet demand would be a small proportion of the 4.8 Tg C yr^{-1} captured residues because the population within the negative mitigation regions is small (0.2 million people, Table 3). Dynamically capturing enough harvest residues to meet local demand and displacing only high-emissions fuels would give a greater mitigation potential, but this is beyond the scope of our study.

Break-even point, or C payback period, is the principal metric identified in a recent meta-analysis of forest bioenergy GHG accounting studies (Buchholz et al., 2016), and hence comparisons were made to other studies of GHG mitigation from use of harvest residues for bioenergy. The average payback period in our study was 6.2 ± 8.6 years, with forest residues from bioenergy predominantly producing heat and replacing fuel oil, coal and petcoke, and natural gas. This compares favourably with the range of values found in the literature, as described below.

An immediate carbon benefit has been found in regions where slash-burning treatment to reduce fuel loads is common practice and is replaced using this material as bioenergy feedstock (Jones et al., 2010). A payback time of 16 years was found in a study by McKechnie et al. (2011) where residues were co-fired in a coal facility to generate electricity. Zanchi et al. (2012) found the break-even time for use of harvest residues ranged from 0 for when they displaced coal, 7 years for oil, and 16 years for natural gas. Guest et al. (2013) found payback times of 13–36 years for natural gas (heat) and 0–12 years for coal (electricity).

Differences in payback times between studies can result from different assumptions about displaced fossil fuels, facility efficiencies, and residue feedstocks. The effectiveness with which fossil C is displaced has a major influence on net GHG balances, and faster climate benefits have been found when bioenergy displaced coal rather than natural gas (Cintas et al., 2015). Shorter break-even times have been found for heat or combined heat and power technologies due to the increased conversion efficiency of woody biomass combustion (Richter et al., 2009). Payback times are also affected by differences in decomposition rates due to biomass type (e.g. capture of stumps) and the overall rate of decomposition (Repo et al., 2011).

Residue extraction can sometimes affect subsequent growth rates of forests and hence C sequestration rate in Europe, but there is little evidence for this in North America (Thiffault *et al.*, 2011) where there is only one known long-term study with growth reductions following nutrient removals in harvest residue (Ponder *et al.*, 2012), and this is on a poor phosphorus-deficient site in the south-eastern United States (Scott & Dean, 2006). On the other hand, residue extraction can assist with site preparation before planting and reduce the fuel hazard (Saarinen, 2006; Helmisaari *et al.*, 2011). In the absence of long-term data to the contrary, we have therefore assumed that there is no reduction in growth rates following harvest residue removal from sites in Canada. Additionally, climate change impacts on forest growth, decomposition, or disturbance regimes were not considered in this analysis.

Future analysis will consider additional impacts on the national mitigation potential, beyond the sensitivity of bioenergy facilities and remote community fuel mix. Refining the spatial allocation of fossil fuel usage to account for the fuel mix used locally could increase the mitigation potential in regions with high fossil fuel emissions intensities. An additional assessment could consider alternate future energy demands, with different trajectories of fossil fuel elimination by decade. The mitigation potential could decrease, relative to the present results, if fossil fuels are more quickly replaced by hydro-electricity, solar, tidal, geothermal, or nuclear energy sources. On the other hand if the current fossil fuel mix is maintained, the mitigation potential would be higher. We did not address the option to transfer residues across FMU boundaries where communities are nearby, but outside the FMU, because these simulations are not spatially explicit. Future analyses could address these issues.

Conclusions

Canada's forests and forest products can contribute to mitigating climate change, and several mitigation options are available for forest management and wood-product use. The results of this study provide estimates of the mitigation potential to 2050 of using the residues for bioenergy that will be useful to strategic decision-makers and planners at regional, provincial, or national scales.

We emphasize the importance of a sound analytical framework for mitigation assessment associated with incremental activities relative to a *Base Case*, and an integrated assessment of harvest residues for bioenergy using a systems approach, and hence we examined C pools in the forest ecosystem, C use and storage in

HWPs, and substitution of wood for other energy sources.

The local use of harvest residues for local production of bioenergy was found to be effective in some locations but counter-productive from a climate change mitigation standpoint in other locations. The total national mitigation potential of 429 $MtCO_2e$ was comparable to some of the other mitigation strategies examined in an earlier analysis (Smyth *et al.*, 2014), but the mitigation potential from this analysis is concentrated in two ecozones, whereas the other strategies had higher mitigation potential in many ecozones for a longer-lived products strategy and a better utilization strategy. Substantial gains could be realized through a portfolio of strategies, both in contributing to Canada's emission reduction targets and in reducing global emissions.

This national-level study estimated the displaced emissions from using bioenergy in place of another energy source by multiplying the captured residue by a displacement factor for each forest management unit. Displacement factors (from Smyth *et al.*, 2016) were based on the local energy demand, strategic displacement of the highest emitting fuels in the fuel mix for heat and electricity, and a selection of the set (size, type and number) of bioenergy facilities that maximized the displaced emissions. Our results show that the use of harvest residues for bioenergy resulted in a reduction in GHG emissions in regions where high-emitting fossil fuels were displaced, mainly from heat production in combined heat and power facilities. Negative mitigation potential was found in regions where harvest residues were used to generate electricity and displace low-emission hydroelectricity. We conclude that national-scale forest sector mitigation options need to be assessed rigorously from a systems perspective to distinguish policies that deliver net benefits to the atmosphere from those policies that do not contribute to climate change mitigation.

Acknowledgements

This study would not have been possible without strong cooperation between provincial, territorial, and federal government agencies. We thank all members of the National Forest Sinks Committee and their colleagues and acknowledge their support in developing model inputs. However, the authors accept full responsibility for the assumptions made in the analysis. Thanks also to the Canadian Forest Service colleagues E. Walsh, M. Fellows, K. Blake, E. Neilson, M. Hafer, M. Magnan, G. Zhang, S. Morken and G. Thandi for their support in model simulations and presentation. We also thank B. Titus and two anonymous reviewers for providing thoughtful comments on the manuscript. Funding for this study was provided by the Government of Canada's Clean Air Agenda, the Program of

Energy Research and Development, and in-kind contributions from provincial and territorial governments.

References

Arena U, Di Gregorio F, Santonastasi M (2010) A techno-economic comparison between two design configurations for a small scale, biomass-to-energy gasification based system. *Chemical Engineering Journal*, **162**, 580–590.

Biopathways (2015) Biopathways database. Available at: www.fpac.ca/bio-pathways and http://www.nrcan.gc.ca/forests/industry/bioproducts/13321 (accessed May 2012).

Böttcher H, Kurz WA, Freibauer A (2008) Accounting of forest carbon sinks and sources under a future climate protocol—factoring out past disturbance and management effects on age-class structure. *Environmental Science & Policy*, **11**, 669–686.

Buchholz T, Hurteau MD, Gunn J, Saah D (2016) A global meta-analysis of forest bioenergy greenhouse gas emission accounting studies. *GCB Bioenergy*, **8**, 281–289.

Cintas O, Berndes G, Cowie AL, Egnell G, Holmström H, Ågren GI (2015) The climate effect of increased forest bioenergy use in Sweden: evaluation at different spatial and temporal scales. *Wiley Interdisciplinary Reviews: Energy and Environment*, **5**, 351–369.

Colombo SJ, Parker WC, Luckai N, Dang Q, Cai T (2005) *The Effects of Forest Management on Carbon Storage in Ontario's Forests, Sault Ste.* Ontario Ministry of Natural Resources, Marie, ON.

DECC (2012) *UK Bioenergy Strategy.* Department of Energy and Climate Change, London, UK.

Domke GM, Becker DR, D'amato AW, Ek AR, Woodall CW (2012) Carbon emissions associated with the procurement and utilization of forest harvest residues for energy, northern Minnesota, USA. *Biomass and Bioenergy*, **36**, 141–150.

Environment Canada (2014) *National Inventory Report: 1990–2012, Greenhouse Gas Sources and Sinks in Canada.* Environment Canada, Greenhouse Gas Division, Ottawa, ON.

Environmental Protection Act Bill C-33 (2008) www.lop.parl.gc.ca/Content/LOP/LegislativeSummaries/39/2/c33-e.pdf (accessed May 2014).

Golden D, Smith MA, Colombo S (2011) Forest Carbon management and carbon trading: a review of Canadian forest options for climate change mitigation. *The Forestry Chronicle*, **87**, 625–635.

Gómez DR, Watterson JD, Americano BB et al. (2006) *2006 IPCC Guidelines for National Greenhouse Gas Inventories.* Intergovernmental Panel on Climate Change, Kamiyamaguchi, Hayama, Japan.

Government of Canada (2013) *Canada's Sixth National Report on Climate Change, 2014: Actions to Meet Commitments Under the United Nations Framework Convention on Climate Change.* Environment Canada, Ottawa, ON.

Government of Canada (2015) Canada's Intended Nationally Determined Contribution Submission to the UNFCCC. Available at: http://www4.unfccc.int/submissions/indc/Submission%20Pages/submissions.aspx (accessed May 2015).

Guest G, Cherubini F, Strømman AH (2013) Climate impact potential of utilizing forest residues for bioenergy in Norway. *Mitigation and Adaptation Strategies for Global Change*, **5**, 459–466.

Guindon L, Bernier PY, Beaudoin A et al. (2014) Annual mapping of large forest disturbances across Canada's forests using 250 m MODIS imagery from 2000 to 2011. *Canadian Journal of Forest Research*, **44**, 1545–1554.

Gustavsson L, Haus S, Ortiz CA, Sathre R, Truong NL (2015) Climate effects of bioenergy from forest residues in comparison to fossil energy. *Applied Energy*, **138**, 36–50.

Helmisaari H-S, Hanssen KH, Jacobson S et al. (2011) Logging residue removal after thinning in Nordic boreal forests: long-term impact on tree growth. *Forest Ecology and Management*, **261**, 1919–1927.

IPCC (2011) *Special Report on Renewable Energy Sources and Climate Change Mitigation.* (eds Edenhofer O, Pichs-Madruga R et al.) Cambridge University Press, Cambridge.

IPCC (2014) *2013 Revised Supplementary Methods and Good Practice Guidance Arising From the Kyoto Protocol.* (eds Hiraishi T, Krug T, Tanabe K et al.). IPCC, Switzerland.

Jones G, Loeffler D, Calkin D, Chung W (2010) Forest treatment residues for thermal energy compared with disposal by onsite burning: emissions and energy return. *Biomass and Bioenergy*, **34**, 737–746.

Kurz WA, Dymond CC, Stinson G et al. (2008) Mountain pine beetle and forest carbon feedback to climate change. *Nature*, **452**, 987–990.

Kurz WA, Dymond CC, White TM et al. (2009) CBM-CFS3: a model of carbon-dynamics in forestry and land-use change implementing IPCC standards. *Ecological Modelling*, **220**, 480–504.

Lemprière TC, Kurz WA, Hogg EH et al. (2013) Canadian boreal forests and climate change mitigation. *Environmental Reviews*, **21**, 293–321.

Lundmark T, Bergh J, Hofer P et al. (2014) Potential roles of Swedish forestry in the context of climate change mitigation. *Forests*, **5**, 557–578.

McKechnie J, Colombo S, Chen J, Mabee W, Maclean HL (2011) Forest bioenergy or forest Carbon? Assessing trade-offs in greenhouse gas mitigation with wood-based fuels. *Environmental Science & Technology*, **45**, 789–795.

Nabuurs GJ, Masera O, Andrasko K et al. (2007) IPCC forestry. In: *Climate Change 2007: Mitigation. Contribution of Working Group III to the Fourth Assessment Report of the Intergovernmental Panel on Climate Change* (eds Metz B, Davidson OR, Bosch PR, Dave R, Meyer LA), pp. 541–584. Cambridge University Press, Cambridge, UK and New York, NY, USA.

National Energy Board (2013) Canada's Energy Future 2013 - Energy Supply and Demand Projections to 2035 – Appendices. Available at: https://www.neb-one.gc.ca/nrg/ntgrtd/ftr/2013/ppndcs/ppndcs-eng.html (accessed May 2013).

Natural Resources Canada (2014) Remote Communities Database. Available at: http://www2.nrcan.gc.ca/eneene/sources/rcd-bce/index.cfm?fuseaction=admin.home1 (accessed May 2014).

Obersteiner M, Böttcher H, Yamagata Y (2010) Terrestrial ecosystem management for climate change mitigation. *Current Opinion in Environmental Sustainability*, **2**, 271–276.

Office of Energy Efficiency (2015) Comprehensive Energy Use Database. Available at: http://oee.nrcan.gc.ca/corporate/statistics/neud/dpa/menus/trends/comprehensive_tables/list.cfm (accessed May 2015).

O'Neill G (2012) 2012 Bioenergy Action Plan, California Energy Commission. CEC-300-2012.

Ponder F, Fleming R, Berch S et al. (2012) Effects of organic matter removal, soil compaction and vegetation control on 10th year biomass and foliar nutrition: LTSP continent-wide comparisons. *Forest Ecology and Management*, **278**, 35–54.

Pröll T, Rauch R, Aichernig C, Hofbauer H (2011) Fluidized bed steam gasification of solid biomass – performance characteristics of an 8 MWth combined heat and power plant. *International Journal of Chemical Reactor Engineering*, **5**, A54.

Ralevic P, Ryans M, Cormier D (2010) Assessing forest biomass for bioenergy: operational challenges and cost considerations. *Forestry Chronicle*, **86**, 43–50.

Rampley GJ, Lemprière TC, Walsh E et al. (2016) The cost of climate change mitigation from bioenergy in Canada's forest sector. In preparation.

Repo A, Tuomi M, Liski J (2011) Indirect carbon dioxide emissions from producing bioenergy from forest harvest residues. *GCB Bioenergy*, **3**, 107–115.

Repo A, Känkänen R, Tuovinen J-P, Antikainen R, Tuomi M, Vanhala P, Liski J (2012) Forest bioenergy climate impact can be improved by allocating forest residue removal. *GCB Bioenergy*, **4**, 202–212.

RETScreen International (2015) RETScreen Project Database. Available at: http://www.retscreen.net/ang/software_and_data.php (accessed May 2015).

Richter D, Jenkins DH, Karakash JT, Knight J, McCreery LR, Nemestothy KP (2009) Wood energy in America. *Science*, **323**, 1432–1433.

Roach J, Berch SM (2014) *A compilation of forest biomass harvesting and related policy in Canada.* BC Technical Report 081. Victoria, BC.

Röder M, Whittaker C, Thornley P (2015) How certain are greenhouse gas reductions from bioenergy? Life cycle assessment and uncertainty analysis of wood pellet-to-electricity supply chains from forest residues. *Biomass and Bioenergy*, **79**, 50–63.

Royer J (2011) Status of remote off-grid communities in Canada. Natural Resources Canada. Available at: http://www.nrcan.gc.ca/energy/publications/sciences-technology/renewable/smart-grid/11916 (accessed May 2016).

Saarinen V-M (2006) The effects of slash and stump removal on productivity and quality of forest regeneration operations—preliminary results. *Biomass and Bioenergy*, **30**, 349–356.

Sathre R, Gustavsson L (2011) Time-dependent climate benefits of using forest residues to substitute fossil fuels. *Biomass and Bioenergy*, **35**, 2506–2516.

Scott DA, Dean TJ (2006) Energy trade-offs between intensive biomass utilization, site productivity loss, and ameliorative treatments in loblolly pine plantations. *Biomass and Bioenergy*, **30**, 1001–1010.

Smyth CE, Stinson G, Neilson E, Lemprière TC, Hafer M, Rampley GJ, Kurz WA (2014) Quantifying the biophysical climate change mitigation potential of Canada's forest sector. *Biogeosciences*, **11**, 3515–3529.

Smyth CE, Rampley GJ, Lemprière TC, Schwab O, Kurz WA (2016) Estimating product and energy substitution benefits in national-scale mitigation analyses. *GCB Bioenergy*, in press.

Statistics Canada (2011) Dissemination Block boundary file. Available at: http://www12.statcan.ca/census-recensement/2011/geo/bound-limit/bound-limit-2011-eng.cfm (accessed May 2014).

Stinson G, Kurz WA, Smyth CE *et al.* (2011) An inventory-based analysis of Canada's managed forest carbon dynamics, 1990 to 2008. *Global Change Biology*, **17**, 2227–2244.

Ter-Mikaelian MT, Colombo SJ, Chen J (2015) The burning question: does forest bioenergy reduce carbon emissions? A review of common misconceptions about forest carbon accounting. *Journal of Forestry*, **113**, 57–68.

Thiffault E, Hannam K, Paré D, Titus B, Hazlett P, Maynard D, Brais S (2011) Effects of forest biomass harvesting on soil productivity in boreal and temperate forests – A review. *Environmental Reviews*, **19**, 278–309.

Valente C, Spinelli R, Hillring BG (2011) LCA of environmental and socio-economic impacts related to wood energy production in alpine conditions: valle di Fiemme (Italy). *Journal of Cleaner Production*, **19**, 1931–1938.

Werner F, Taverna R, Hofer P, Thürig E, Kaufmann E (2010) National and global greenhouse gas dynamics of different forest management and wood use scenarios: a model-based assessment. *Environmental Science & Policy*, **13**, 72–85.

White EM (2010) *Woody Biomass for Bioenergy and Biofuels in the United States – A Briefing Paper*. (ed. Department of Agriculture). Pacific Northwest Research Station, Portland, OR.

Whittaker C, Mortimer N, Murphy R, Matthews R (2011) Energy and greenhouse gas balance of the use of forest residues for bioenergy production in the UK. *Biomass and Bioenergy*, **35**, 4581–4594.

Wood SR, Rowley PN (2011) A techno-economic analysis of small-scale, biomass-fuelled combined heat and power for community housing. *Biomass and Bioenergy*, **35**, 3849–3858.

Zanchi G, Pena N, Bird N (2012) Is woody bioenergy carbon neutral? A comparative assessment of emissions from consumption of woody bioenergy and fossil fuel. *GCB Bioenergy*, **4**, 761–772.

5

Transcriptomic characterization of candidate genes responsive to salt tolerance of *Miscanthus* energy crops

ZHIHONG SONG[1,2,*], QIN XU[1,*], CONG LIN[1], CHENGCHENG TAO[1,2], CAIYUN ZHU[2,3], SHILAI XING[2,3], YANGYANG FAN[1,2], WEI LIU[3], JUAN YAN[4], JIANQIANG LI[4] and TAO SANG[1,2,3]

[1]Key Laboratory of Plant Resources and Beijing Botanical Garden, Institute of Botany, Chinese Academy of Sciences, Beijing 100093, China, [2]University of Chinese Academy of Sciences, Beijing 100049, China, [3]State Key Laboratory of Systematic and Evolutionary Botany, Institute of Botany, Chinese Academy of Sciences, Beijing 100093, China, [4]Key Laboratory of Plant Germplasm Enhancement and Speciality Agriculture, Wuhan Botanical Garden, Chinese Academy of Sciences, Wuhan 430074, China

Abstract

Given the growing need for biofuel production but the lack of suitable land for producing biomass feedstock, development of stress-tolerant energy crops will be increasingly important. We used comparative transcriptomics to reveal differential responses to long-term salt stress among five populations of *Miscanthus lutarioriparius* grown in the natural habitats and salinity experimental site. A total of 59 genes were found to be potentially responsive to the high-salinity conditions shared by the five populations, including those involved in detoxification, plant defense, photosynthesis, and signal transduction. Of these genes, about 70% were related to abiotic stress response. Among five populations, the most contrasting performance between relatively high survival rates and the relatively weak growing traits was in accordance with the down-regulation of genes involved in growth and up-regulation of genes related to plant stress tolerance in one of the populations. These results might reveal a potential tolerance-productivity trade-off, where resources were allocated from growth to stress resistance. The comparative transcriptomics of different populations among different environments will provide a basis for breeding and domestication of energy crops.

Keywords: bioenergy, energy crop domestication, long-term salt tolerance, marginal land, *Miscanthus lutarioriparius*, resource allocation

Introduction

With the increasing demand for fuel production from renewable resources, the development of second-generation energy crops capable of growing on marginal land becomes increasingly urgent (Sang & Zhu, 2011; Allwright & Taylor, 2016). Given that salinity is a major adverse environmental factor affecting plant growth and productivity (Boyer, 1982), it is important to improve salt tolerance of energy crops. Considerable efforts have been undertaken to elucidate salt-responsive mechanisms of plants (Flowers & Yeo, 1995; Zhu, 2001; Zhang et al., 2004; Brinker et al., 2010; Cherel et al., 2014; Bushman et al., 2016). However, understanding of plant responding to long-term salt stress in the field conditions that could facilitate energy crop development remained limited.

Over the past decades, intense studies were devoted to understanding the mechanisms of plant responding to salt stress using model plants under short-term (several hours) salt stress in controlled laboratory or greenhouse conditions (Munns & Termaat, 1986; Zhu et al., 1998; Taji et al., 2004; Fujii & Zhu, 2009; Sun et al., 2010; Wang et al., 2015). Although this has been a powerful approach for revealing detailed molecular mechanisms, mechanistic study of salt tolerance of plants growing in salinity soil under long-term field conditions is needed to close the gap for the development of salt-tolerant crops (Zhu et al., 1998; Brosché et al., 2005; Vicente et al., 2016). In such efforts, plants grown in the field conditions do represent a valuable resource for elucidating the tolerance mechanisms (Brosché et al., 2005). The importance of acclimation process to long-term salt stress in the field is even more crucial for perennial grasses, as they face repeated episodes of abiotic stresses during their life cycles (Brosché et al., 2005; Moinuddin et al., 2014). Thus, a better understanding of long-term salt tolerance mechanisms under field

*These two authors contributed equally to the work.

Correspondence: Tao Sang
e-mail: sang@ibcas.ac.cn

conditions should benefit breeding salt-tolerant perennial grasses.

Generally, plants responding to various abiotic stresses often show a notable alteration in gene expression at the transcriptional level (Kreps *et al.*, 2002). High-throughput RNA sequencing has been increasingly adopted to monitor global gene expression changes and identify candidate genes by comparative analyses of transcriptional profiles between normal and stressed conditions (Walia *et al.*, 2005; Wang *et al.*, 2009; Beritognolo *et al.*, 2011; Alvarez *et al.*, 2015). It is particularly useful in studying nonmodel plants whose reference genome sequences are not available (Trick *et al.*, 2009; Libault *et al.*, 2010a,b). Moreover, according to recent researches exploring the performance of native plants in adverse environments, transcriptomic analysis has been proven to be an effective approach to discover candidate genes with drastically altered expressional levels (Morozova & Marra, 2008; Wang *et al.*, 2009; Champigny *et al.*, 2013).

Several studies aiming to determine which physiological parameters best reflecting the response of *Miscanthus* to salt stress showed that salinity resulted in a reduction in plant growth and photosynthetic rates (Plazek *et al.*, 2014). Nevertheless, *Miscanthus* could still maintain a relatively high growth rate and biomass accumulation compared to other plants (Plazek *et al.*, 2014). *Miscanthus lutarioriparius*, an endemic species in central China, is considered to be a promising candidate of second-generation energy crops due to its capability of producing high biomass on the semiarid marginal land (Sang & Zhu, 2011; Yan *et al.*, 2012; Liu & Sang, 2013; Mi *et al.*, 2014; Fan *et al.*, 2015; Xu *et al.*, 2015). It is self-incompatible and reproduces through both cross pollination and clonal growth by rhizomes (Chen & Renvoize, 2006). The adaptation of *M. lutarioriparius* to the semiarid environment could be related to the change of expressional patterns of candidate genes responsible for abiotic stress resistance and the improvement of water use efficiency (Fan *et al.*, 2015; Xu *et al.*, 2015).

In this study, we sequenced transcriptomes of 50 one-year-old *M. lutarioriparius* individuals grown in the salinity experimental field in Dongying, Shandong Province of China (DS). These individuals were collected from five natural populations (NP), and seeds were planted in the experimental field. The gene expression of 50 individuals was compared with those of 40 individuals whose leaves were directly sampled from these five natural populations for transcriptomic studies. The comparative transcriptomic study allowed for the identification of candidate genes responsive to long-term salt tolerance and one underlying mechanism of tolerance-productivity trade-off.

Materials and methods

Sample collection

A total of 40 individuals of *M. lutarioriparius* from five populations across the natural habitats of the species (NP) and 50 individuals from same populations grown at the salinity experimental site (DS) were sampled in June and July of 2013, respectively. Our previous study had predicted a high degree of adaptability in *M. lutarioriparius* based on the genetic analysis of 644 individuals from 25 populations along the Yangtze River, which span the whole geographic range of the species (Yan *et al.*, 2016). To investigate whether these populations have salt tolerance and reveal the underlying mechanism of salt tolerance at the transcriptomic level, we focused on five of these populations of *M. lutarioriparius* which represent different habitats in the middle reaches of Yangtze River in this study. They are LU5, LU7, LU10, LU14, and LU9, which are from Hekou, Jianli, Honghu, and Jiayu of Hubei Province and Junshan of Hunan Province, respectively. The population identities used here are corresponding to codes in Yan *et al.* (2016). Seeds of the five populations of *M. lutarioriparius* were collected in NP and planted in the experimental field in Dongying, Shandong Province (DS), with high salt content in April 2012 (Fig. 1). The five populations of *M. lutarioriparius* were planted with randomized block design to reduce the variety of salinity in the saline field. We investigated the survival rate, plant height, and net photosynthetic rate of each sampled individual in DS. To reveal the intrinsic expression patterns of underlying adaptation in natural populations, the five populations in NP had been sampled for transcriptome sequencing around noon of 18–22 June, 2013, with eight individuals for each of the populations (J. Yan, Z.H. Song, J. Greimler, J.Q. Li, T. Sang, unpublished). To obtain samples of DS harvested at a similar growing phase with that of NP, the collecting time was accordingly adopted. Due to the seasonal climate conditions, the same temperature was found about one month later in DS than in NP. Ten individuals from each of the populations grown in DS were sampled for transcriptome sequencing around noon on 24 July, 2013. For all individuals, the fourth leaf from the top of each individual was cut and immediately placed in liquid nitrogen and stored at −80 °C for further analysis.

RNA-Seq, preprocessing, and annotating RNA-Seq data

Total RNA of leaves was extracted using Qiagen Plant Mini Kit (Qiagen, Stanford, CA, USA) and purified using the RNase-free DNaseI (TaKaRa, Otsu, Shiga, Japan) following the manufacturer's protocol. The concentration of purified RNA was quantified by NanoDrop ND-1000 spectrophotometer (Thermo Scientific, Wilmington, DE, USA) and stored at −80 °C until next step. The mRNA was isolated from 5 µg purified total RNA using one round of purification with oligo d (T) beads Dynabeads® mRNA Purification Kit (Invitrogen, Carlsbad, CA, USA). The cDNA libraries were prepared with the NEB-Next Ultra RNA Library Prep Kit for Illumina (New England

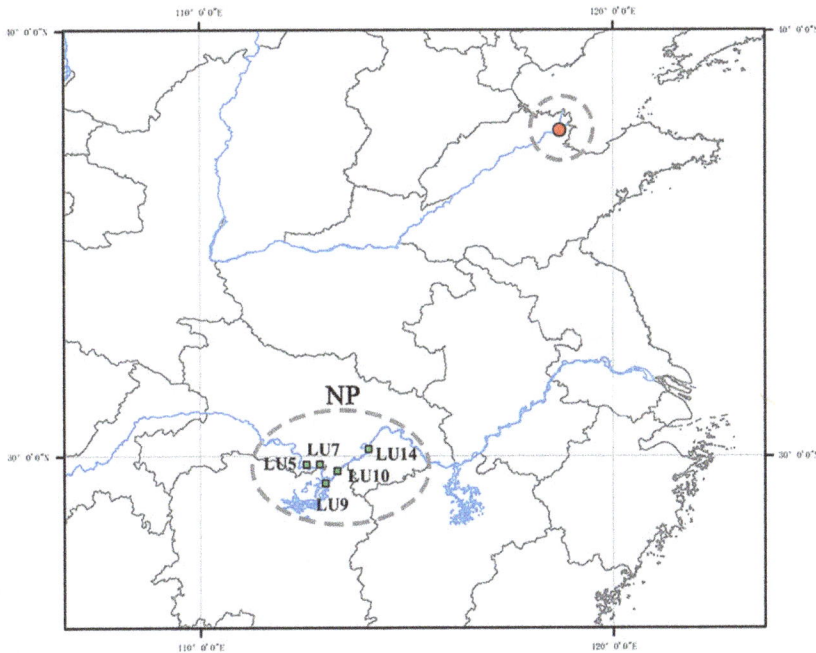

Fig. 1 Sampling map in the transplanted site and the native habitats. Map of the locations indicating the populations sampling sites with dotted circle: Dongying, Shandong Province (DS), natural populations (NP). The detailed site for the collection in DS is indicated with red circle, while the five detailed sites for the collection of each population in NP (LU5, LU7, LU9, LU10, and LU14) are indicated with green squares.

Biolabs, Ipswich, MA, USA). The first strand cDNA was synthesized using random hexamer-primed reverse transcription. Following the second-strand cDNA synthesis and adaptor ligation, approximately 450-bp cDNA fragments were isolated by the selection of Ampure XP beads (Beckman Coulter, Brea, CA, USA). The isolated cDNA fragments were amplified by 10 cycles of PCR. Library integrity and quality were estimated with Agilent 2100 Bioanalyzer (Agilent Technologies, Palo Alto, CA, USA) and Qubit 2.0 fluorometer (Life Technologies, Grand Island, NY, USA), respectively. The libraries were subsequently sequenced on the Illumina HiSeq 2500 system as 2×100-bp paired-end reads.

Raw data were sorted into the corresponding individual according to their indexed nucleotides using bcl2fastq-1. 8.4 (http://support.illumina.com/downloads/bcl2fastq_conversion_software_184.html). The unstable sequences content in the first nine bases of the 100-bp reads were trimmed in all samples. To control the quality, raw reads were filtered based on quality scores (Q = 20) and trimmed the indexed using FASTQC (http://www.bioinformatics.bbsrc.ac.uk/projects/fastqc/) and FASTX (http://hannonlab.cshl.edu/fastx_toolkit). Although there was no reference genome sequence of *M. luparioriparius*, Xu *et al.* (2015) assembled a high-quality reference transcriptome (TSA accession no. GEDE00000000). To measure the expression level of each gene in each sample, trimmed reads of each individual were mapped to the Bowtie-build indexed reference transcriptome using TopHat v2.0.0 with default settings (Langmead *et al.*, 2009; Trapnell *et al.*, 2009; Xu *et al.*, 2015). Expression level of each sample was calculated using FPKM standing for fragments per kilobase of exon per million

fragments mapped with Cufflinks v2.0.2 (Trapnell *et al.*, 2010). The trimmed sequence data are available at Sequence Read Archive (SRA) at NCBI under Project ID (SRP068901). By mapping reads from each sample onto the reference transcriptome, we obtained the number of genes expressed in each sample and the expression level for each gene.

SAMtools was used to identify single-nucleotide polymorphisms (SNPs) with default settings (Li *et al.*, 2009). We eliminated SNPs with quality score ≤ 10 and minor allele frequency (MAF) ≤ 0.05 to ensure the accuracy of SNPs. After the low-quality reads were discarded, the remaining SNPs were retained for analysis. Genetic differentiation (F_{ST}) was estimated based on those SNPs (Wright, 1978).

Expression pattern analysis

We used Xu *et al.*'s method (Xu *et al.*, 2015) to describe the level and diversity of gene expression at the population level. Population expression level (E_p) was calculated as the mean of FPKM values of the given individuals to descript the mean level of gene expression in each population. Considering the unequal number of samples collected from two sites, the extent of variability in relation to the mean of the population of gene expression in each population (expression diversity, E_d) was calculated using the formula ($E_d = \frac{\sum_{i=1}^{n} |E_i - E_p|}{(n-1)E_p}$). The above n represents the number of individuals sampled and E_i represents the FPKM of a given gene of the ith individual. The change of E_p was further examined by the ratios of E_ps (the E_p in DS divided by the E_p in NP).

Identification of population-specific and shared responsive genes

To more accurately detect population gene expression either shared or responding differently to long-term salt stress, we used both t-test and a fold change ranking (Cui & Churchill, 2003; Lee *et al.*, 2012; Parker *et al.*, 2016). The differentially expressed genes between the two sites were identified using the parametric t-test (normal bimodal distribution) or the non-parametric Wilcoxon test (nonnormal unimodal distribution). Gene expression change with $P < 0.05$ and larger than twofold change was considered to be statistically significant. Those genes significant differentially expressed in all of the five populations were considered as shared responsive genes. To further estimate the pattern of gene expression of shared responsive genes (down-regulated and up-regulated), we compared the E_ps and E_ds of these genes with those of all genes that expressed in both sites, respectively.

Functional annotation of population shared and differently responsive genes

Hierarchical cluster analysis of shared responsive genes in all individuals was carried out using MULTI EXPERIMENT VIEWER (MEV) v4.9 software using Spearman's rank correlation (Saeed *et al.*, 2003). These normalized genes' FPKM values of individuals of two sites were put together as input data for MEV v4.9. The sequences of shared responsive genes were searched against the *Arabidopsis* database, using the defaulted TAIR BLASTn parameters to analyze the detailed functional categorization. The expected value (e-value) threshold was set at 10^{-10} when we did the BLAST search. Functional classifications of genes expressed significantly different in each population were achieved using AgriGO (Du *et al.*, 2010). KOBAS 2.0 was used to analyze the pathway annotation and enrichment of the differentially expressed genes in each population and chose false discovery rate (FDR) <0.05 as the threshold (Xie *et al.*, 2011).

Results

Sequencing quality, gene expression of five M. lutarioriparius populations

Based on RNA-seq, 50 *M. lutarioriparius* individuals collected from five populations in the experimental field DS generated approximately 1 369 882 768 raw reads and 128.9 Gb of 100-bp paired-end reads after quality control (Table S1). The detail data size information for each individual in DS is shown in Table S1. After excluding genes with average FPKM value of zero in each population, the number of expressed genes for five populations in DS ranged from 17 315 to 17 560.

Then, we compared the data with that of 40 individuals of the five populations in NP. The further population transcriptome analyses were based on 16 678 genes expressed in five populations in both environments. We compared the E_ps for all genes from NP to DS, and we found that E_ps in DS shifted significantly toward to higher levels, suggesting an overall increase in expression from NP to DS (Wilcoxon's test, $P < 0.01$; Fig. S1a). We also compared the change of E_ps for each of the five different populations from NP to DS separately. From NP to DS, the proportion of genes that up-regulated was highest in LU5 and lowest in LU9 (67.4%, 62.8%, 62.1%, 61.4%, and 48.2% genes were up-regulated in LU5, LU10, LU7, LU14, and LU9, respectively; Fig. S1b–f), suggesting the majority decrease in expression from NP to DS in LU9.

Responsive genes shared in five populations of M. lutarioriparius

The salinity of NP ranges from 0.55‰ to 1.65‰ (Harmonized World Soil Database v1.2, FAO/IIASA/ISRIC/ISSCAS/JRC 2012), while the salinity of DS is around 7‰, about seven times higher than that of NP on average. To accurately identify salt stress-responsive genes shared by different populations of *M. lutarioriparius*, we conducted t-test/Wilcoxon's test and a fold change method to screen genes significantly up or down-regulated in five populations (Fig. S2). As a result, 9945 genes were significantly differentially expressed in only one of the five populations ($P < 0.05$ and twofold change; Fig. 2a), while 375, 172, and 89 genes were significantly differentially expressed together in two, three, and four of the five populations, respectively ($P < 0.05$ and twofold change; Fig. 2a). Moreover, there were 59 differentially expressed genes shared by five populations of *M. lutarioriparius*, in which 46 genes were up-regulated and 13 genes were down-regulated. The E_ps of the 59 responsive genes shared in five populations of *M. lutarioriparius* tended to be significantly higher than that of all genes in both sites (Wilcoxon's test, $P < 0.01$; Fig. 2b, c). E_ds of 59 responsive genes shared in five populations of *M. lutarioriparius* tended to be lower compared with that of the all genes in both environments (Fig. 2d, e). The ratio of E_ds between DS and NP for the 59 responsive genes shared in five populations was significantly lower than that of all genes ($P < 0.01$; Fig. S3). Together these results demonstrated the conserved expression of these shared genes in response to salt stress.

Three distinct clusters were obtained based on hierarchical cluster analysis of those genes using Spearman's rank correlation (Fig. 3a; Table 1). In addition, all 59 responsive genes shared in five populations were placed into several main functional categories based on the results of BLASTn against the *Arabidopsis* database. The major functional category was involved in plant defense (23.73%), followed by photosynthesis (15.25%),

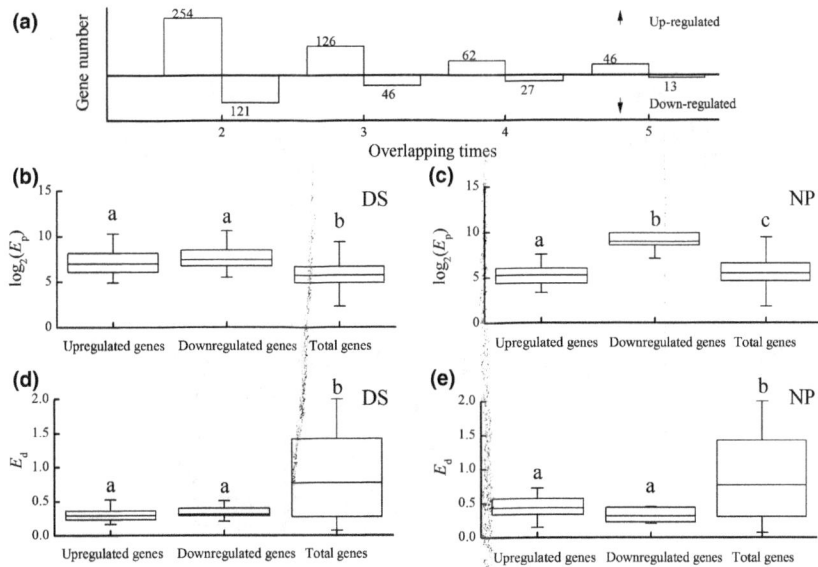

Fig. 2 E_ps and E_ds of 59 responsive genes shared in five populations of *Miscanthus lutarioriparius*. (a) The histogram shows the total number of transcripts up- or down-regulated in different times in response to salinity stress at a level of twofold or more and a P value <0.05. (b) Box plot demonstrating the $\log_2(E_p)$s of 46 up-regulated genes and 13 down-regulated genes compared with that of total genes in DS. (c) Box plot demonstrating the $\log_2(E_p)$s of 46 up-regulated genes and 13 down-regulated genes compared with that of total genes in NP. (d) Box plot demonstrating the E_ds of 46 up-regulated genes and 13 down-regulated genes compared with that of total genes in DS. (e) Box plot demonstrating the E_ds of 46 up-regulated genes and 13 down-regulated genes compared with that of total genes in NP.

cellular metabolism (13.56%), signal transduction (8.47%), and detoxification (8.47%). Some of the responsive genes shared in five populations (5.08%) were unable to assign functions (Fig. 3b). A detailed functional categorization for 59 responsive genes shared by five populations is shown in Table 1. Of these genes, about 61%, 69%, and 90% genes were reported abiotic stress-responsive genes in clusters I, II, and III, respectively, with an average of about 70% (Table 1). Interestingly, of these genes, about 33%, 11%, and 56% of reported photosynthesis genes were within clusters I, II, and III, respectively (Table 1).

Population-different phenotype and candidate genes responding to salt stress

Pair-wise F_{ST} estimated with SNPs from the transcriptomes between the five populations ranged from 0.028 to 0.049 (Table S2). Then, we monitored the survival rate and trait performances of each population respectively in response to salt stress to screen the variation of salt stress tolerance. The average survival rate for the majority of populations was smaller than 50% (Fig. 4). And there was significant difference for their survival rates among the five populations, with the survival rate 45.9, 38.3, 31.6, 29.4, and 26% in LU14, LU9, LU5, LU7, and LU10, respectively (Fig. 4). Moreover, plant height

and photosynthetic rate of five populations were significantly different from each other ($P < 0.05$). The average plant height of LU10, LU5, LU14, LU7, and LU9 was 255, 243.3, 218.3, 209, and 201.7 cm, respectively (Fig. 4). The average photosynthetic rates of LU7, LU10, LU14, LU5, and LU9 were 35.0, 33.7, 33.0, 31.7, and 29.6 μm m^{-1} s^{-1}, respectively (Fig. 4). Therefore, among the five populations, LU9 showed the most contrasting performance between relatively high survival rates and relatively weak growing traits including plant height and photosynthetic rates in salinity soil. LU10 showed the contrary performance with relatively lower survival rates and relatively high growing traits for survival individuals in salinity soil.

We further monitored the expression performance of each population respectively in response to salt stress to screen the candidate genes. Among the 16 678 shared expressed genes in all five populations in two environments, 1774 (10.6%), 239 (1.4%), 981 (5.9%), 2015 (12.1%), and 912 (5.5%) population-specific responsive genes ($P < 0.05$ and twofold change) were detected in LU5, LU7, LU9, LU10, and LU14, respectively. The E_ps and E_ds of these genes were quite different among populations ($P < 0.05$; Fig. S4a–d). The E_ps of LU9 exhibited the widest range in both two environments. What's more, for LU9 the E_ps were largest in NP and the smallest in DS ($P < 0.05$; Fig. S4a, b), and the E_ds were largest

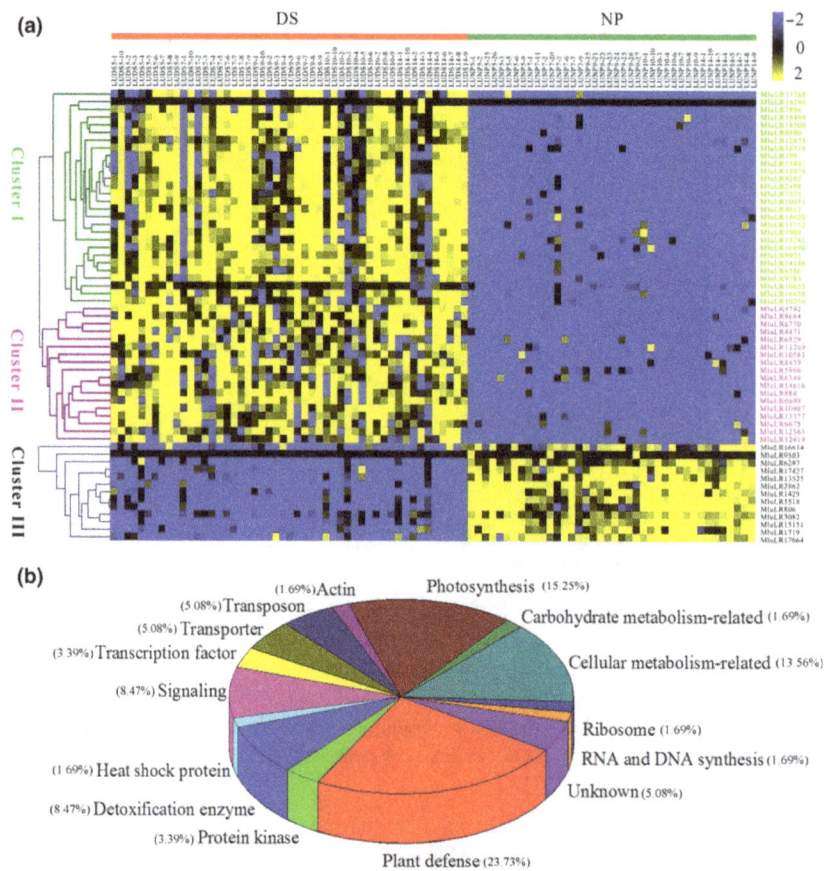

Fig. 3 Clustering of expression and function in shared responsive genes, respectively. (a) Hierarchical clustering of 59 shared responsive genes of all individuals in DS (left; red) and NP (right; green). Normalized gene expression values (FPKM) of those genes of each individual are used for the cluster display. Values are given for all shared responsive genes both differentially expressed among five populations between the two sites ($P < 0.05$, twofold change). Sample clustering is displayed at the top, and the three gene clusters are displayed on the left. The blue-to-yellow color scale representing up- to down-regulation is shown at the top. The genes that share similar expression patterns are divided into three groups. (b) The functional categorization of top-hit distribution of BLAST matches of 59 shared responsive genes. Areas on the pie chart are proportional to the total number of shared responsive genes. A detailed list of the function for each gene can be found in Table 1.

in NP and smallest in DS ($P < 0.05$; Fig. S4c, d). LU9, with most of the genes down-regulated, was found to be with the significantly widest range of E_p ratio from NP to DS ($P < 0.05$; Fig. 5a). LU9 was also found to be with the significantly lowest E_d ratio from NP to DS ($P < 0.05$; Fig. 5b).

The enriched GO terms in population-specific responsive genes were listed in Table S3. The significantly enriched GO terms of population-specific responsive genes in LU9 were involved in the up-regulation of defense response, the down-regulation of photosystem, the down-regulation of isopentenyl diphosphate biosynthetic process, mevalonate-independent pathway, and the pentose-phosphate shunt in the category of biological process (Fig. 6; Table S3). Enriched GO terms of genes between the up- and down-regulated genes were similar among other four populations (Table S3). Thus,

LU9 showed the most typical expression coordination, in which the decreased expression of genes involved in normal growth was accompanied by the increased expression of genes related to stress tolerance. To improve our understanding of the important biochemical pathways of population-specific responsive genes, we further analyzed the KEGG pathway of those genes in each population. With the KO annotation of expressed genes in each population as background, four pathways that were photosynthesis (ko00195), protein processing in endoplasmic reticulum (ko04141), phenylpropanoid biosynthesis (ko00940), and circadian rhythm plant (ko04712) had significant enrichment between two environments in LU9 (Table 2). Two pathways that were ribosome (ko03010) and aminoacyl-tRNA biosynthesis (ko00970) had significant enrichment between two sites in LU5 and LU10 (Table 2). One pathway that

Table 1 The accession, annotation, and potential functional groups of responsive genes shared in five populations was performed using blastn of TAIR. The clusters were grouped using hierarchical clustering method. The direction of expression regulation is shown by plus sign (up) and minus sign (down), respectively. Different stress types and corresponding reference of the reported abiotic stress-responsive genes in those genes were also listed

Gene	Cluster	Function category	Accession	TAIR description	Regulation	Reference	Stress type
MluLR16374	I	Cellular metabolism-related	AT5G01820.1	Pigment Defective 312 (PDE312)	+		
MluLR4612	I	Cellular metabolism-related	AT2G24200.1	Chloroplast-targeting Obg GTPase (ObgC)	+	Chen et al. (2014)	Wounding, salt stress
MluLR1909	I	Cellular metabolism-related	AT2G37220.1	Myristoyl-CoA:protein N-myristoyltransferase	+		
MluLR9800	I	Cellular metabolism-related	AT2G07686.1	Embryo sac development arrest 41 (EDA41)	+		
MluLR199	I	Cellular metabolism-related	AT4G32551.2	ATP-binding microtubule motor family protein	+	Singh et al. (2011)	Salt stress, hormones
MluLR10256	I	Cellular metabolism-related	AT5G19855.1	FAD/NAD (P)-binding oxidoreductase family protein	+	Yu et al. (2014)	Saline–alkaline stress
MluLR12673	I	Detoxification enzyme	AT4G33510.2	A homolog of animal DJ-1 superfamily protein	+	Park et al. (2005)	Oxidative stress
MluLR9783	I	HSP90	AT3G27190.1	Heat Shock Protein 90.5 (HSP90.5)	+	Lockwood et al. (2010)	Heat stress
MluLR16658	I	Photosynthesis	AT2G42490.1	Heme oxygenase-like protein	+	Pham et al. (2015)	Oxidative stress
MluLR10655	I	Photosynthesis	AT3G47450.1	Chaperonin-60 alpha	+	Shi et al. (2011)	Salt stress
MluLR10490	I	Photosynthesis	AT5G26345.1	DegP2 protease (DEGP2)	+	Kley et al. (2011)	Light
MluLR14620	I	Plant defense	AT3G04650.1	Nitric Oxide Synthase 1 (NOS1)	+	Zhao et al. (2007)	Salt stress
MluLR18500	I	Plant defense	AT2G47940.2	SNF1-related protein kinases (SnRK2)	+	Pillai et al. (2012)	Drought, hyperosmotic stress
MluLR7896	I	Plant defense	AT1G11430.1	Activated Disease Susceptibility 1 (ADS1)	+	Huang et al. (2003)	Salt stress
MluLR14296	I	Plant defense	AT5G52470.1	Acetohydroxy Acid Synthase (AHAS)	+	Lee et al. (2011)	Herbicide
MluLR10051	I	Plant defense	AT5G52460.1	Arginosuccinate Synthase (ASS)	+	Kalamaki et al. (2009)	Drought, salt stress
MluLR13497	I	Plant defense	AT1G71220.2	A substrate of the type III effector HopU1	+	Jeong et al. (2011)	Plant immunity
MluLR15976	I	Plant defense	AT1G20620.5	Glycine-Rich RNA-Binding Protein 8 (GR-RBP8)	+	Lorkovic (2009)	Cold stress
MluLR2498	I	Plant defense	AT4G35090.2	A substrate of the type III effector HopU1	+	Jeong et al. (2011)	Plant immunity
MluLR7575	I	Ribosome	AT3G48530.1	Fibrillarin 1 (FBR1)	+		
MluLR9202	I	RNA and DNA synthesis	AT1G51410.1	Multiple Organellar Rna Editing Factor 9 (MORF9)	+		
MluLR9925	I	Signaling	AT4G25990.2	Chloroplast Signal Recognition Particle 54 Kda Subunit (cpSRP54)	+		
MluLR14186	I	Transporter	AT5G09810.1	Heavy Metal Atpase 2 (HMA2)	+	Tan et al. (2013)	Heavy metal stresses

Table 1 (continued)

Gene	Cluster	Function category	Accession	TAIR description	Regulation	Reference	Stress type
MfuLR6586	I	Transporter	AT1G79040.1	Copper uptake transmembrane transporter	+		
MfuLR18499	I	Transposon	AT2G04030.2	Non-LTR retrotransposon family	+		
MfuLR11152	I	Transposon	AT3G14940	Mutator-like transposase family	+		
MfuLR13768	I	Transposon	AT3G48000.1	Gypsy-like retrotransposon family	+		
MfuLR15241	I	Unknown	AT5G03940.1	SWIB/MDM2 domain superfamily protein	+		
MfuLR9684	II	Carbohydrate metabolism-related	AT4G07970.1	Uridine Kinase-Like 2 (UKL2)	+		
MfuLR9792	II	Cellular metabolism-related	AT1G18550.1	3-Deoxy-D-Arabino-Heptulosonate 7-Phosphate Synthase	+	Keith et al. (1991)	Wounding, pathogenic attack
MfuLR4471	II	Cellular metabolism-related	AT2G37220.1	Alanine Aminotransferase 2 (ALAAT2)	+	Limami et al. (2008)	Hypoxic stress
MfuLR6349	II	Detoxification enzyme	AT2G28000.1	Catalase 3	+	Engel et al. (2006)	Salt stress
MfuLR14616	II	Detoxification enzyme	AT3G48560.1	Catalase 2	+	Engel et al. (2006)	Salt stress
MfuLR12619	II	Detoxification enzyme	AT5G20550.1	Soluble Epoxide Hydrolase (SEH)	+	Kiyosue et al. (1994)	Auxin, water Stress
MfuLR8439	II	Detoxification enzyme	AT1G56340.2	Copper Amine Oxidase Family Protein	+	Wimalasekera et al. (2011)	Abscisic acid
MfuLR6699	II	Photosynthesis	AT2G14880.1	Phosphoenolpyruvate carboxylase (PEPC)	+	Monreal et al. (2013)	Salt stress
MfuLR884	II	Plant defense	AT4G34020.2	Aldehyde Dehydrogenase 2 (ALDH2)	+	Sunkar et al. (2003)	Oxidative stress
MfuLR12563	II	Plant defense	AT2G26740.1	2-oxoglutarate (2OG) and Fe (II)-dependent oxygenase superfamily protein	+	Kumchai et al., (2013)	Molybdenum
MfuLR6829	II	Plant defense	AT3G31395.1	Cytosol Aminopeptidase Family Protein (CAP)	+	Zhou et al. (2009)	Aluminum stress
MfuLR5906	II	Plant defense	AT2G19170.1	UDP-glucose:glycoprotein glucosyltransferase (UGGT)	+	Howell (2013)	Endoplasmic reticulum stress
MfuLR13269	II	Protein kinase	AT2G33170.1	SNF1-related protein kinase regulatory subunit gamma 1 (KING1)	+	Akkasaeng et al. (2007)	Water stress
MfuLR6675	II	Protein kinase	AT_G08450.3	Leucine-rich repeat receptor-like protein kinase family protein	+	Wu et al. (2009)	Salt stress, cold, wounding
MfuLR10987	II	Signaling	AT2G26550.3	Calreticulin 3 (CRT3)	+	Jia et al. (2008)	Salt stress, tunicamycin
MfuLR13377	II	Signaling	AT2G15290.1	Calreticulin 1 (CRT1)	+	Jia et al. (2008)	Salt stress, tunicamycin
MfuLR10583	II	Signaling	AT5G52310.1	BRI1-LIKE 3 (BRL3)	+	Cano-Delgado et al. (2004)	Brassinosteroid regulation
MfuLR6770	II	Unknown	AT4G39260.4	Similar to Eucalyptus gunnii alcohol dehydrogenase of unknown physiological function	+		
MfuLR806	III	Actin	AT4G29140.1	Member of Actin gene family	-	Wang et al. (2012)	Salt stress

Table 1 (continued)

Gene	Cluster	Function category	Accession	TAIR description	Regulation	Reference	Stress type
MluLR15151	III	Cellular metabolism-related	AT2G23030.1	Subtilisin-Like Serine Protease 3 (SLP3)	-	Liu et al. (2007)	Salt stress
MluLR13525	III	Photosynthesis	AT5G18570.1	Photosystem II encoding the Light-harvesting chlorophyll a/b-binding protein CP26	-		
MluLR16614	III	Photosynthesis	AT3G48500.2	PSI type III chlorophyll a/b-binding protein (Lhca3*1)	-	Jung et al. (2003)	Cold stress
MluLR17427	III	Photosynthesis	AT4G24830.2	Encodes the only subunit of photosystem I located entirely in the thylakoid lumen	-		
MluLR1429	III	Photosynthesis	AT5G64040.2	A chloroplast stromal localized RbcX protein	-	Kolesinski et al. (2011)	Salt stress, oxidative stress and so on
MluLR6297	III	Photosynthesis	AT1G04445.1	The 10-kDa PsbR subunit of photosystem II	-	Suja & Parida (2008)	H2O2 stress, polyethylene glycol stress
MluLR2862	III	Plant defense	AT4G10340.1	LEUNIG	-	Shrestha et al. (2014)	Salt stress, osmotic stress
MluLR9503	III	Signaling	AT1G72330.3	CBL-Interacting Protein Kinase 14 (CIPK14)	-	Zhang et al. (2008)	Cold, drought, salt stress
MluLR1719	III	Transcription factor	AT4G30110.1	Chloroplast Import Apparatus Cia2-Like	-		
MluLR5518	III	Transcription factor	AT1G61520.3	C2H2-like zinc finger protein	-	Plavcova et al. (2013)	Nitrogen stress
MluLR3082	III	Transporter	AT5G57020.1	ABC-2 type transporter family protein	-	Gu et al. (2004)	Salt stress
MluLR17664	III	Unknown	AT3G13380.1	Hypothetical protein	-		

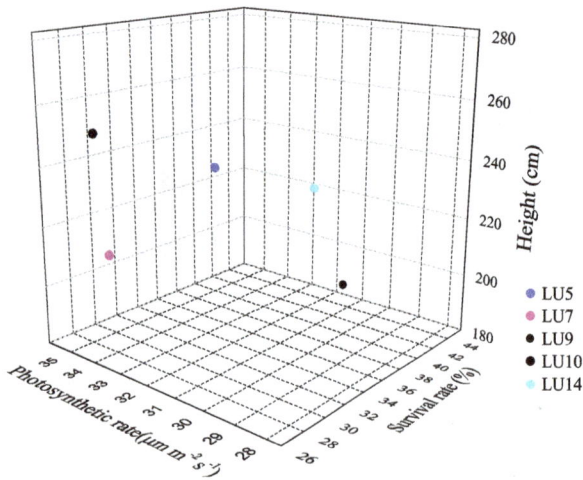

Fig. 4 Contrasting performance between mean survival rate and growing traits (plant height, photosynthetic rate) in populations collected in DS. The survival rate, plant height, and net photosynthetic rate of each individual were investigated when harvested the samples. Each dot represents the mean value of each population.

was phenylpropanoid biosynthesis (ko00940) had significant enrichment between two environments in LU7 (Table 2). One pathway that was ribosome (ko03010) had significant enrichment between two sites in LU14 (Table 2).

Discussion

The important role of shared responsive genes in stress responses

To avoid getting false-positive results of responsive genes, only those showing significantly differential

expression in all the five populations ($P < 0.05$ and two-fold change) were identified as candidate genes. This is necessary for studying plants responding to stress in the field conditions where variables are relatively difficult to control. The finding that responsive genes shared by five populations tended to have lower E_ds and higher E_ps compared with all genes (Fig. 2b–e) may indicate that those shared responsive genes tend to be more conserved among individuals than others. They are expected to have experienced strong positive or purifying selective constraint (Hannah *et al.*, 2006; Des Marais *et al.*, 2012; Rengel *et al.*, 2012; Lasky *et al.*, 2014). This result was also in agreement with previous reports that E_ds of genes with higher E_ps tended to be more conserved in changing environment (Xu *et al.*, 2016), which had been explained that highly expressed genes experienced stronger purifying selection than those at lower level. The results that the decrease of E_ds of shared responsive genes following *M. lutarioriparius* being transplanted from its native habitats NP to DS in high saline field reflected the dominant trend of narrowed range of gene expression in the new environment. Thus, these shared responsive genes may be considered as consistently expressed salt-responsive genes and likely represent adaptive responses to common saline conditions (Lasky *et al.*, 2014).

Indeed, these shared genes are also found to play an important role in stress response in the related species. As we found that the functions of those genes were mainly involved in detoxification enzymes, plant defense and photosynthesis, and up to 70% of these genes were reported to be stress-responsive genes. For example, detoxification enzymes can reduce excessive reactive oxygen species (ROS) accumulation as various environmental stresses such that salinity universally causes oxidative stress with an excessive accumulation

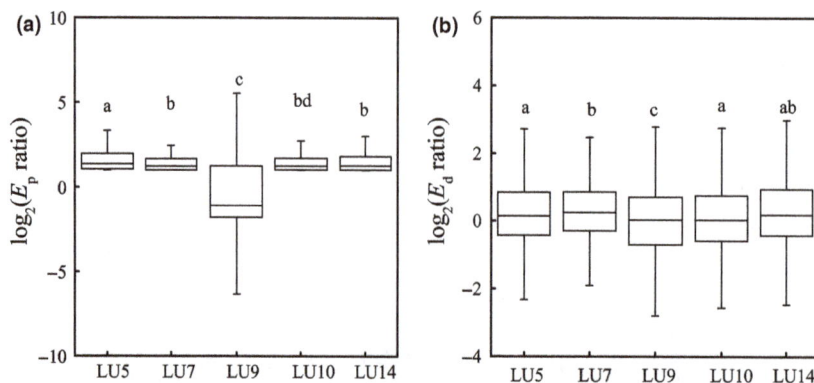

Fig. 5 Different expression variation of population-specific responsive genes for each population between two environments. (a) The $\log_2 E_p$ ratio of population-specific responsive genes in each population of *Miscanthus lutarioriparius* from DS to NP. (b) The $\log_2 E_d$ ratio of population-specific responsive genes in each population of *M. lutarioriparius* from DS to NP. Each bar represents the mean ± SD. Means without the same letter are significantly different ($P < 0.05$).

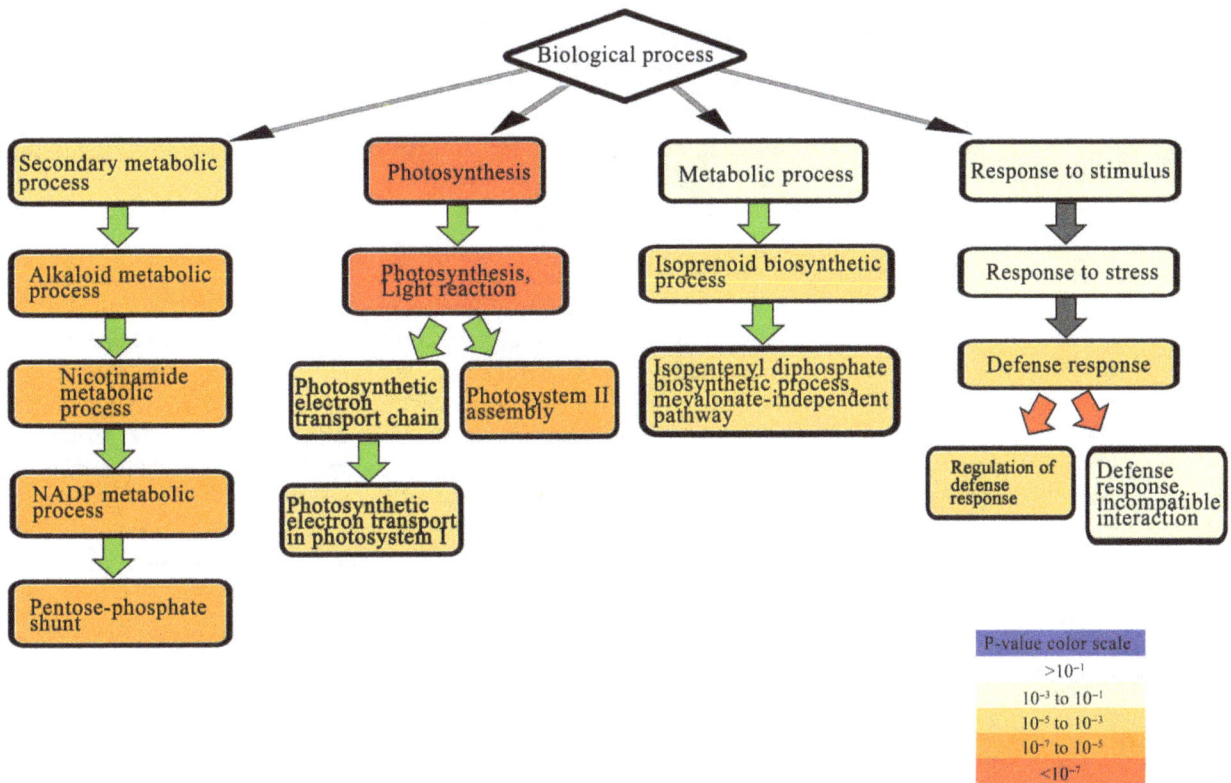

Fig. 6 Gene Ontology (GO) term enrichment result for biological process based on GO terms for population-specific responsive genes in LU9. The directed graph shows the relationship of the GO terms over-represented. All colored boxes are enriched with adjusted P value < 0.05. The degree of enrichment was shown from red (adjusted P value < 10^{-7}) to white (10^{-1} < adjusted P value). Arrows stand for relationships between different GO terms. Red arrow stands for the up-regulated expression, and green arrow stands for the down-regulated expression.

Table 2 Significantly enriched pathway of genes expressed significantly different in each population. KOBAS 2.0 was used to analyze the pathway enrichment of genes expressed significantly different in each population and chose FDR (false discovery rate) <0.05 as the threshold

Pathways	LU5	LU7	LU9	LU10	LU14
Ribosome	√			√	√
Aminoacyl-tRNA biosynthesis	√			√	
Phenylpropanoid biosynthesis		√	√		
Photosynthesis			√		
Circadian rhythm plant			√		
Protein process in endoplasmic reticulum			√		

of ROS in plants such as *Soldanella alpina* and *Arabidopsis thaliana* (Kiyosue *et al.*, 1994; Sunkar *et al.*, 2003; Engel *et al.*, 2006). Therefore, the complex regulation of the mRNA of catalase (CAT) (Engel *et al.*, 2006), aldehyde dehydrogenase (ALDH) (Sunkar *et al.*, 2003), and epoxide hydrolase (EH) (Kiyosue *et al.*, 1994) seemed to constitute a detoxification mechanism that limits excessive ROS accumulation in *M. lutarioriparius* leaves. Plant defense was found to frequently respond to various stress environments in other plants (Park *et al.*, 2005; Lockwood *et al.*, 2010), which is according to our result that genes encoding DJ-1/PfpI proteins family and heat shock proteins (HSPs) were up-regulated in DS. Photosynthetic response to salinity stress is extremely complex because salinity stress indirectly affects photosynthesis and includes expression of those genes connected to growth inhibition or leaf shedding. Through limiting water consumption, plant could help to maintain the carbon assimilation (Chaves *et al.*, 2009). In this case, down-regulated genes in *M. lutarioriparius* encoding photosynthesis-related proteins such as light-harvesting chlorophyll a/b-binding protein 5 (LHCP5) may function in the same way. Consistently, the previous studies also showed that plants responding to salt stress displayed complex molecular responses including the production of ion transporters and detoxification enzymes to re-establish homeostasis and resume growth in *Arabidopsis* (Zhu, 2001).

The potential tolerance-productivity trade-off mechanism in response to salt stress

The different performances existing among different populations or species may be used for dissecting different salt tolerance mechanisms (Walia *et al.*, 2005; Platten *et al.*, 2013). This study analyzed the expression pattern of population-specific responsive genes in five populations. The $E_d s'$ range of these genes narrowed down only in LU9 when *M. lutarioriparius* was transplanted from NP to DS (Fig. S4c, d), suggesting that the expression for these genes in LU9 experienced a purifying selection (Brawand *et al.*, 2011) or responded to long-term salt stress. Another possible explanation for this might be that these genes would be under a more systematic and strict regulation when encountered with salt stress environment (Zhu, 2001; Chen *et al.*, 2002).

In addition, the different performances of phenotype were also observed among the five populations (Fig. 4). The strongest growth traits and lowest survival rate of LU10 might be a result of selection under which a relatively small portion of individuals with outstanding phenotype was able to survive (Randerson & Hurst, 2001). Therefore, the surviving individuals with such phenotype could have a great potential for energy crop development in salinity soil. In further studies, salt tolerance and growth traits of *M. lutarioriparius* can be selected separately for pyramiding breeding (Handa *et al.*, 2014). The detailed response mechanisms could be revealed more clearly by associating the expression pattern with phenotype (Atwell *et al.*, 2010). First of all, the responsive genes were analyzed for GO enrichment in each population separately, and the results showed that the up-regulation of stress response, the down-regulation of photosystem II assembly, isopentenyl diphosphate biosynthetic process, mevalonate-independent pathway, and the pentose-phosphate shunt were enriched in LU9 (Fig. 6; Table S3). Moreover, pathway enrichment analysis showed that photosynthesis, protein processing in endoplasmic reticulum, and phenylpropanoid biosynthesis were enriched in LU9 (Table 2). Phenylpropanoid biosynthesis (ko00940) (Vogt, 2010) and the up-regulation of defense response might have contributed to the salt tolerance of LU9. The pentose-phosphate shunt pathway that was well known as the fundamental metabolic pathway was down-regulated (Wood, 1986). And the down-regulation of photosystem II assembly, isopentenyl diphosphate biosynthetic process, mevalonate-independent pathway, and photosynthesis (Lichtenthaler *et al.*, 1997) might have contributed to the lowest photosynthetic rate in LU9 (Fig. 4). The down-regulation of function in photosynthesis and cellular metabolism that consequently limited carbon dioxide uptake and compromised plant growth might be

one cause of the weak phenotype in LU9 under salt stress (Chaves *et al.*, 2009).

The contrasting performance of LU9 between relatively high survival rate and relatively weak growing traits including plant height and photosynthetic rate was in accordance with gene expression patterns in such a way that the decreased expression of genes involved in normal growth was accompanied by the increased expression of genes related to plant stress tolerance. These results revealed the underlying mechanism of tolerance-productivity trade-off, that is, the molecular basis for redirecting resources from growth to stress resistance (Dupont-Prinet *et al.*, 2010; Muller-Landau, 2010). The coordination of decreasing the expression of genes related to normal growth and increasing the expression of genes involved in plant stress-responsive mechanisms could have contributed to allocation of resources from rapid growth to stress protection (Sabreen & Sugiyama, 2008; Lopez-Maury *et al.*, 2009; Zakrzewska *et al.*, 2011) and resulted in higher survival rates at a cost of other physiological attributes (Dupont-Prinet *et al.*, 2010).

The findings of response of M. lutarioriparius to long-term salt stress in natural environment

We compared the findings of our study with those studies of poplar and *Arabidopsis* under short-term controlled environment salt stress. The *CAT* gene family involved in antioxidant defense in poplar (Ding *et al.*, 2010) and SOS pathway that salt stress induced calcium-signaling pathway in *Arabidopsis* (Chinnusamy *et al.*, 2004) were the well-known salt stress-associated genes which were revealed under short-term (0–48 h) controlled environment salt stress. We also found genes such as *CAT* involved in antioxidant defense and *CBL* that was identified as one of the calcium sensors in calcium-signaling pathway, which is in coincidence with the findings in short-term controlled environment (Engel *et al.*, 2006; Zhang *et al.*, 2008). And the expression of catalase was only up-regulated under long-term salt stress treatments, which was consistent with the previous study (Hernandez *et al.*, 2010). Compared with findings of studies under short-term salt stress, the long-term study could also have a better observation of survival rates and growing traits.

We also compared the findings of our study with those studies of *Miscanthus*, pea, and wheat under controlled long-term salt stress. We found in *M. lutarioriparius* leaves, genes involved in the function of detoxification, plant defense, photosynthesis, and signal transduction were potentially responsive to the high-salinity conditions. One of the *M. lutarioriparius* populations had the potential tolerance-productivity trade-off, in which down-regulation of cellular metabolisms, photosynthesis genes, and the up-regulation of stress-related genes were coordinated with the contrasting performance between relatively high survival rate in salinity soil and relatively weak growing traits including plant height and photosynthetic rates (Fig. 4; Fig. 6). The previous studies found that *Miscanthus* x *giganteus* and *Miscanthus sinensis* tended to have a reduction in plant growth and photosynthetic rate in response to salt stress under controlled hydroponic conditions (Plazek *et al.*, 2014), which were in accordance with the relatively weak growing traits. In the expression profile, the induction of antioxidant defense (Hernandez *et al.*, 2000) and the reduction in photosynthesis-related genes (Kiani-Pouya, 2015) were also found in pea and wheat responding to long-term (15–60 days) salt stress under controlled environment. Those findings focused on investigating the different responses between salt-tolerant and salt-sensitive genotypes facing long-term salt stress. Salt-tolerant genotypes had stronger growing traits or higher expression of antioxidant defense-related genes than salt-sensitive genotypes. However, only our study did find the link between gene expression and growth traits.

In this study, we aimed to understand how plants respond to long-term salt stress by conducting comparative population transcriptomic analyses between native habitat and salinity field. Through identifying candidate stress-responsive genes shared by populations, we demonstrated that consistently expressed responsive genes were likely to represent adaptive responses to common stress conditions and genes shared for stress responses appeared to have been subject to purifying selection. Based on the population which had the most contrasting performance between survival rates and growing traits, we found this contrasting performance was in accordance with the up-regulation of genes involved in stress tolerance and down-regulation of certain growth-related genes. The finding suggested that there might be a tolerance-productivity trade-off when plants were exposed to long-term salt stress in the field conditions, where resource allocation from growth to stress resistance might be the mechanism at the expressional level for stress tolerance. Fine tuning of the tolerance-productivity trade-off could potentially contribute to the development of second-generation energy crops that have satisfactory establishment rates and productivity in salinity land.

Acknowledgements

This work was supported by grants from the National Natural Science Foundation of China [31500186 and 31400284] and the Science and Technology Service Network Initiative of the Chinese Academy of Sciences [KFJ-EW-STS-061 and KFJ-EW-STS-119]. We thank the Beijing Center for Physical and Chemical Analysis for generating the sequencing data, the Beijing Computing Center for providing computational infrastructure for data analysis. The government of Kenli County and government of Dongying City of the Shandong Province provide assistance and support for this work.

References

Akkasaeng C, Tantisuwichwrong N, Chairam I, Prakrongrak N, Jogloy S, Pathanothai A (2007) Isolation and identification of peanut leaf proteins regulated by water stress. *Pakistan Journal of Biological Sciences: PJBS*, **10**, 1611–1617.

Allwright MR, Taylor G (2016) Molecular breeding for improved second generation bioenergy crops. *Trends Plant Science*, **21**, 43–54.

Alvarez M, Schrey AW, Richards CL (2015) Ten years of transcriptomics in wild populations: what have we learned about their ecology and evolution? *Molecular Ecology*, **24**, 710–725.

Atwell S, Huang YS, Vilhjalmsson BJ et al. (2010) Genome-wide association study of 107 phenotypes in *Arabidopsis thaliana* inbred lines. *Nature*, **465**, 627–631.

Beritognolo I, Harfouche A, Brilli F et al. (2011) Comparative study of transcriptional and physiological responses to salinity stress in two contrasting *Populus alba* L. genotypes. *Tree Physiology*, **31**, 1335–1355.

Boyer JS (1982) Plant productivity and environment. *Science*, **218**, 443–448.

Brawand D, Soumillon M, Necsulea A et al. (2011) The evolution of gene expression levels in mammalian organs. *Nature*, **478**, 343–348.

Brinker M, Brosche M, Vinocur B et al. (2010) Linking the salt transcriptome with physiological responses of a salt-resistant populus species as a strategy to identify genes important for stress acclimation. *Plant Physiology*, **154**, 1697–1709.

Brosché M, Vinocur B, Alatalo ER et al. (2005) Gene expression and metabolite profiling of *Populus euphratica* growing in the Negev desert. *Genome Biology*, **6**, R101.

Bushman BS, Amundsen KL, Warnke SE, Robins JG, Johnson PG (2016) Transcriptome profiling of Kentucky bluegrass (*Poa pratensis* L.) accessions in response to salt stress. *BMC Genomics*, **17**, 48.

Cano-Delgado A, Yin YH, Yu C et al. (2004) BRL1 and BRL3 are novel brassinosteroid receptors that function in vascular differentiation in *Arabidopsis*. *Development*, **131**, 5341–5351.

Champigny MJ, Sung WWL, Catana V et al. (2013) RNA-seq effectively monitors gene expression in *Eutrema salsugineum* plants growing in an extreme natural habitat and in controlled growth cabinet conditions. *BMC Genomics*, **14**, 578.

Chaves MM, Flexas J, Pinheiro C (2009) Photosynthesis under drought and salt stress: regulation mechanisms from whole plant to cell. *Annals of Botany*, **103**, 551–560.

Chen SL, Renvoize SA (2006) *Miscanthus*. In: *Flora of China* (eds Wu ZY, Raven PH, Hong DY), pp. 581–583. Science Press, Beijing, China, Missouri Botanical Garden Press, St. Louis, MO, USA.

Chen WQ, Provart NJ, Glazebrook J et al. (2002) Expression profile matrix of *Arabidopsis* transcription factor genes suggests their putative functions in response to environmental stresses. *Plant Cell*, **14**, 559–574.

Chen J, Bang WY, Lee Y et al. (2014) AtObgC-AtRSH1 interaction may play a vital role in stress response signal transduction in *Arabidopsis*. *Plant Physiology and Biochemistry*, **74**, 176–184.

Cherel I, Lefoulon C, Boeglin M, Sentenac H (2014) Molecular mechanisms involved in plant adaptation to low K$^+$ availability. *Journal of Experimental Botany*, **65**, 833–848.

Chinnusamy V, Schumaker K, Zhu JK (2004) Molecular genetic perspectives on cross-talk and specificity in abiotic stress signalling in plants. *Journal of Experimental Botany*, **55**, 225–236.

Cui XQ, Churchill GA (2003) Statistical tests for differential expression in cDNA microarray experiments. *Genome Biology*, **4**, 210.

Des Marais DL, Mckay JK, Richards JH, Sen S, Wayne T, Juenger TE (2012) Physiological genomics of response to soil drying in diverse *Arabidopsis* accessions. *Plant Cell*, **24**, 893–914.

Ding MQ, Hou PC, Shen X et al. (2010) Salt-induced expression of genes related to Na$^+$/K$^+$ and ROS homeostasis in leaves of salt-resistant and salt-sensitive poplar species. *Plant Molecular Biology*, **73**, 251–269.

Du Z, Zhou X, Ling Y, Zhang ZH, Su Z (2010) Agrigo: A go analysis toolkit for the agricultural community. *Nucleic Acids Research*, **38**, W64–W70.

Dupont-Prinet A, Chatain B, Grima L, Vandeputte M, Claireaux G, Mckenzie DJ (2010) Physiological mechanisms underlying a trade-off between growth rate and tolerance of feed deprivation in the European sea bass (*Dicentrarchus labrax*). *Journal of Experimental Biology*, **213**, 1143–1152.

Engel N, Schmidt M, Lutz C, Feierabend J (2006) Molecular identification, heterologous expression and properties of light-insensitive plant catalases. *Plant Cell and Environment*, **29**, 593–607.

Fan YY, Wang Q, Kang LF *et al.* (2015) Transcriptome-wide characterization of candidate genes for improving the water use efficiency of energy crops grown on semiarid land. *Journal of Experimental Botany*, **66**, 6415–6429.

Flowers TJ, Yeo AR (1995) Breeding for salinity resistance in crop plants: where next? *Australian Journal of Plant Physiology*, **22**, 875–884.

Fujii H, Zhu JK (2009) An autophosphorylation site of the protein kinase SOS2 is important for salt tolerance in *Arabidopsis*. *Molecular Plant*, **2**, 183–190.

Gu RS, Fonseca S, Puskas LG, Hackler L, Zvara A, Dudits D, Pais MS (2004) Transcript identification and profiling during salt stress and recovery of *Populus euphratica*. *Tree Physiology*, **24**, 265–276.

Handa N, Bhardwaj R, Thukral AK, Arora S, Kohli SK, Gautam V, Kaur T (2014) Gene pyramiding and omics approaches for stress tolerance in leguminous plants. In: *Legumes Under Environmental Stress: Yield, Improvement and Adaptations*, 265. John Wiley & Sons, Chichester, UK.

Hannah MA, Wiese D, Freund S, Fiehn O, Heyer AG, Hincha DK (2006) Natural genetic variation of freezing tolerance in *Arabidopsis*. *Plant Physiology*, **142**, 98–112.

Hernandez JA, Jimenez A, Mullineaux P, Sevilla F (2000) Tolerance of pea (*Pisum sativum l.*) to long-term salt stress is associated with induction of antioxidant defences. *Plant Cell and Environment*, **23**, 853–862.

Hernandez M, Fernandez-Garcia N, Diaz-Vivancos P, Olmos E (2010) A different role for hydrogen peroxide and the antioxidative system under short and long salt stress in *Brassica oleracea* roots. *Journal of Experimental Botany*, **61**, 521–535.

Howell SH (2013) Endoplasmic reticulum stress responses in plants. *Annual Review of Plant Biology*, **64**, 477–499.

Huang J, Zhang HS, Wang JF, Yang JS (2003) Molecular cloning and characterization of rice 6-phosphogluconate dehydrogenase gene that is up-regulated by salt stress. *Molecular Biology Reports*, **30**, 223–227.

Jeong BR, Lin Y, Joe A *et al.* (2011) Structure function analysis of an ADP-ribosyl-transferase type III effector and its RNA-binding target in plant immunity. *Journal of Biological Chemistry*, **286**, 43272–43281.

Jia XY, Xu CY, Jing RL, Li RZ, Mao XG, Wang JP, Chang XP (2008) Molecular cloning and characterization of wheat calreticulin (CRT) gene involved in drought-stressed responses. *Journal of Experimental Botany*, **59**, 739–751.

Jung SH, Lee JY, Lee DH (2003) Use of SAGE technology to reveal changes in gene expression in *Arabidopsis* leaves undergoing cold stress. *Plant Molecular Biology*, **52**, 553–567.

Kalamaki MS, Alexandrou D, Lazari D *et al.* (2009) Over-expression of a tomato N-acetyl-L-glutamate synthase gene (SlNAGS1) in *Arabidopsis thaliana* results in high ornithine levels and increased tolerance in salt and drought stresses. *Journal of Experimental Botany*, **60**, 1859–1871.

Keith B, Dong XN, Ausubel FM, Fink GR (1991) Differential induction of 3-deoxy-D-arabino-heptulosonate 7-phosphate synthase genes in *Arabidopsis thaliana* by wounding and pathogenic attack. *Proceedings of the National Academy of Sciences*, **88**, 8821–8825.

Kiani-Pouya A (2015) Changes in activities of antioxidant enzymes and photosynthetic attributes in *Triticale* (x *Triticosecale wittmack*) genotypes in response to long-term salt stress at two distinct growth stages. *Acta Physiologiae Plantarum*, **37**, 1–11.

Kiyosue T, Beetham JK, Pinot F, Hammock BD, Yamaguchishinozaki K, Shinozaki K (1994) Characterization of an *Arabidopsis* cDNA for a soluble epoxide hydrolase gene that is inducible by auxin and water-stress. *Plant Journal*, **6**, 259–269.

Kley J, Schmidt B, Boyanov B *et al.* (2011) Structural adaptation of the plant protease Deg1 to repair photosystem II during light exposure. *Nature Structural and Molecular Biology*, **18**, 728–731.

Kolesinski P, Piechota J, Szczepaniak A (2011) Initial characteristics of RbcX proteins from *Arabidopsis thaliana*. *Plant Molecular Biology*, **77**, 447–459.

Kreps JA, Wu YJ, Chang HS, Zhu T, Wang X, Harper JF (2002) Transcriptome changes for *Arabidopsis* in response to salt, osmotic, and cold stress. *Plant Physiology*, **130**, 2129–2141.

Kumchai J, Huang JZ, Lee CY, Chen FC, Chin SW (2013) Proline partially overcomes excess molybdenum toxicity in cabbage seedlings grown in vitro. *Genetics and Molecular Research*, **12**, 5589–5601.

Langmead B, Trapnell C, Pop M, Salzberg SL (2009) Ultrafast and memory-efficient alignment of short DNA sequences to the human genome. *Genome Biology*, **10**, R25.

Lasky JR, Marais DLD, Lowry DB *et al.* (2014) Natural variation in abiotic stress responsive gene expression and local adaptation to climate in *Arabidopsis thaliana*. *Molecular Biology and Evolution*, **31**, 2283–2296.

Lee H, Rustgi S, Kumar N *et al.* (2011) Single nucleotide mutation in the barley acetohydroxy acid synthase (AHAS) gene confers resistance to imidazolinone herbicides. *Proceedings of the National Academy of Sciences of the United States of America*, **108**, 8909–8913.

Lee ST, Xiao YY, Muench MO *et al.* (2012) A global DNA methylation and gene expression analysis of early human B-cell development reveals a demethylation signature and transcription factor network. *Nucleic Acids Research*, **40**, 11339–11351.

Li H, Handsaker B, Wysoker A *et al.* (2009) The sequence alignment/map format and SAMtools. *Bioinformatics*, **25**, 2078–2079.

Libault M, Farmer A, Brechenmacher L *et al.* (2010a) Complete transcriptome of the soybean root hair cell, a single-cell model, and its alteration in response to *Bradyrhizobium japonicum* infection. *Plant Physiology*, **152**, 541–552.

Libault M, Farmer A, Joshi T *et al.* (2010b) An integrated transcriptome atlas of the crop model glycine max, and its use in comparative analyses in plants. *Plant Journal*, **63**, 86–99.

Lichtenthaler HK, Rohmer M, Schwender J (1997) Two independent biochemical pathways for isopentenyl diphosphate and isoprenoid biosynthesis in higher plants. *Physiologia Plantarum*, **101**, 643–652.

Limami AM, Glevarec G, Ricoult C, Cliquet JB, Planchet E (2008) Concerted modulation of alanine and glutamate metabolism in young *Medicago truncatula* seedlings under hypoxic stress. *Journal of Experimental Botany*, **59**, 2325–2335.

Liu W, Sang T (2013) Potential productivity of the *Miscanthus* energy crop in the loess plateau of china under climate change. *Environmental Research Letters*, **8**, 044003.

Liu JX, Srivastava R, Che P, Howell SH (2007) Salt stress responses in *Arabidopsis* utilize a signal transduction pathway related to endoplasmic reticulum stress signaling. *Plant Journal*, **51**, 897–909.

Lockwood BL, Sanders JG, Somero GN (2010) Transcriptomic responses to heat stress in invasive and native blue mussels (genus *Mytilus*): molecular correlates of invasive success. *Journal of Experimental Botany*, **213**, 3548–3558.

Lopez-Maury L, Marguerat S, Bahler J (2009) Tuning gene expression to changing environments: from rapid responses to evolutionary adaptation. *Nature Reviews Genetics*, **9**, 583–593.

Lorkovic ZJ (2009) Role of plant RNA-binding proteins in development, stress response and genome organization. *Trends Plant Science*, **14**, 229–236.

Mi J, Liu W, Yang WH, Yan J, Li JQ, Sang T (2014) Carbon sequestration by *Miscanthus* energy crops plantations in a broad range semi-arid marginal land in china. *Science of the Total Environment*, **496**, 373–380.

Moinuddin M, Gulzar S, Ahmed MZ, Gul B, Koyro HW, Khan MA (2014) Excreting and non-excreting grasses exhibit different salt resistance strategies. *Aob Plants*, **6**, plu038.

Montreal JA, Arias-Baldrich C, Tossi V *et al.* (2013) Nitric oxide regulation of leaf phosphoenolpyruvate carboxylase-kinase activity: implication in sorghum responses to salinity. *Planta*, **238**, 859–869.

Morozova O, Marra MA (2008) From cytogenetics to next-generation sequencing technologies: advances in the detection of genome rearrangements in tumors. *Biochemistry and Cell Biology-Biochimie et Biologie Cellulaire*, **86**, 81–91.

Muller-Landau HC (2010) The tolerance-fecundity trade-off and the maintenance of diversity in seed size. *Proceedings of the National Academy of Sciences of the United States of America*, **107**, 4242–4247.

Munns R, Termaat A (1986) Whole-plant responses to salinity. *Australian Journal of Plant Physiology*, **13**, 143–160.

Park J, Kim SY, Cha GH, Lee SB, Kim S, Chung J (2005) Drosophila DJ-1 mutants show oxidative stress-sensitive locomotive dysfunction. *Gene*, **361**, 133–139.

Parker BL, Thaysen-Andersen M, Fazakerley DJ, Holliday M, Packer NH, James DE (2016) Terminal galactosylation and sialylation switching on membrane glycoproteins upon TNF-alpha-induced insulin resistance in adipocytes. *Molecular and Cellular Proteomics*, **15**, 141–153.

Pham NT, Kim JG, Jung S (2015) Differential antioxidant responses and perturbed porphyrin biosynthesis after exposure to oxyfluorfen and methyl viologen in *oryza sativa*. *International Journal of Molecular Sciences*, **16**, 16529–16544.

Pillai BVS, Kagale S, Chellamma S (2012) Enhancing productivity and performance of oil seed crops under environmental stresses. In: *Crop Stress and its Management:*

Perspectives and Strategies (ed. Venkateswarlu B), pp. 139–161. Springer Science + Media, Mumbai, India.

Platten JD, Egdane JA, Ismail AM (2013) Salinity tolerance, Na⁺ exclusion and allele mining of HKT1;5 in *Oryza sativa* and *O. glaberrima*: many sources, many genes, one mechanism? *BMC Plant Biology*, **13**, 32.

Plavcova L, Hacke UG, Almeida-Rodriguez AM, Li EY, Douglas CJ (2013) Gene expression patterns underlying changes in xylem structure and function in response to increased nitrogen availability in hybrid poplar. *Plant Cell and Environment*, **36**, 186–199.

Plazek A, Dubert F, Koscielniak J, Tatrzanska M, Maciejewski M, Gondek K, Zurek G (2014) Tolerance of *Miscanthus x giganteus* to salinity depends on initial weight of rhizomes as well as high accumulation of potassium and proline in leaves. *Industrial Crops and Products*, **52**, 278–285.

Randerson JP, Hurst LD (2001) A comparative test of a theory for the evolution of anisogamy. *Proceedings of the Royal Society B-Biological Sciences*, **268**, 879–884.

Rengel D, Arribat S, Maury P et al. (2012) A gene-phenotype network based on genetic variability for drought responses reveals key physiological processes in controlled and natural environments. *PLoS ONE*, **7**, e45249.

Sabreen S, Sugiyama SI (2008) Trade-off between cadmium tolerance and relative growth rate in 10 grass species. *Environmental and Experimental Botany*, **63**, 327–332.

Saeed AI, Sharov V, White J et al. (2003) TM4: a free, open-source system for microarray data management and analysis. *BioTechniques*, **34**, 374–378.

Sang T, Zhu WX (2011) China's bioenergy potential. *Global Change Biology Bioenergy*, **3**, 79–90.

Shi SS, Chen W, Sun WN (2011) Comparative proteomic analysis of the *Arabidopsis* CBL1 mutant in response to salt stress. *Proteomics*, **11**, 4712–4725.

Shrestha B, Guragain B, Sridhar VV (2014) Involvement of co-repressor LUH and the adapter proteins SLK1 and SLK2 in the regulation of abiotic stress response genes in *Arabidopsis*. *BMC Plant Biology*, **14**, 54.

Singh K, Singla-Pareek SL, Pareek A (2011) Dissecting out the crosstalk between salinity and hormones in roots of *Arabidopsis*. *Omics - A Journal of Integrative Biology*, **15**, 913–924.

Suja G, Parida A (2008) Isolation and characterization of photosystem 2 PsbR gene and its promoter from drought-tolerant plant *Prosopis juliflora*. *Photosynthetica*, **46**, 525–530.

Sun W, Xu XN, Zhu HS, Liu AH, Liu L, Li JM, Hua XJ (2010) Comparative transcriptomic profiling of a salt-tolerant wild tomato species and a salt-sensitive tomato cultivar. *Plant and Cell Physiology*, **51**, 997–1006.

Sunkar R, Bartels D, Kirch HH (2003) Overexpression of a stress-inducible aldehyde dehydrogenase gene from *Arabidopsis thaliana* in transgenic plants improves stress tolerance. *Plant Journal*, **35**, 452–464.

Taji T, Seki M, Satou M et al. (2004) Comparative genomics in salt tolerance between *Arabidopsis* and *Arabidopsis*-related halophyte salt cress using *Arabidopsis* microarray. *Plant Physiology*, **135**, 1697–1709.

Tan JJ, Wang JW, Chai TY et al. (2013) Functional analyses of TaHMA2, a P1B-type ATPase in wheat. *Plant Biotechnology Journal*, **11**, 420–431.

Trapnell C, Pachter L, Salzberg SL (2009) Tophat: discovering splice junctions with RNA-seq. *Bioinformatics*, **25**, 1105–1111.

Trapnell C, Williams BA, Pertea G et al. (2010) Transcript assembly and quantification by RNA-seq reveals unannotated transcripts and isoform switching during cell differentiation. *Nature Biotechnology*, **28**, 511–515.

Trick M, Kwon SJ, Choi SR et al. (2009) Complexity of genome evolution by segmental rearrangement in *Brassica rapa* revealed by sequence-level analysis. *BMC Genomics*, **10**, 539.

Vicente O, Al Hassan M, Boscaiu M (2016) Contribution of osmolyte accumulation to abiotic stress tolerance in wild plants adapted to different stressful environments. In: *Osmolytes and Plants Acclimation to Changing Environment: Emerging Omics Technologies.* (eds Iqbal N, Nazar R, Khan NA), pp. 13–25. Springer, New Delhi, India.

Vogt T (2010) Phenylpropanoid biosynthesis. *Molecular Plant*, **3**, 2–20.

Walia H, Wilson C, Condamine P et al. (2005) Comparative transcriptional profiling of two contrasting rice genotypes under salinity stress during the vegetative growth stage. *Plant Physiology*, **139**, 822–835.

Wang Z, Gerstein M, Snyder M (2009) RNA-seq: a revolutionary tool for transcriptomics. *Nature Reviews Genetics*, **10**, 57–63.

Wang F, Yang CL, Wang LL, Zhong NQ, Wu XM, Han LB, Xia GX (2012) Heterologous expression of a chloroplast outer envelope protein from *Suaeda salsa* confers oxidative stress tolerance and induces chloroplast aggregation in transgenic *Arabidopsis* plants. *Plant Cell and Environment*, **35**, 588–600.

Wang T, Tohge T, Ivakov A et al. (2015) Salt-related MYB1 (SRM1) coordinates abscisic acid biosynthesis and signaling during salt stress in *Arabidopsis*. *Plant Physiology*, **169**, 1027–1041.

Wimalasekera R, Villar C, Begum T, Scherer GFE (2011) COPPER AMINE OXIDASE1 (CuAO1) of *Arabidopsis thaliana* contributes to abscisic acid- and polyamine-induced nitric oxide biosynthesis and abscisic acid signal transduction. *Molecular Plant*, **4**, 663–678.

Wood T (1986) Physiological functions of the pentose-phosphate pathway. *Cell Biochemistry and Function*, **4**, 241–247.

Wright S (1978) *Evolution and the Genetics of Population, Variability within and among Natural Populations*. University of Chicago press, Chicago, IL, USA.

Wu T, Tian ZD, Liu J, Xie CH (2009) A novel leucine-rich repeat receptor-like kinase gene in potato, StLRPK1, is involved in response to diverse stresses. *Molecular Biology Reports*, **36**, 2365–2374.

Xie C, Mao XZ, Huang JJ et al. (2011) KOBAS 2.0: a web server for annotation and identification of enriched pathways and diseases. *Nucleic Acids Research*, **39**, W316–W322.

Xu Q, Xing SL, Zhu CY et al. (2015) Population transcriptomics reveals a potentially positive role of expression diversity in adaptation. *Journal of Integrative Plant Biology*, **57**, 284–299.

Xu Q, Zhu CY, Fan YY et al. (2016) Population transcriptomics uncovers the regulation of gene expression variation in adaptation to changing environment. *Scientific Reports*, **6**, 25536.

Yan J, Chen WL, Luo F et al. (2012) Variability and adaptability of *Miscanthus* species evaluated for energy crop domestication. *Global Change Biology Bioenergy*, **4**, 49–60.

Yan J, Zhu MD, Liu W, Xu Q, Zhu CY, Li JQ, Sang T (2016) Genetic variation and bidirectional gene flow in the riparian plant *Miscanthus lutarioriparius*, across its endemic range: Implications for adaptive potential. *Global Change Biology Bioenergy*, **8**, 764–776.

Yu Y, Huang WG, Chen HY et al. (2014) Identification of differentially expressed genes in flax (*Linum usitatissimum* L.) under saline-alkaline stress by digital gene expression. *Gene*, **549**, 113–122.

Zakrzewska A, Van Eikenhorst G, Burggraaff JEC et al. (2011) Genome-wide analysis of yeast stress survival and tolerance acquisition to analyze the central trade-off between growth rate and cellular robustness. *Molecular Biology of the Cell*, **22**, 4435–4446.

Zhang JZ, Creelman RA, Zhu JK (2004) From laboratory to field. Using information from *Arabidopsis* to engineer salt, cold, and drought tolerance in crops. *Plant Physiology*, **135**, 615–621.

Zhang HC, Yin WL, Xia XL (2008) Calcineurin B-Like family in *Populus*: comparative genome analysis and expression pattern under cold, drought and salt stress treatment. *Plant Growth Regulation*, **56**, 129–140.

Zhao MG, Tian QY, Zhang WH (2007) Nitric oxide synthase-dependent nitric oxide production is associated with salt tolerance in *Arabidopsis*. *Plant Physiology*, **144**, 206–217.

Zhou SP, Sauve R, Thannhauser TW (2009) Proteome changes induced by aluminium stress in tomato roots. *Journal of Experimental Botany*, **60**, 1849–1857.

Zhu JK (2001) Plant salt tolerance. *Trends Plant Science*, **6**, 66–71.

Zhu JK, Liu JP, Xiong LM (1998) Genetic analysis of salt tolerance in *Arabidopsis*: evidence for a critical role of potassium nutrition. *Plant Cell*, **10**, 1181–1191.

6

Belowground impacts of perennial grass cultivation for sustainable biofuel feedstock production in the tropics

YUDAI SUMIYOSHI[1], SUSAN E. CROW[1], CREIGHTON M. LITTON[1], JONATHAN L. DEENIK[2], ANDREW D. TAYLOR[3], BRIAN TURANO[2] and RICHARD OGOSHI[2]

[1]Department of Natural Resources and Environmental Management, University of Hawaii Manoa, Honolulu, HI 96822, USA, [2]Department of Tropical Plant and Soil Sciences, University of Hawaii Manoa, Honolulu, HI 96822, USA, [3]Department of Biology, University of Hawaii Manoa, Honolulu, HI 96822, USA

Abstract

Perennial grasses can sequester soil organic carbon (SOC) in sustainably managed biofuel systems, directly mitigating atmospheric CO_2 concentrations while simultaneously generating biomass for renewable energy. The objective of this study was to quantify SOC accumulation and identify the primary drivers of belowground C dynamics in a zero-tillage production system of tropical perennial C4 grasses grown for biofuel feedstock in Hawaii. Specifically, the quantity, quality, and fate of soil C inputs were determined for eight grass accessions – four varieties each of napier grass and guinea grass. Carbon fluxes (soil CO_2 efflux, aboveground net primary productivity, litterfall, total belowground carbon flux, root decay constant), C pools (SOC pool and root biomass), and C quality (root chemistry, C and nitrogen concentrations, and ratios) were measured through three harvest cycles following conversion of a fallow field to cultivated perennial grasses. A wide range of SOC accumulation occurred, with both significant species and accession effects. Aboveground biomass yield was greater, and root lignin concentration was lower for napier grass than guinea grass. Structural equation modeling revealed that root lignin concentration was the most important driver of SOC pool: varieties with low root lignin concentration, which was significantly related to rapid root decomposition, accumulated the greatest amount of SOC. Roots with low lignin concentration decomposed rapidly, but the residue and associated microbial biomass/by-products accumulated as SOC. In general, napier grass was better suited for promoting soil C sequestration in this system. Further, high-yielding varieties with low root lignin concentration provided the greatest climate change mitigation potential in a ratoon system. Understanding the factors affecting SOC accumulation and the net greenhouse gas trade-offs within a biofuel production system will aid in crop selection to meet multiple goals toward environmental and economic sustainability.

Keywords: carbon sequestration, guinea grass, napier grass, root decomposition, soil carbon, structural equation modeling, total belowground carbon flux

Introduction

Global interest in the development of renewable energy systems stems largely from combined concerns about climate change, energy security, and environmental sustainability (Tilman *et al.*, 2009). Biofuel production from plant feedstocks is a promising component of diverse, renewable energy plans (Liska & Perrin, 2009; Werling *et al.*, 2014). However, biofuel production can result in net CO_2 emissions through the use of fossil fuels during land-use conversion and the production and processing of biomass (Gibbs *et al.*, 2008; Cherubini *et al.*, 2009). The use of conservation management practices during biofuel cultivation could sequester soil organic carbon

Correspondence: Susan E. Crow
e-mail: crows@hawaii.edu

(SOC) (Lal, 2013), providing an offset to some or all of the CO_2 emissions associated with biofuel feedstock production (Adler *et al.*, 2007; Davis *et al.*, 2013). However, considerable uncertainty exists regarding the magnitude and direction of changes in SOC under various biofuel feedstock production systems (Stockmann *et al.*, 2013).

The SOC pool is a function of the dynamic balance between inputs and outputs of C belowground, with SOC increasing when inputs exceed outputs (Six & Jastrow, 2002; Cotrufo *et al.*, 2015). Inputs of C to belowground include aboveground and belowground litter and total belowground carbon flux (TBCF), which is the sum of autotrophic C input to belowground to support root production and respiration, exudates, and microbial symbioses (Giardina & Ryan, 2002). The quantity of belowground C input is influenced by overall stand productivity and the partitioning of photosynthetically fixed C to aboveground vs. belowground (Litton *et al.*, 2007). Primary C outputs from belowground include soil CO_2 efflux ('soil respiration') and, in some systems, leaching. Soil CO_2 efflux, the primary belowground output of C in most systems, can be decreased through reduced tillage or increased chemical recalcitrance of inputs (Davidson & Janssens, 2006).

The use of perennial grasses for biofuel production has the potential to mitigate SOC loss and greenhouse gas (GHG) emissions within the production system due to high yield and low requirements for fertilizer, pesticides, and irrigation (Liebig *et al.*, 2008; DeLucia, 2016). In tropical and subtropical regions, perennial C4 grasses such as *Saccharum officinarum* (sugarcane), *Saccharum spontaneum* (energy cane), *Pennisetum purpureum* (napier grass), *Megathyrsus maximus* (guinea grass), and *P. purpureum* × *Pennisetum glaucum* (sterile napier grass hybrids) have been identified as promising candidates for biofuel feedstocks, due primarily to their potential for high biomass yields (Tran *et al.*, 2011; Hashimoto *et al.*, 2012; Meki *et al.*, 2014; Mochizuki *et al.*, 2014). Most assessments of tropical perennial C4 grasses to date have focused on aboveground yields and agronomic requirements (Kinoshita *et al.*, 1995; Keffer *et al.*, 2009). In turn, very little attention has been given to belowground C dynamics associated with these grasses even though SOC accrual could potentially provide a strong C sink to offset nonrenewable GHG emissions associated with their production (Gelfand *et al.*, 2013; DeLucia, 2016).

In addition to the potential for high yield and low agricultural inputs, perennial grasses under zero-tillage cultivation have the potential to store a large amount of SOC compared to annual crops due to their extensive belowground root system (Powlson *et al.*, 2011; Anderson-Teixeira *et al.*, 2012). Napier grass and guinea grass species produce large stocks of root biomass, both as fine roots in the surface root zone, and at depths of up to 4.5 m (Khanal *et al.*, 2010). Perennial grasses can be harvested by ratooning, a form of zero-tillage harvest that leaves the lower part of the plant and living roots and soil undisturbed, thereby potentially contributing to the sustainability of the production system via SOC accumulation, erosion control, and improved soil fertility (Anderson-Teixeira *et al.*, 2013). Compared to annual crops such as corn and soybeans, perennial species such as napier grass and guinea grass can be ratooned for as many as 4 years without reduction in production (Samson *et al.*, 2005). In temperate ecosystems, the capacity for perennial grass feedstocks to mitigate climate change is fairly well documented (Anderson-Teixeira *et al.*, 2012), but whether this will be the case in tropical regions is unknown.

Both the quantity and quality of plant inputs affect the balance of SOC. A recent meta-analysis revealed that differences in SOC pools between tillage and zero-tillage soils were due primarily to differences in the quantity of C inputs (Virto *et al.*, 2012). However, in other cases, input quantity itself has no impact on the SOC pool (Al-Kaisi & Grote, 2007; Sanderson, 2008), suggesting a predominant influence by other factors such as input quality (i.e., decomposition dynamics). Traditionally, inputs with lower overall chemical quality and therefore lower decomposability were thought to contribute to the accumulation of SOC more than inputs with higher chemical quality due to the complex structure of low-quality molecules (Jastrow & Miller, 1996). Grandy & Neff (2008) reported that the majority of recent, plant-derived SOC was comprised of lignin-related compounds. Recent concepts of litter transformation to SOC include rapid and slow modes of transport of different plant compounds into the mineral soil (Cotrufo *et al.*, 2013, 2015), and multiple mechanisms of stabilization involving the protection of inputs from microbial degradation (Schmidt *et al.*, 2011; Lehmann & Kleber, 2015). Ample evidence suggests that plant tissue decomposition rate, determined by the decay constant (*k*), is related to tissue chemical characteristics such as C : N and lignin : N (Melillo *et al.*, 1982; Silver & Miya, 2001; Johnson *et al.*, 2007). And, in at least one study on switchgrass, SOC positively correlated to root C : N, a ratio often used as a predictor of root decomposition (Ma *et al.*, 2000).

The overarching objectives of this research were to quantify change in SOC over time and to identify the primary drivers of belowground C dynamics in a ratoon production system of eight perennial grass accessions under consideration as biofuel feedstocks in Hawaii. It was expected that SOC would increase following cultivation as a result of high belowground organic matter inputs. A series of hypotheses were developed to determine the primary drivers of increased SOC. First, it was hypothesized that the selected perennial grass systems would have a wide range of ecosystem C pools, fluxes, and input quality indicators that contribute collectively to the soil C balance (H1). Measured ecosystem C pools and fluxes included aboveground biomass yield, soil and root C, soil CO_2 efflux, total belowground C flux, and changes in pools over time. Input quality indicators included concentrations of cellulose, hemicellulose, lignin, C, N, and C/lignin to N ratios in roots. A negative relationship between decomposition rate constant (*k*) and litter lignin concentration (Johnson *et al.*, 2007; Zhang *et al.*, 2008) and lignin : N (Melillo *et al.*, 1982) has been widely reported. Therefore, it was hypothesized that grasses with higher root lignin, C : N ratio, and lignin : N ratio would result in lower decomposition rates (H2). Finally, it was hypothesized that high aboveground biomass yield and high root lignin concentration would be the primary, direct drivers of SOC accumulation in this zero-tillage, tropical perennial grass system. Further, root biomass, root decay constant, and soil CO_2 efflux were expected to mediate indirect controls of these factors on SOC accumulation (H3). H3 was assessed using structural equations modeling (SEM), a multivariate approach that develops a conceptual model of causal relationships among the quantity and quality of C pools and fluxes that ultimately drive SOC accumulation.

Materials and methods

Field site and experimental design

The study was conducted at the University of Hawaii's Experimental Research Station at Waimanalo, on the Island of Oahu, Hawaii (21°20′15″N, 157°43′30″W). The plots were located on alluvial fans at 30 m elevation with a mean annual temperature of 24.6 °C and mean annual precipitation of 938 mm, 72% of which occurs between November and April (Giambelluca et al., 2013). The soil is classified as silty clay with smectitic and halloysitic mineralogy (Waialua series, very fine, mixed, superactive, isohyperthermic Pachic Haplustolls) (Soil Survey staff, accessed 11/1/14). For 24 years prior to the start of this experiment, the field site was a fallow, mechanically maintained grassy field. Information before 24 years is limited, but the presence of a dense plow layer suggests past extended periods of heavy cultivation.

The field plot design was established in October 2009 and consisted of a randomized complete block design with four replicates of eight grass accessions. The accessions studied were three napier grasses, one pearl millet × dwarf napier grass (PM×D) cross hybrid, and four guinea grass accessions (Table S1). These accessions were selected to encompass a wide range of aboveground yield potential based on preliminary site data. Each grass plot consisted of four rows planted in a 2 × 3 m area, with a nonplanted buffer of 0.6 m between plots. Inter rows and buffers were covered with nylon mats to suppress weed growth. Plants were produced from stem cuttings and irrigated 3 days a week from 9 : 00 to 14 : 00 with drip irrigation for the duration of the study at ~3.6 L m^{-2} day^{-1}. Fertilization with granular triple-16 formulation at the rate of 53 kg N, 23 kg phosphorus, and 44 kg potassium ha^{-1} occurred when the grasses were first planted in November 2009, and after the second ratooning in November 2010. The grasses were ratoon harvested in March and November 2010, and July 2011. At each harvest, grasses were cut at 10 cm above the soil surface using a brush-cutter, and all aboveground biomass was removed from the plot. Material harvested from the inner two rows, excluding the outer 0.5 m of those rows on each end, was considered the experimental unit for analysis.

Ecosystem carbon pools and fluxes

Total wet weight of the harvested material was recorded, and a subset of the material was dried at 105 °C to determine

moisture content, from which dry weights of the total harvested material (i.e., aboveground biomass yield) were calculated. Subsamples of leaf tissue were oven-dried at 75 °C, homogenized with a ball mill (Retsch MM200 mixer mill; Retsch GmbH, Haan, Germany) to pass through a 250-μm sieve, and the C and N concentrations of leaf samples were determined by oxidative combustion on an elemental analyzer (Costech ECS 4010 CHNSO Analyzer; Costech Analytical Technologies Inc., Valencia, CA, USA). Aboveground biomass yields from the three ratoons were combined, multiplied by C concentrations of leaf materials, and averaged across years for an annualized biomass C value (g C m^{-2} yr^{-1}).

Two perforated aluminum pans (16 × 26 × 2 cm) were placed in the inter-rows in each plot to measure aboveground litterfall (F_A). Litterfall measurements were made monthly from February to July 2011. It was assumed thereafter that, because yields over time were steady, the litterfall rate during the study period remained constant. Thus, the average monthly litterfall rate from the five measurements was multiplied by 12 to determine annual cumulative F_A (g C m^{-2} yr^{-1}). Collected litter was oven-dried at 75 °C to a constant mass and weighed. Elemental C and N composition of litter was determined as described above.

Soil CO_2 efflux (F_S) was measured monthly for 1 year from August 2010 to July 2011 between 9 : 00 and 18 : 00 from each plot using a LI-6400XT portable photosynthesis system with a soil respiration chamber (LI-COR Inc., Lincoln, NE, USA). In each plot, F_S was measured on five 10-cm-diameter polyvinyl chloride (PVC) collars that were inserted 2 cm into the soil ~1 month prior to initial measurements. This was reduced to three collars $plot^{-1}$ after November 2010 based on a lack of within plot variability in F_S. Living vegetation inside the collars was clipped and removed 1 day prior to measurements. Soil temperature at 10 cm depth was measured adjacent to each collar at the time of F_S measurements with a temperature probe. Volumetric soil moisture was also measured at 5 cm depths adjacent to each collar using an impedance probe calibrated to the study site soil (Hydra Soil Moisture Probe; Stevens Water Monitoring Systems Inc., Beaverton, OR, USA).

Annual cumulative F_S (g C m^{-2} yr^{-1}) was calculated using linear interpolation between monthly measurements (Litton et al., 2008). Mathematically, the interpolation can be expressed as:

$$F_S = \frac{360\left[\sum_1^i (D_i - D_{(i-1)})\left(\frac{F_i + F_{(i+1)}}{2}\right)\right]}{(D_i - D_1)} \quad (1)$$

where F_i is monthly soil CO_2 efflux rates (g C m^{-2} day^{-1}) and D_i is date of the efflux measurement, starting at 1 in July 2010 and ending as i in August 2011. Daily F_i values were not corrected for diel variability because flux measurements taken every 2 h from 11 : 00 hours to 3 : 00 hours in one replicate of NG1 revealed no diel variability in F_S (Tukey multiple comparison with 95% confidence) and no relationship between F_S and soil temperature ($r = -0.030$; $P = 0.704$; $n = 160$). Previous studies have also documented lack of diel patterns in F_S in tropical dry forest (Litton et al., 2008) and tropical wet forest in Hawaii (Giardina & Ryan, 2002; Litton et al., 2011). Gomez-Casanovas et al. (2013) found that linear interpolation between survey measurements was the second best method overall for estimating cumulative F_S, with only linear interpolation supplemented by site-specific soil temperature dependence of F_S outperforming this approach. In our study, there was no temperature dependence of F_S ($P = 0.704$), making linear interpolation between monthly measurements the most robust approach to annual estimates of F_S. Litton et al. (2008) did document an increase in F_S following a large pulse precipitation event (95 mm in a system receiving ~750 mm annually) that, if ignored, resulted in an underestimate of F_S when using linear interpolation between monthly measurements. However, this underestimate was very small on an annual basis (0.5–7%). In addition, the focus herein is on comparison between treatments, and not with other studies, and any bias in F_S resulting from the use of linear interpolations between monthly measurements would be the same across treatments.

Samples for SOC (C_S) and root biomass C (C_R) pools were collected during two sampling periods: (i) after the first ratooning in April 2010 and (ii) after the 3rd ratooning in August 2011. In April 2010, two cored (5-cm-diameter) samples for C_R and two augered (6-cm-diameter) samples for C_S were collected at two depths (0–15 and 15–30 cm) from each plot. For August 2011, four cored (5-cm-diameter) samples (two for C_R and two for C_S) were collected at the same depth increments. Soil samples were air-dried at 25 °C for a week until no mass change, sieved to 2 mm, and weighed, and elemental C and N concentration was determined on a subsample as described above. Measurements were limited to 30 cm depths due to the presence of a dense, clay-rich plow layer.

The SOC pool was quantified using the equivalent mass of soil method (Ellert & Bettany, 1995; Gifford & Roderick, 2003), which negates the issue of core compaction during sampling and the effect of land management on bulk density. Equivalent mass of soil (g C m^{-2}) at target depth (t) was calculated for each plot as:

$$C_S(t) = C_S(Z_a) + \frac{C_S(Z_b) - C_S(Z_a)}{M_S(Z_b) - M_S(Z_a)}(M_S(t) - M_S(Z_a)) \quad (2)$$

where $M_S(Z_a)$ and $M_S(Z_b)$ are the masses of soil for the first and second increments at depth Z_a and Z_b, $C_S(Z_a)$ and $C_S(Z_b)$ are the masses of C of both depth increments, and $M_S(t)$ is the target mass of soil that all samples are compared against. This linear interpolation allows comparison of SOC pools without the need of an accurate core volume. The reference soil mass selected to make comparisons within this study was 300 kg m^{-2} because it was approximately the mean mass of soil samples collected to 30 cm depth across all cores. As such, the SOC pools were expressed as g C m^{-2} in 300 kg of soil m^{-2}.

For C_R, cored soil samples were split into four parts, inserted into 250 ml Nalgene polypropylene bottles and shaken for 16 h with 100 ml of 10% sodium hexametaphosphate. Dispersed soils were wet-sieved through a 0.5 mm sieve, and collected roots were washed with deionized water, ground to pass through a 250-μm sieve using a UDY cyclone mill (Tecator Inc., Boulder, CO, USA), and root C concentrations were determined as above. No separation of live and dead roots was conducted. The root biomass in soil cores was expressed in terms

of C by multiplying root C concentration by mass of root biomass extracted from wet sieving (g C m^{-2} to 30 cm depth). Average values across cores of C_S and C_R were calculated for each plot.

The belowground autotrophic input of C was estimated in each plot using the TBCF mass balance approach for nonsteady state conditions (Giardina & Ryan, 2002). The mass balance approach quantifies the sum of all pertinent C inputs that plants send belowground including root production and respiration, root exudation, and C flow to symbionts, all of which are difficult to measure independently. Use of this method for quantifying autotrophic flux of C to belowground outside forest ecosystems has been limited (Adair et al., 2009; Ford et al., 2012), but as long as all important C fluxes (inputs and outputs) and changes in C pools are quantified accurately, the mass balance approach for estimating TBCF should be applicable in any environment and is calculated as:

$$\text{TBCF} = F_S - F_A + \Delta(C_S + C_R + C_L) \quad (3)$$

where TBCF is the total flux of C that plants send belowground, F_S is soil CO_2 efflux (g C m^{-2} yr^{-1}), F_A is aboveground litterfall (g C m^{-2} yr^{-1}), and C_S, C_R, and C_L are soil, root, and litter layer C pools (g C m^{-2}), respectively. This approach assumes there are negligible losses of C via erosion and leaching. Soil erosion in the study site was nonexistent due to a flat topography and controlled water input. Leaching loss of C was also assumed to be negligible with <1 m precipitation annually. In addition, leaching of C has been shown to account for <2% of F_S in a variety of ecosystems including grasslands and croplands (Kindler et al., 2011). It was also assumed that changes in soil and root biomass C pools below 30 cm were negligible during the measurement period. Trenches adjacent to the study plots demonstrated that minimal root growth extended below 30 cm depth in this system. Annualized changes in C_S (ΔC_S) and C_R (ΔC_R) were calculated as the differences between the 2010 and 2011 annualized pools. Change in litter layer C was assumed to be zero from 2010 to 2011 because all aboveground biomass was removed regularly for feedstock conversion.

Root quality and decay

Grass stalks with roots attached were excavated from soil pits (30 cm × 30 cm, 15 cm in depth) in each plot, washed with deionized water, and air-dried for 1 week. Approximately 20 g of subsampled root tissue was analyzed for neutral detergent fiber (NDF), acid detergent fiber (ADF), cellulose, and acid unhydrolyzable compounds (lignin and other recalcitrant compounds hereafter referred to as lignin) using standard reagents (Van Soest, 1963) on a Fiber Analyser (Ankom, Macedon, NY, USA) at the Agricultural Diagnostic Service Center of the University of Hawaii Manoa. The quantities of nonfiber carbohydrates (NFC) such as organic acids, sugars, and starch were determined as 100 – NDF, and hemicellulose was calculated as NDF – ADF using results of the fiber analysis (Hall, 2003). The lignin value estimated using the sequential digestion method is a coarse estimate of lignin as polyphenolic and other unsaturated substances such as tannins and suberin may be included in the value (Van Soest & Wine, 1968). Carbon and N concentrations were determined on subsamples as described above.

A common area root decomposition experiment was conducted from December 2010 to August 2011 to determine the root decomposition constant (k) using the litterbag method (Ostertag, 2001). For each replicate of each accession (four replicates × eight accessions = 32 experimental units), a set of five bags was made with 0.5 g subsamples of air-dried root materials 2–3 cm in length. Each bag was constructed of 5 cm × 5 cm 0.132 mm nylon mesh. Bags were buried in a 1.5 × 7 m area adjacent to each accession plot, in a line (to facilitate future destructive sampling events) in random order such that 5-cm borders separated 32 lines of bag sets. Individual bags were buried at a 45° angle to the surface and inserted 3.5 cm into the soil. Bags from each replicate set were collected randomly at 1, 2, 3, 5, and 8 months after deployment. The site was covered with a weed mat and irrigated three times a week to simulate conditions similar to the accession plots. Collected bags were rinsed with deionized water and dried to a constant mass at 75 °C, and residue C and N were determined following the procedure described above. For each experimental unit, a first-order single-pool exponential function was applied for the root C decline over time to characterize decomposition (Wider & Lang, 1982):

$$L_t = e^{(-kt)} \tag{4}$$

where L_t represents the proportion of original mass at time t (year), and k represents the decay constant (t^{-1}). Both first-order single- and double-pool exponential functions were considered, but a single-pool exponential function was best suited for the data based on r^2 values (mean r^2 was 0.90 with a range of 0.69–0.99) and visual analysis of residuals.

Statistical analysis

Comparison of C pools and fluxes, root composition, and decay constant. In the statistical comparisons of C pools and fluxes (H1) and root elemental/chemical composition and decay constant (H2), three levels of comparisons were made. First, an accession effect was determined by treating each of the eight accessions independently. Analysis of variance (ANOVA) was performed to compare the effect of accessions on ecosystem C pools, fluxes, and input quality indicators using the software R 2.15 (R Development Core Team, 2012). Prior to analysis, homogeneity of variances for block and accession effects was tested with Levene's test on all response variables. Graphical assessment of normality and outliers revealed no severe nonnormality or outliers in any of the variables. A wrong accession was planted in one of the replicates of GG2 and therefore was omitted from the data analysis, giving GG2 only three replicates (Table 1). The unorthogonality created by the missing observation on sum of squares of accession effect was negligible, and therefore, type II sum of squares was used to calculate F values. When the effect of accession on response variables was significant at $P \leq 0.05$, Tukey *post hoc* comparisons were performed to determine significant differences between accessions.

Next, two preplanned custom contrasts comparing (i) guinea grass vs. napier grass + PM×D and (ii) napier grass vs. PM×D were tested using a multcomp package in R (Hothorn *et al.*, 2008) to determine species effect (guinea grass vs. napier grass hybrid + PM×D). Further, because PM×D is a napier grass hybrid with pearl millet, a variety effect was also determined (napier grass vs. PM×D). The two custom contrasts are independent of each other, negating the risk of inflating type I experiment-wise error rate (Seltman, 2012).

A correlation analysis was used to identify factors controlling the decay constant (k) (H2). Pearson correlation coefficients (r) were calculated ($n = 32$) for compositional factors (C, N, NFC, hemicellulose, cellulose, lignin, C : N ratio, and lignin : N ratio) and the root decay constant (k) calculated from the litterbag experiment.

Functional belowground carbon dynamics. Structural equation modeling (SEM), a multivariate statistical method, was used to disentangle the effect of the multiple explanatory variables into hypothesized causal pathways (Grace, 2006; Tabachnick & Fidell, 2007) (H3). Several recent ecological studies have used SEM to investigate the causal relationships among multicollinear predicting variables and their effect on soil CO_2 efflux (F_S) (Geng *et al.*, 2012; Matias *et al.*, 2012), the SOC pool (Jonsson & Wardle, 2010; Brahim *et al.*, 2011), and soil microbial community composition (Eisenhauer *et al.*, 2012). The overall purpose of SEM in this study was to delineate patterns of direct and indirect effects of explanatory variables by formulating a causal model based on both results of this study and prior knowledge of SOC accumulation. Structural equation modeling was performed using the 'sem' package in R. Variables are connected with one-way arrows indicating the flow of causal relationships. Regression coefficients are then parameterized simultaneously using a maximum-likelihood method for each arrow (Grace, 2006). The variance and covariance matrices from the parameterized coefficients were tested against the matrix from the data to determine overall fit of the hypothesized models to the data. Unstandardized and standardized path coefficients and associated standard error, Z-score, and P-value are reported for each parameter estimate.

All plots were treated as independent, and thus, all 32 plots were included in this analysis. Given the small sample size ($n = 32$), only observed variables and their relationships were considered; no latent (i.e., unobserved, underlying) variables were included in the model. Also, variables were limited to six to minimize estimated parameters (Grace, 2006). The six variables chosen were root lignin concentration, aboveground biomass yield, k, F_S, C_R 2010, and C_S 2011. Because SEM relies on variance and covariance matrices of the variables, assessment of univariate and multivariate distribution of variables for outliers, linearity, and normality is crucial for subsequent inferences (Tabachnick & Fidell, 2007). Bivariate scatterplots and residual plots were used to assess linearity and normality and to check for outliers using Minitab 16 (Minitab Inc., State College, PA, USA).

The acceptability of the final model was first determined by chi-square (χ^2) tests ($P > 0.05$). Nonsignificant chi-square test indicates that the variance and covariance matrices of the

Table 1 Mean values ± 1 SE for pools (g C m^{-2}) and fluxes (g C m^{-2} yr^{-1}) of soil carbon for grass accessions ($n = 4$ except for GG2 where $n = 3$). C_S and C_R were expressed in terms of equivalent mass of 300 g m^{-2}, which was approximately equal to 0–30 cm depth. ANOVA F and P values of the random effect of blocks (as replicates), fixed effect of accessions, and custom contrasts of species on pools and fluxes

| | Napier | | | Hybrid | Guinea | | | | Accession | GG vs. | NG vs. PM×D |
	NG1	NG2	NG3	PM×D	GG1	GG2	GG3	GG4	effect	NG + PM×D	
Pools											
C_S 2010	5156 ± 168	4908 ± 20	5062 ± 143	5393 ± 258	5273 ± 187	5144 ± 184	5002 ± 253	5334 ± 300	ns	ns	ns
C_S 2011	5534 ± 221	5424 ± 64	5409 ± 110	5483 ± 221	5391 ± 123	5327 ± 148	5262 ± 115	5617 ± 133	ns	ns	ns
C_R 2010	50 ± 6	39 ± 8	34 ± 2	39 ± 6	40 ± 7	41 ± 5	21 ± 2	35 ± 5	ns	ns	ns
C_R 2011	154 ± 10	106 ± 11	110 ± 16	91 ± 14	111 ± 14	96 ± 15	83 ± 25	85 ± 29	ns	ns	ns
Fluxes											
Yield	1805 ± 300	1351 ± 87	931 ± 227	1590 ± 230	1007 ± 170	1135 ± 250	983 ± 198	1085 ± 168	$P = 0.07$, $F = 2.29$	$P = 0.026$, $F = 5.77$	ns
ΔC_S	284 ± 120	387 ± 37	260 ± 45	68 ± 70	88 ± 62	137 ± 49	195 ± 123	213 ± 159	$P = 0.031$, $F = 2.86$	$P = 0.034$, $F = 5.16$	$P = 0.005$, $F = 9.94$
ΔC_R	79 ± 11	50 ± 14	57 ± 13	39 ± 14	54 ± 8	41 ± 8	47 ± 18	37 ± 23	ns	ns	ns
F_S	1784 ± 216	1325 ± 114	1788 ± 107	1536 ± 195	1489 ± 173	1565 ± 235	1555 ± 182	1518 ± 105	ns	ns	ns
F_A	229 ± 59	215 ± 17	213 ± 19	128 ± 6	120 ± 8	145 ± 45	170 ± 35	110 ± 15	$P = 0.006$, $F = 4.07$	$P = 0.002$, $F = 11.99$	$P = 0.005$, $F = 11.62$
TBCF	1917 ± 223	1546 ± 99	1892 ± 163	1516 ± 157	1510 ± 166	1599 ± 332	1627 ± 144	1658 ± 233	ns	ns	ns

SE, standard error; Yield, aboveground yield (g C m^{-2} yr^{-1}); F_S, soil CO_2 efflux (g C m^{-2} yr^{-1}); F_A, litter fall (g C m^{-2} yr^{-1}); ΔC_S, difference in soil C pools from 2010 to 2011 sampling dates (g C m^{-2} yr^{-1}); ΔC_R, difference in root C pools from 2010 to 2011 sampling dates (g C m^{-2} yr^{-1}); C_S, soil C pool (g C m^{-2}); C_R, root C pool (g C m^{-2}); TBCF, total belowground carbon flux (g C m^{-2} yr^{-1}).

hypothesized model have no difference from observed data. Model fit was further ensured by various fit indicators such as low root-mean-square error of approximation (RMSEA) (<0.05), high Tucker–Lewis index (TLI) (>0.95), standardized root-mean-square residual (SRMR) (<0.06), and low Akaike information criterion (AIC) (Hu & Bentler, 1999). The modification indices >5, which suggests that removing a variable would improve the model fit substantially, were considered when the paths had theoretical support and estimated parameters could be interpreted (Jöreskog & Sorbom, 1989; Grace, 2006). Additionally, a test of indirect effects through intermediate variables, called mediation, was carried out using the delta method (Sobel, 1982).

Results

Ecosystem C pools and fluxes

In 2011, SOC pools (mean of 5434 ± 51 g C m^{-2}) were, on average, 5% higher than in 2010 (mean of 5160 ± 70 g C m^{-2}) (Table 1), resulting in positive ΔC_S for all grass accessions (Fig. 1a). Mean SOC ranged from 4908 to 5617 g C m^{-2} in 2010; PM×D

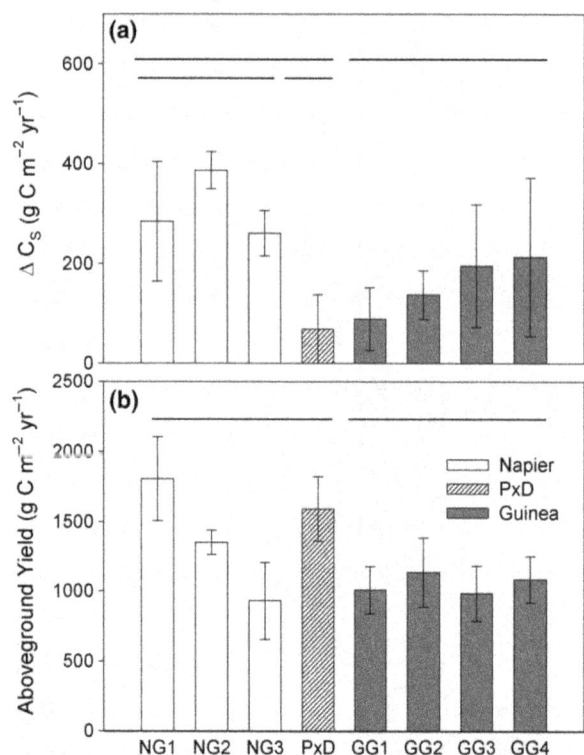

(5394 $\pm$ 257 g C m^{-2}) was significantly greater than napier grass (5021 $\pm$ 73 g C m^{-2}); but neither napier grass nor PM×D differed significantly from guinea grass (5191 $\pm$ 114 g C m^{-2}). No significant differences were observed across accessions, species, or variety within napier grass in 2011. The increase in SOC pool from 2010 to 2011 (ΔC_S) was greater for napier grass (249 $\pm$ 45 g C m^{-2}) compared to guinea grass (160 $\pm$ 53 g C m^{-2}) and for napier grass (without PM×D) (285 $\pm$ 43 g C m^{-2}) compared to PM×D (67 $\pm$ 69 g C m^{-2}) (Table 1).

In 2010, root C pools (C_R) ranged from 21 to 50 g C m^{-2} with an average of 42 ± 2.5 g C m^{-2} across accessions and species, and no significant differences among accessions (Table 1). In 2011, average C_R increased to 123 ± 0.8 g C m^{-2} across accessions, again with no significant differences among accessions. Accrual of root biomass (ΔC_R) from 2010 to 2011 averaged 61 ± 5.5 g C m^{-2} yr^{-1} across accessions and was highly variable across replicates within accessions (coefficient of variations as high as 120%), which resulted in no significant differences across accessions (Table 1).

Average aboveground biomass C ranged from 931 to 1805 g C m^{-2} yr^{-1} across accessions (Table 1) and was significantly greater for napier grasses and PM×D (1419 $\pm$ 136 g C m^{-2} yr^{-1}) than guinea grasses (1047 $\pm$ 87 g C m^{-2} yr^{-1}) (Fig. 1b). Aboveground production values (931–1805 g C m^{-2} yr^{-1}) were consistent with other reported values and yields (1020–1347 g C m^{-2} yr^{-1}) in the literature for perennial grasses in temperate and tropical regions assuming that 48% of reported yields was C (Bremer et al., 1998; Tufekcioglu et al., 2001; Dornbush & Raich, 2006). Yields of irrigated and rainfed napier grass in Hawaii were estimated to be 2352 and 1902 g C m^{-2} yr^{-1}, respectively (Kinoshita et al., 1995). Although no trials have been carried out on guinea grass in Hawai'i prior to this study, yield of guinea grass reported in Mexico reaches up to 1152 g C m^{-2} yr^{-1} (Reynoso et al., 2009).

Annualized litterfall (F_A) ranged from 110 to 229 g C m^{-2} yr^{-1} (Table 1) and was also greater for napier grasses and PM×D (128 $\pm$ 6 g C m^{-2} yr^{-1}) than for guinea grass accessions (159 $\pm$ 20 g C m^{-2} yr^{-1}). Within the napier grasses and PM×D, napier grass F_A was greater than PM×D. Litterfall C values were 1.6–3.3 times greater than root C values, but nearly 10 times less than the aboveground C that was removed as biomass for fuel production. In addition, although some dissolved C from litter may have passed, weed mats prevented the incorporation of aboveground litter into the soil profile. Litterfall therefore was unlikely to be a substantial contributor to SOC pools.

Annualized soil CO_2 efflux (F_S) ranged from 1325 to 1788 g C m^{-2} yr^{-1} and averaged 1570 ± 58 g C m^{-2}

Fig. 1 Change in soil organic carbon (SOC) pool from 2010 to 2011 (ΔC_S) (a) and annualized aboveground biomass yield (b). White vertical bars indicate napier grass accessions, gray striped bar indicates PM×D, and gray vertical bars indicate guinea accessions. Unconnected horizontal bars indicate significant differences at $P \leq 0.05$; species (guinea grass vs. napier grass) and/or variety (Napier vs. PM×D within napier grass) differences are reported as described in the Methods section.

yr^{-1} with no significant differences across accessions (Table 1). Significant differences in ΔC_S and F_A canceled each other out, resulting in no significant differences in TBCF among accessions. The mean TBCF was 1614 ± 62 g C m^{-2} yr^{-1}, which was positively correlated with aboveground biomass C across accessions ($r = 0.433$, $P = 0.015$; $n = 31$). The range of TBCF was between 1510 and 1917 g C m^{-2} yr^{-1}, which was nearly twice as much as values reported in a nutrient poor sandy soil grassland in Minnesota (Adair et al., 2009) and in line with values for a highly productive Eucalyptus saligna plantation in Hawai'i (1400–1900 g C m^{-2} yr^{-1}; Ryan et al., 2004).

Root quality and decay

Multiple aspects of root chemical characteristics varied significantly among accessions and by species (Table 2). The mean concentration of root NFC, which contains soluble sugars and amino acids, varied from 21.3% to 26.8% across accessions, and root NFC concentrations were significantly greater for NG1 and NG3 than GG3 ($P = 0.008$). Averaged over the different accessions, napier grasses and PM×D ($25.9 \pm 0.8\%$) had significantly greater root NFC concentrations than guinea grass ($22.7 \pm 0.7\%$) ($P \leq 0.001$). Mean root cellulose and hemicellulose concentrations for all accession were $33.4 \pm 0.2\%$ and $22.6 \pm 0.3\%$, respectively, with no accession or species effects.

Root lignin concentration ranged from 17.9% to 20.9% across accessions, with GG2, GG3, and GG4 significantly greater than NG1, NG2, and PM×D (Table 2). Comparisons of species revealed a significant and distinct pattern of root lignin concentrations in the order of PM×D ($17.9 \pm 0.7\%$) < napier grass ($19.0 \pm 0.4\%$) < guinea grass ($20.8 \pm 0.4\%$). Root C concentrations also differed significantly across accessions (Table 2), guinea grass GG1 was greater than any napier grass accessions and PM×D, and guinea grass accessions on average had significantly greater root C ($44.6 \pm 0.3\%$) compared to the average of napier grass accessions and PM×D ($42.6 \pm 0.2\%$). Root N concentrations did not vary across accessions (Table 2). However, napier grass and PM×D ($0.72 \pm 0.04\%$) had significantly greater root N concentrations than guinea grass (0.64 ± 0.04).

Root C : N ranged widely from 57 to 82 with significant differences across accessions (Table 2). On average, napier grass accessions and PM×D together had significantly lower C : N (62 ± 4) compared to guinea grass accessions (74 ± 6). Root lignin : N ratios were less variable at 24–39, but still varied significantly across accessions. GG1 had significantly greater lignin : N compared to PM×D. Similar to C : N, lignin : N was

significantly lower for napier and PM×D compared to guinea grass accessions. Root chemical characteristics such as root N and lignin concentrations, and lignin : N ratio observed in this study were similar to values reported for temperate C4 grasses where % N ranged from 0.76 to 1.19, % lignin ranged from 9.3 to 29.3, and lignin : N ratio was between 11.0 and 25.8 (Vivanco & Austin, 2006).

The decay constant, k, ranged from 0.95 to 1.69, similar to the range observed for temperate grass root decomposition (0.51–1.82 yr^{-1}; Vivanco & Austin, 2006). On average, roots of napier grass accessions and PxD decomposed significantly faster (1.59 ± 0.09) compared to guinea grass accessions (1.19 ± 0.09), while napier grasses and PxD showed a similar decay constant. Pearson correlation analysis determined that k was significantly correlated with all root components except cellulose and hemicellulose (Table 3).

High negative and significant correlations with k were observed in root C and lignin concentration, C : N, and lignin : N. The k value was also significantly and positively correlated with root N and NFC concentrations. Root lignin concentration was the single best predictor of k, and adding N as a second predictor did not significantly improve the fit, which is consistent with a recent litter decomposition review that manipulating a single chemical characteristic would not contribute to drastic differences in decomposition rates (Prescott, 2010).

Functional belowground C dynamics

In the initial model structure developed around our hypotheses regarding the primary drivers of SOC accumulation under cultivation, aboveground biomass yield and root lignin were assumed independent of other variables and to have a direct effect on C_S 2011 and C_R 2010. Fs and k served as both explanatory and response variables in the initial model, and C_S 2011 did not affect any other variables (Fig. 2a). Ad hoc model modification was not considered because the value of the highest modification index was <5, suggesting that the hypothesized initial model fit the data well enough to not require major changes (Jöreskog & Sorbom, 1989). In the final model, nonsignificant paths from lignin and yield to C_S 2011 were removed because the low probability associated with these paths suggested limited chance that such an effect was present in the study (Table 4). Hypothesized paths from k to C_R 2010, C_R 2010 to F_S, and C_R 2010 to C_S 2011 were retained in the model because such effects are likely to be real. As a result of these adjustments to the conceptual model, the final model fit the data well with $\chi^2 = 3.15$, df = 6, $P = 0.79$, RMSEA = 0, TLI = 1.18, and SRMR = 0.05 (Fig. 2b).

Table 2 Mean values ± 1 SE for chemical composition (%) and decay constant (yr^{-1}) of root biomass for grass accessions ($n = 4$ except for GG2 where $n = 3$). ANOVA F and P values of the random effect of blocks (as replicates), fixed effect of accessions, and custom contrasts of species on chemical composition of root biomass

	Napier			Hybrid	Guinea				Accession effect	GG vs. NG + P×D	NG vs. PM×D
	NG1	NG2	NG3	PM×D	GG1	GG2	GG3	GG4			
Fiber analysis											
NFC	26.8 ± 1.6	25.2 ± 0.79	26.8 ± 2.8	24.8 ± 0.6	23.5 ± 1.8	23.5 ± 1.2	21.3 ± 0.9	22.7 ± 1.4	$P = 0.008$, $F = 3.87$	$P < 0.001$, $F = 18.75$	ns
Hemicellulose	21.7 ± 1.1	22.5 ± 0.9	21.1 ± 1.4	23.9 ± 1.2	21.8 ± 0.5	21.6 ± 0.5	24.6 ± 0.4	23.3 ± 0.5	ns	ns	ns
Cellulose	32.9 ± 0.7	33.6 ± 0.3	32.6 ± 0.4	33.5 ± 0.8	34.3 ± 0.4	34.1 ± 1.0	33.2 ± 0.4	33.2 ± 0.9	ns	ns	ns
Lignin	18.6 ± 0.3	18.7 ± 0.5	19.5 ± 1.0	17.9 ± 0.7	20.5 ± 0.9	20.9 ± 0.9	20.9 ± 0.8	20.8 ± 0.9	$P < 0.001$, $F = 8.66$	$P < 0.001$, $F = 52.4$	$P = 0.042$, $F = 4.75$
Elemental concentrations and ratios											
C	42.5 ± 0.2	42.6 ± 0.4	42.5 ± 0.6	42.8 ± 0.47	45.3 ± 0.2	44.8 ± 1.0	43.7 ± 0.5	44.8 ± 0.3	$P = 0.001$, $F = 5.74$	$P < 0.001$, $F = 34.39$	ns
N	0.65 ± 0.08	0.71 ± 0.08	0.73 ± 0.08	0.77 ± 0.09	0.59 ± 0.06	0.55 ± 0.03	0.69 ± 0.08	0.7 ± 0.11	ns	$P = 0.024$, $F = 5.92$	ns
C : N	68 ± 9	63 ± 8	60 ± 7	57 ± 6	80 ± 11	82 ± 4	66 ± 9	69 ± 11	$P = 0.038$, $F = 2.71$	$P = 0.004$, $F = 10.91$	ns
Lignin : N	30 ± 4	28 ± 4	28 ± 4	24 ± 3	37 ± 6	39 ± 4	32 ± 5	33 ± 6	$P = 0.032$, $F = 2.83$	$P = 0.001$, $F = 14.92$	ns
Decay constant (k)	1.64 ± 0.13	1.64 ± 0.13	1.38 ± 0.1	1.69 ± 0.3	1.23 ± 0.18	0.95 ± 0.21	1.14 ± 0.12	1.38 ± 0.22	ns	$P = 0.004$, $F = 10.85$	ns

SE, standard error; NFC, nonfiber carbohydrates (%); HC, hemicellulose (%); CEL, cellulose (%); % N, root nitrogen concentration; C : N, the C to N ratio; Lignin : N, the lignin to N ratio; k, root decay constant (yr^{-1}).

Table 3 Pearson correlation coefficients (r) and significance (P) for each chemical composition factor and the decay constant (k) from litterbag experiment ($n = 31$)

Root composition factor	r	P
C	−0.390	0.030
N	**0.497**	**0.004**
NFC	0.483	0.006
Hemicellulose	0.021	0.911
Cellulose	−0.286	0.118
Lignin	**−0.709**	**<0.001**
C : N	−0.520	0.003
Lignin : N	−0.597	<0.001

Bolded factors indicate the largest r for positive and negative correlations.

In the final model, root lignin concentration had a highly significant, negative effect on k, as expected from correlation analysis of root chemical compositions (Table 4). The standardized path coefficient of −0.680 in the lignin to k path implies that as lignin decreases by one standard deviation, it is predicted that k increases by 68% of its standard deviation (Table 4; Fig. 2b). Expressed in terms of an unstandardized coefficient, a one unit decrease in lignin concentration would result in an increase of 0.159 in k. Expressed in terms of relevant ranges, as lignin concentrations increase its effective range from 15.88 to 22.53, k was predicted to decrease 52.8% of its effective range from 0.565 to 2.569

(Fig. 2b). Almost half (46%) of the variation in k was accounted for by root lignin concentrations.

Aboveground yield had significant positive effects on C_R 2010 and F_S (Table 4), explaining 24% and 23% of variation in each variable, respectively (Fig. 2b). Aboveground yield did not have a direct effect on C_S 2011, and therefore whatever effect the yield had on C_S 2011 was mediated through F_S and C_R 2010. Soil CO_2 efflux, a pathway for loss of SOC, had a marginal negative effect on C_S 2011 (Fig. 2b).

Contrary to expectation, C_R 2010 was a poor predictor of both F_S and C_S 2011. Due to the weak effects of C_R 2010 on F_S or C_S 2011, the indirect effect of yield on C_S 2011 through F_S and C_R 2010 also were not significant (Table 4). Root k was the only variable in the model with a significant direct effect on C_S 2011 (Fig. 2b). Together with F_S, k explained 40% of the variation in C_S 2011. However, k had a positive effect on C_S 2011, opposite what was originally hypothesized. Moreover, a direct effect of lignin on C_S 2011 was not observed in the model, while the indirect negative effect of lignin on C_S 2011 through k was significant (Table 4). Therefore, the effect of lignin on C_S 2011 was negatively mediated through the intervening variable k indicating that greater root lignin concentration would result in less SOC, contrary to the initial expectation. Furthermore, root k had a marginally negative effect on F_S (Table 4). Contrary to initial expectations, root lignin indirectly affected F_S positively at a marginal level (Table 4). The

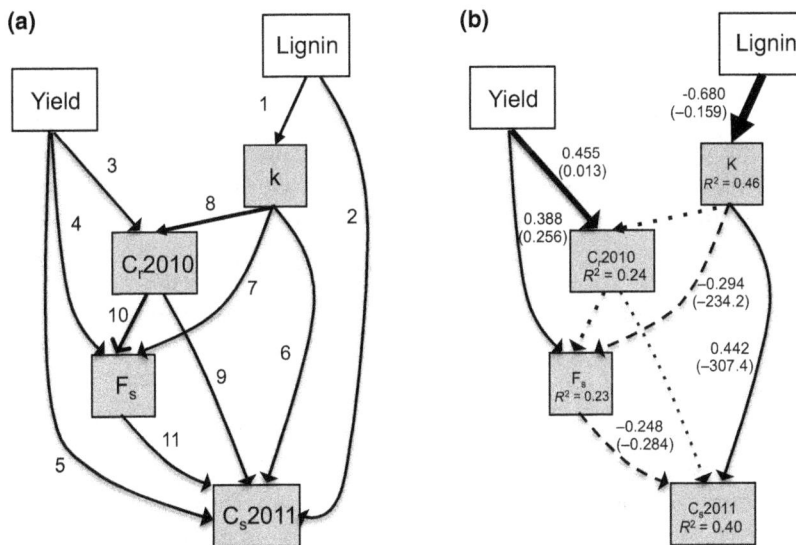

Fig. 2 Conceptual model of hypothesized relationships between soil organic carbon (SOC) after 20 months of grass growth and quantity and quality of grass C inputs and outputs (a) and best-fit model resulting from structural equation modeling (SEM) (b). Values on arrows are standardized coefficients, and values in parentheses are unstandardized coefficients. Only values with significant paths were reported. The solid arrows indicate statistically significant paths at $P \leq 0.05$. The bold arrows indicate paths at $P \leq 0.01$. Thick bold arrows indicate $P \leq 0.001$. Dashed arrows indicate marginally significant paths ($0.1 > P > 0.05$). Dotted arrows indicate nonsignificant paths. The final model fit the data well ($\chi^2 = 3.15$, df = 6, $P = 0.79$, RMSEA = 0, TLI = 1.18, and SRMR = 0.05).

Table 4 Summary of direct and indirect effects in the structural equation modeling ($n = 32$)

	Unstd. coeff.	SE	Z	P	Std. coeff.	Std. relevant
Direct effects						
k						
Lignin → k	**−0.159**	**0.031**	**−5.158**	**<0.001**	**−0.680**	**−0.528**
C_R 2010						
Yield → C_R 2010	**0.013**	**0.005**	**2.911**	**0.004**	**0.455**	**0.358**
k → C_R 2010	6.464	5.487	1.178	0.239	0.184	0.200
F_S						
Yield → F_S	**0.256**	**0.118**	**2.176**	**0.030**	**0.388**	**0.409**
C_R 2010 → F_S	−1.178	4.12	−0.286	0.775	−0.052	−0.068
k → F_S	−234.186	128.64	−1.82	0.069	−0.294	−0.421
C_S 2011						
k → C_S 2011	**307.350**	**103.95**	**2.957**	**0.031**	**0.442**	**0.548**
C_R 2010 → C_S 2011	3.291	2.831	1.162	0.245	0.166	0.190
F_S → C_S 2011	−0.248	0.129	−1.926	0.054	−0.284	−0.246
Indirect effects						
C_R 2010						
Lignin → k → C_R 2010	−1.031	0.897	−1.148	0.251	–	–
F_S						
Lignin → k → F_S	37.338	21.750	1.717	0.086	–	–
Yield → C_R 2010 → F_S	−0.016	0.055	−0.285	0.776	–	–
C_S 2011						
Lignin → k → C_S 2011	**−49.004**	**19.104**	**−2.565**	**0.010**	–	–
k → C_R 2010 → C_S 2011	21.271	25.708	0.827	0.408	–	–
k → F_S → C_S 2011	57.987	43.825	1.323	0.186	–	–
Yield → C_R 2010 → C_S 2011	0.043	0.040	1.080	0.280	–	–
Yield → F_S → C_S 2011	−0.063	0.044	−1.442	0.149	–	–
C_R 2010 → F_S → C_S 2011	0.292	1.031	0.283	0.777	–	–

Unstd. coeff., unstandardized path coefficient; Std. coeff., standardized path coefficient; Std relevant, standardized relevant range; Lignin, root lignin concentration (%); k, root decay constant (yr^{-1}); F_S, annualized soil CO_2 efflux (g C m^{-2} yr^{-1}); C_S, soil C pool (g C m^{-2}); C_R, root C pool (g C m^{-2}).

Significant effects ($P < 0.05$) are given in bold. Standard errors (SE) of indirect effects were estimated using the delta method (see Methods).

final model indicated that the indirect effect of belowground input quality, specifically, low lignin concentration, was a more important determinant of the SOC pool size than the indirect effect of aboveground biomass yield.

Discussion

Ecosystem C pools and fluxes

Cultivation of fallow soil typically results in rapid decreases in SOC pools (Davidson & Ackerman, 1993), particularly if long-term, intensive tillage and planting are practiced (Dalal & Mayer, 1986). In contrast, when conservation practices such as zero-tillage or ratoon harvest are used with crops known to have extensive belowground root systems such as perennial grasses, C loss can be minimized and even accumulation of SOC is possible (Lal & Kimble, 1997; Franzluebbers, 2012; Anderson-Teixeira et al., 2009; Powlson et al., 2011;

Anderson-Teixeira et al., 2013). In this tropical study, an accumulation of SOC occurred in the initial years of cultivation of a fallow field for perennial C4 grasses for biofuel production. Napier grass varieties, in particular, accumulated greater SOC during a 1-year period than guinea grass varieties, suggesting that cultivation of napier grass instead of guinea grass may have a greater benefit to the greenhouse warming potential of the production system in the form of C sequestration in soil.

In 2010, the SOC pool in adjacent fallow soils was 5423 g C m^{-2} (Sumiyoshi, 2011) compared to 5174 and 5442 g C m^{-2} for the accession plots for 2010 and 2011, respectively. Thus, it appears that any initial decrease in SOC after preparation of the field recovered under the cultivation of ratoon-harvested perennial grasses within 2 years and, in the interim, this dataset captured a dynamic window of postcultivation soil carbon dynamics. Root biomass builds rapidly after the first planting and then stabilizes following the first ratoon harvest as it comes into equilibrium with root mortality during subsequent harvests. This suggests that tilling and replanting a field that has been in zero-tillage cultivation may not lose accumulated C, as is often assumed in simulation models and life cycle analyses.

Soil CO_2 efflux and TBCF were substantial, and annual C fluxes were greater than aboveground production, supporting the hypothesis that a wide range of ecosystem C pools and fluxes would contribute collectively to the soil C balance of our study systems. Consistent with the observed pattern between TBCF and aboveground net primary productivity in forest ecosystems (Litton et al., 2007), TBCF was positively related to aboveground yield in this study. This is the first study, to our knowledge, to use this mass balance approach to estimate the autotrophic flux of C to belowground in a tropical agricultural system, and suggests that coupling of aboveground and belowground C fluxes occurs universally across terrestrial ecosystems.

That total SOC pool size did not differ among the grasses was not surprising given the large residual stock of C in relation to the comparatively small change in SOC pool (<5% of total SOC) over 1 year of cultivation. Differences in the component pools and fluxes among the grasses, however, suggest that over time SOC pools may change substantially in response to cultivation of different grasses. For example, aboveground biomass yield and the annual increase in soil C differed among the accessions and more generally by species. If high-yielding grasses (such as two of the tested napier grass accessions and the napier grass hybrid) also have high root biomass, and these inputs result in a net gain of soil C over time, then selecting grasses with high aboveground biomass yield may have a net greenhouse warming benefit to the production system.

Root quality and decay

The link between high-yielding grasses and soil C accumulation is dependent in part on the fate of the root biomass as an input to the belowground soil system. The fate (i.e., whether C is released from the system or retained), in turn, is in part dependent on the chemical composition of detrital inputs (Cotrufo et al., 2015). Here, the hypothesis that the selected perennial grasses would have a wide range of root input quality indicators was supported. The concentration of carbohydrates, lignin, C, N, and their relative abundance varied among grasses, and often by species. Specifically, many indicators of low C quality, such as high lignin concentration, C : N ratio, and lignin : N ratio, were greatest for the guinea grasses compared to napier grasses and the napier grass hybrid. A view emerges of complex interactions among the potential factors controlling soil C accumulation when considering variation within these

chemical indicators of input quality in combination with variation described above in ecosystem C pools and fluxes.

Root decay was strongly related to root lignin concentration, which is consistent with a recent review by Prescott (2010) that identified lignin as the most reliable predictor of root decomposition rate. Mass loss during the litterbag study, and the resultant decay constant, was interpreted to represent losses of potential C inputs from the soil system through microbial respiration or leaching. Accordingly, we expected root lignin concentration to have a direct, positive effect on SOC pool size or an indirect effect via the pathway of high lignin concentration driving low root decay (and therefore low losses of C outputs via respiration and leaching) resulting in the accumulation of root fragments or residues as SOC.

Functional belowground C dynamics

Linking ecosystem C pools and fluxes to accurately represent soil function remains a challenge, yet is critical for predicting the effect of land-use change and management on the net C balance of agroecosystems. In a renewable energy system, environmental sustainability is often assessed from the perspective of the net global warming potential for the collective feedstock production system, fuel conversion, and delivery in a life cycle analysis. Understanding functional belowground C dynamics and linking soil C accumulation to measureable indices can be important in guiding feedstock and management choices during the planning and implementation of a production system where climate change mitigation is a desirable outcome. Based on the results from H1 and H2, we developed a conceptual model for the factors controlling soil C accumulation in the study system. However, hypothesis H3, that aboveground biomass and root lignin concentration (both easily measured factors) provide direct positive control on SOC pool size, was not supported. Further, the roles of root C, soil CO_2 efflux, and decay constant as drivers of the indirect effects of yield and root lignin on SOC pool size were confirmed, but also were not as expected.

The marginal and negative correlation between root decay rate (k) and soil CO_2 efflux (F_S) was unexpected. However, F_S is the cumulative flux of CO_2 from soils to the overlying atmosphere that is comprised of both autotrophic and heterotrophic components, and k is only one component of heterotrophic respiration contributing to F_S in addition to decomposition of aboveground litter and SOM. Prior studies have shown that F_S is largely driven by recent photosynthetic C input to belowground (see Kuzyakov & Gavrichkova, 2010) and that the root respiration contribution to F_S is ~30–50% and increases asymptotically with increasing F_S (Bond-

Lamberty et al., 2004). As such, k is likely to be a small component of overall F_S (~10–15%) and it is entirely feasible that k could be marginally and negatively correlated with F_S. In this study, dissolved organic matter transport and fragmentation contributed to k, in addition to biochemical oxidation.

In the final model describing the functional belowground C dynamics of this system, high aboveground biomass yields did result in high root biomass C, but did not directly result in high SOC pools. Instead of high root biomass C resulting in high SOC, the chemistry of roots, primarily lignin concentration, was the critical factor in controlling root decomposition and subsequent SOC accumulation. Specifically, low root lignin concentration drove high root decomposition rates, as hypothesized. High decomposition rates, however, resulted not in greater soil CO_2 losses as expected, but rather in greater SOC accumulation. Thus, the hypothesis that grasses with higher aboveground productivity and higher lignin concentration (and lower decomposition rate) would accumulate the greatest amount of SOC over the study period was not supported. Instead, biofuel grasses cultivated with zero-tillage that had high productivity and roots with low lignin (i.e., high quality) and high decomposition rates accumulated the most SOC in this study. In temperate switchgrass systems, Ontl et al. (2015) reported that crops that maximize root productivity would lead to the largest gains in protected soil C. On a longer time frame, as roots die and decompose with repeated ratooning, it seems likely that accessions with high root biomass, high proportion of root death upon ratoon, low total lignin concentration, and high proportion of readily degradable lignin monomers would accumulate the greatest amount of SOC.

Soil OC formation occurs via both a rapid dissolved organic matter–microbial path and a slower physical transfer path as litter fragments enter the soil (Cotrufo et al., 2013, 2015). A likely mechanism for SOC accumulation may be through incorporation of root fragments and microbial biomass C, necromass, and by-products into aggregates in the absence of tillage (Six et al., 2000; Ontl et al., 2015; Tiemann & Grandy, 2015). Huang et al. (2011) showed in a forest environment that inputs with lower lignin concentration resulted in more SOC accumulation due to greater inputs through fragments, similar to results from this study. Once within aggregates, fragmented tissue, soluble organic matter, and microbial biomass by-products are physically protected from further decomposition. Taken together, the results from this study agree with the recent understanding that microbial transformation of plant residue and physical protection within aggregates can be a more important driver of SOC stabilization than chemical recalcitrance per se (Schmidt et al., 2011; Lehmann & Kleber, 2015).

The effect of biofuel biomass cultivation on SOC may be positive or negative depending on the type of cropping system. Perennial crops and low- or zero-tillage systems with little residue removal are the most likely to result in a net increase or maintenance of SOC (Lewandowski, 2013). Much of the research on C sequestration in bioenergy crop cultivation has focused on temperate and subtropical regions (Gelfand et al., 2013; Agostini et al., 2015), but increasingly tropical perennial grasses such as sugarcane and napier grass are being utilized (Lemus & Lal, 2005; Anderson-Teixeira et al., 2009). Recent work completed on Miscanthus (switchgrass) cultivation in Europe showed that the accumulation of SOC was dependent on the genotype-specific root distribution of each variety (Richter et al., 2015), further highlighting the importance of understanding the underlying mechanisms driving SOC dynamics in a given system. On similar systems in the US Midwest, soil type affected the efficacy of aggregates to form and protect C derived from giant Miscanthus inputs under cultivation for biofuel (Tiemann & Grandy, 2015). Identifying characteristics of plants (such as biomass production, rooting morphology, and tissue chemistry) and of soil (such as texture, mineralogy, and organic matter dynamics) and how they will interact under cultivation is critical to predicting the capacity of a production system to sequester C in soil.

The most beneficial, sustainable biofuel systems should focus on feedstocks that complement food production, do not result in direct or indirect losses of C through land-use change, and do not cause net greenhouse gas emissions (Tilman et al., 2009; Anderson-Teixeira et al., 2012). Productivity and sustainability of systems cultivated for biofuel feedstocks often are positively associated with SOC content (Lehmann, 2012; Anderson-Teixeira et al., 2013), and increasing SOC in agricultural systems simultaneously restores soil quality through improvements in the biological, physical, and chemical functions provided by organic matter (Powlson et al., 2011; Franzluebbers, 2012; Murphy, 2014) while sequestering atmospheric CO_2 (Lal, 2013; Werling et al., 2014; DeLucia, 2016). In this study of tropical perennial C4 grasses, high-yielding varieties with low root lignin cultivated in zero-tillage systems were the most promising biomass feedstock to meet the complex needs of society for renewable fuels, food, and climate change mitigation through net carbon sequestration during feedstock production. Genetic improvements are possible through technology and breeding to maximize climate change mitigation of renewable fuels systems, and these results reveal plant traits that may be the best predictors of soil carbon accumulation in the tropics. Identifying soil management practices that ensure productivity and profitability for farmers while building SOC remains a challenge (Hill *et al.*, 2006; Meki *et al.*, 2014), but isolating readily measured indicators of potential SOC accumulation such as yield, root morphology, and tissue chemistry improves our ability to make system-level decisions that ensure both environmental and economic sustainability.

Acknowledgements

Funding for this research was provided by Department of Energy award number DE-FG36-08GO88037 to Andrew G. Hashimoto, Office of Naval Research Grant N00014-12-1-0496 to Hawaii Natural Energy Institute, and a Specific Cooperative Agreement between Susan E. Crow and the United States Department of Agriculture-Agricultural Research Service (award number 003232-00001) with funds provided by the Office of Naval Research. This work was further supported by the USDA National Institute of Food and Agriculture, Hatch project (project HAW01130-H), managed by the College of Tropical Agriculture and Human Resources. Roger Corrales and the field crew at the Waimanalo Agricultural Research Station, College of Tropical Agriculture and Human Resources, University of Hawaii at Manoa assisted with the field trials. We thank Heather Kikkawa, Mataia Reeves, Meghan Pawlowski, Mariko Panzella, Jon Wells, Anne Quidez, Nate Hunter, Alisa Davis, and Hironao Yamasaki for their assistance with field and laboratory work. Adel Youkhana and three anonymous reviewers provided helpful comments on the manuscript. Drs Goro Uehara, Hue Nguyen, and Russell Yost provided invaluable mentoring and advice.

References

Adair EC, Reich PB, Hobbie SE, Knops JMH (2009) Interactive effects of time, CO_2, N, and diversity on total belowground carbon allocation and ecosystem carbon storage in a grassland community. *Ecosystems*, 12, 1037–1052.

Adler PR, Del Grosso SJ, Parton WJ (2007) Life-cycle assessment of net greenhouse-gas flux for bioenergy cropping systems. *Ecological Applications*, 17, 675–691.

Agostini F, Gregory AS, Richter GM (2015) Carbon sequestration by perennial energy crops: is the jury still out? *Bioenergy Research*, 8, 1057–1080.

Al-Kaisi MM, Grote JB (2007) Cropping systems effects on improving soil carbon stocks of exposed subsoil. *Soil Science Society of America Journal*, 71, 1381–1388.

Anderson-Teixeira KJ, Davis SC, Masters MD, DeLucia EH (2009) Changes in soil organic carbon under biofuel crops. *GCB Bioenergy*, 1, 75–96.

Anderson-Teixeira KJ, Snyder PK, Twin TE, Cuadra CV, Costa MH, DeLucia EH (2012) Climate-regulation services of natural and agricultural ecoregions of the Americas. *Nature Climate Change*, 2, 177–181.

Anderson-Teixeira KJ, Masters MD, Black CK, Zeri M, Hussain MZ, Bernacchi CJ, DeLucia EH (2013) Altered belowground carbon cycling following land-use change to perennial bioenergy crops. *Ecosystems*, 16, 508–520.

Bond-Lamberty B, Wang C, Gower ST (2004) A global relationship between the heterotrophic and autotrophic components of soil respiration? *Global Change Biology*, 10, 1756–1766.

Brahim N, Blavet D, Gallali T, Bernoux M (2011) Application of structural equation modeling for assessing relationships between organic carbon and soil properties in semiarid Mediterranean region. *International Journal of Environmental Science and Technology*, 8, 305–320.

Bremer DJ, Ham JM, Owensby CE, Knapp AK (1998) Responses of soil respiration to clipping and grazing in a tallgrass prairie. *Journal of Environmental Quality*, 27, 1539–1548.

Cherubini F, Bird ND, Cowie A, Jungmeier G, Schlamadinger B, Woess-Gallasch S (2009) Energy- and greenhouse gas-based LCA of biofuel and bioenergy systems: key issues, ranges and recommendations. *Resources, Conservation and Recycling*, 53, 434–447.

Cotrufo MF, Wallenstein MD, Boot CM, Denef K, Paul E (2013) The Microbial Efficiency-Matrix Stabilization (MEMS) framework integrates plant litter decomposition with soil organic matter stabilization: do labile plant inputs form stable soil organic matter? *Global Change Biology*, 19, 988–995.

Cotrufo MF, Soong JL, Horton AJ, Campbell EE, Haddix ML, Wall DH, Parton WJ (2015) Formation of soil organic matter via biochemical and physical pathways of litter mass loss. *Nature Geoscience*, 8, 776–779.

Dalal RC, Mayer RJ (1986) Long-term trends in fertility of soils under continuous cultivation and cereal. Cropping in southern Queensland. 2. Total organic carbon and its rate of loss from the soil profile. *Australian Journal of Soil Research*, 24, 281–292.

Davidson E, Ackerman I (1993) Changes in soil carbon inventory following cultivation of previously untilled soils. *Biogeochemistry*, 20, 161–193.

Davidson EA, Janssens IA (2006) Temperature sensitivity of soil carbon decomposition and feedbacks to climate change. *Nature*, 440, 165–173.

Davis SC, Boddey RM, Alves BJR et al. (2013) Management swing potential for bioenergy crops. *GCB Bioenergy*, 5, 623–638.

DeLucia EH (2016) How biofuels can cool our climate and strengthen our ecosystems. *Eos*, 97, 14–19.

Dornbush M, Raich J (2006) Soil temperature, not aboveground plant productivity, best predicts intra-annual variations in soil respiration in central Iowa grasslands. *Ecosystems*, 9, 909–920.

Eisenhauer N, Cesarz S, Koller R, Worm K, Reich PB (2012) Global change belowground: impacts of elevated CO_2, nitrogen, and summer drought on soil food webs and biodiversity. *Global Change Biology*, 18, 435–447.

Ellert BH, Bettany JR (1995) Calculation of organic matter and nutrients stored in soils under contrasting management regimes. *Canadian Journal of Soil Science*, 75, 529–538.

Ford CR, Mcgee J, Scandellari F, Hobbie EA, Mitchell RJ (2012) Long- and short-term precipitation effects on soil CO_2 efflux and total belowground carbon allocation. *Agricultural and Forest Meteorology*, 156, 54–64.

Franzluebbers A (2012) Grass roots of soil carbon sequestration. *Carbon Management*, 3, 9–11.

Gelfand I, Sahajpal R, Zhang X, Izaurralde RC, Gross KL, Robertson GP (2013) Sustainable bioenergy production from marginal lands in the US Midwest. *Nature*, 493, 514–517.

Geng Y, Wang Y, Yang K et al. (2012) Soil respiration in Tibetan alpine grasslands: belowground biomass and soil moisture, but not soil temperature, best explain the large-scale patterns. *PLoS One*, 7, e34968.

Giambelluca TW, Chen Q, Frazier AG et al. (2013) Online rainfall atlas of Hawai'i. *Bulletin of the American Meteorology Society*, 94, 313–316 (and online tool).

Giardina CP, Ryan MG (2002) Total belowground carbon allocation in a fast-growing eucalyptus plantation estimated. *Ecosystems*, 5, 487–499.

Gibbs HK, Johnston M, Foley JA, Holloway T, Monfreda C, Ramankutty N, Zaks D (2008) Carbon payback times for crop-based biofuel expansion in the tropics: the effects of changing yield and technology. *Environmental Research Letters*, 3, 034001.

Gifford RM, Roderick ML (2003) Soil carbon stocks and bulk density: spatial or cumulative mass coordinates as a basis of expression? *Global Change Biology*, 9, 1507–1514.

Gomez-Casanovas N, Anderson-Teixeira K, Zeri M, Bernacchi CJ, DeLucia EH (2013) Gap filling strategies and error in estimating annual soil respiration. *Global Change Biology*, 19, 1941–1952.

Grace J (2006) *Structural Equation Modeling and Natural Systems*. Cambridge University Press, Cambridge.

Grandy AS, Neff JC (2008) Molecular C dynamics downstream: the biochemical decomposition sequence and its impact on soil organic matter structure and function. *Science of the Total Environment*, 404, 297–307.

Hall MB (2003) Challenges with nonfiber carbohydrate methods. *Journal of Animal Science*, 81, 3226–3232.

Hashimoto A, Arnold J, Ayars J et al. (2012) High-yield tropical biomass for advanced biofuels. Proceedings from Sun Grant National Conference: Science for Biomass Feedstock Production and Utilization, New Orleans, LA. Available at: www.sungrant.tennessee.edu/NatConference/ (accessed 23 June 2016).

Hill J, Nelson E, Tilman D, Polasky S, Tiffany D (2006) Environmental, economic, and energetic costs and benefits of biodiesel and ethanol biofuels. *Proceedings of the National Academy of Science*, 103, 11206–11210.

Hothorn T, Bretz F, Westfall P (2008) Simultaneous inference in general parametric models. *Biometrical Journal*, 50, 346–363.

Hu LT, Bentler PM (1999) Cutoff criteria for fit indexes in covariance structure analysis: conventional criteria versus new alternatives. *Structural Equation Modeling*, 6, 1–55.

Huang YH, Li YL, Xiao Y et al. (2011) Controls of litter quality on the carbon sink in soils through partitioning the products of decomposing litter in a forest succession series in South China. Forest Ecology and Management, 261, 1170–1177.

Jastrow JD, Miller RM (1996) Soil aggregate stabilization and carbon sequestration: feedbacks through organomineral associations. In: Soil Processes and the Carbon Cycle (eds Lal R, Kimble J, Follett R, Stewart BA), pp. 207–223. CRC Press, Boca Raton.

Johnson JMF, Barbour NW, Weyers SL (2007) Chemical composition of crop biomass impacts its decomposition. Soil Science Society of America Journal, 71, 155–162.

Jonsson M, Wardle DA (2010) Structural equation modelling reveals plant-community drivers of carbon storage in boreal forest ecosystems. Biology Letters, 6, 116–119.

Jöreskog KG, Sorbom D (1989) Lisrel 7. A Guide to the Program and Applications, 2nd edn. SPSS Inc., Chicago, IL.

Keffer VI, Turn SQ, Kinoshita CM, Evans DE (2009) Ethanol technical potential in Hawaii based on sugarcane, banagrass, Eucalyptus, and Leucaena. Biomass and Bioenergy, 33, 247–254.

Khanal SK, Surampalli RY, Zhang TC, Lamsal BP, Tyagi RD, Kao CM (eds.) (2010) Bioenergy and Biofuel from Biowastes and Biomass. American Society of Civil Engineers, Reston, VA.

Kindler R, Siemens J, Kaiser K et al. (2011) Dissolved carbon leaching from soil is a crucial component of the net ecosystem carbon balance. Global Change Biology, 17, 1167–1185.

Kinoshita C, Ishimura D, Jakeway L, Osgood R (1995) Production of Biomass for Electricity Generation on the Island of Oahu. Hawaii Natural Energy Institute, Honolulu, HI.

Kuzyakov Y, Gavrichkova O (2010) Time lag between photosynthesis and carbon dioxide efflux from soil: a review of mechanisms and controls. Global Change Biology, 16, 3386–3406.

Lal R (2013) Soil carbon management and climate change. Carbon Management, 4, 439–462.

Lal R, Kimble JM (1997) Conservation tillage for carbon sequestration. Nutrient Cycling in Agroecosystems, 49, 243–253.

Lehmann J (2012) Keeping carbon down. Carbon Management, 3, 21–22.

Lehmann J, Kleber M (2015) The contentious nature of soil organic matter. Nature, 528, 60–68.

Lemus R, Lal R (2005) Bioenergy crops and carbon sequestration. Critical Reviews in Plant Sciences, 24, 1–21.

Lewandowski I (2013) Soil carbon and biofuels: multifunctionality of ecosystem services. In Ecosystem Services and Carbon Sequestration in the Biosphere (eds Lal R, Lorenz K, Huttl RF, Schneider BU, von Braun J), pp. 333–356. Springer, Dordrecht.

Liebig MA, Schmer MR, Vogel KP, Mitchell RB (2008) Soil carbon storage by switchgrass grown for bioenergy. BioEnergy Research, 1, 215–222.

Liska AJ, Perrin RK (2009) Indirect land use emission in the life cycle of biofuels: regulations vs. science. Biological Systems Engineering: Papers and Publications, Paper 144. Biofuels, Bioproducts and Biorefining, 3, 318–328.

Litton CM, Ryan MG, Raich JW (2007) Carbon allocation in forest ecosystems. Global Change Biology, 13, 2089–2109.

Litton CM, Cordell S, Sandquist DR (2008) A non-native invasive grass increases soil carbon flux in a Hawaiian tropical dry forest. Global Change Biology, 14, 726–739.

Litton CM, Giardina CP, Albano JK, Long MS, Asner GP (2011) The magnitude and variability of soil-surface CO_2 efflux increase with mean annual temperature in Hawaiian tropical montane wet forests. Soil Biology & Biochemistry, 43, 2315–2323.

Ma Z, Wood CW, Bransby DI (2000) Carbon dynamics subsequent to establishment of switchgrass. Biomass and Bioenergy, 18, 93–104.

Matias L, Castro J, Zamora R (2012) Effect of simulated climate change on soil respiration in a Mediterranean-type ecosystem: rainfall and habitat type are more important than temperature or the soil carbon pool. Ecosystems, 15, 299–310.

Meki MN, Kiniry JR, Behrman KD, Pawlowski MN, Crow SE (2014) The role of simulation models in monitoring soil organic carbon storage and greenhouse gas mitigation potential in bioenergy cropping systems. In: CO_2 Sequestration and Valorization (ed. Esteves V), InTech, Croatia.

Melillo JM, Aber JD, Muratore JF (1982) Nitrogen and lignin control of hardwood leaf litter decomposition dynamics. Ecology, 63, 621–626.

Mochizuki J, Yanagida JF, Kumar D, Takara D, Murthy GF (2014). Life cycle assessment of ethanol production from tropical banagrass (Pennisetum purpureum) using green and dry processing technologies in Hawaii. Journal of Renewable Sustainable Energy, 6, 043128.

Murphy BW (2014) Soil Organic Matter and Soil Function – Review of the Literature and Underlying Data. Department of the Environment, Canberra, Australia.

Ontl TF, Cambardella CA, Schulte LA, Kolka RK (2015) Factors influencing soil aggregation and particulate organic matter responses to bioenergy crops across a topographic gradient. Geoderma, 255–256, 1–11.

Ostertag R (2001) The effects of nitrogen and phosphorus availability on fine root dynamics in Hawaiian montane forests. Ecology, 82, 485–499.

Powlson DS, Whitmore AP, Goulding WT (2011) Soil carbon sequestration to mitigate climate change: a critical re-examination to identify the true and the false. European Journal of Soil Science, 62, 42–55.

Prescott CE (2010) Litter decomposition: what controls it and how can we alter it to sequester more carbon in forest soils? Biogeochemistry, 101, 133–149.

R Development Core Team (2012) R: A Language and Environment for Statistical Computing.

Reynoso OR, Garay AH, Da Silva SC et al. (2009) Herbage accumulation, growth and structural characteristics of Mombasa grass (Panicum maximum Jacq.) harvested at different cutting intervals. Acumulación de forraje, crecimiento y características estructurales del pasto Mombaza (Panicum maximum Jacq.) cosechado a diferentes intervalos de corte, 47, 203–213.

Richter GM, Agostini F, Redmile-Gordon M, White R, Goulding KWT (2015) Sequestration of C in soils under Miscanthus can be marginal and is affected by genotype-specific root distribution. Agriculture Ecosystems & Environment, 200, 169–177.

Ryan MG, Binkley D, Fownes JH, Giardina CP, Senock RS (2004) An experimental test of the causes of forest growth decline with stand age. Ecological Monographs, 74, 393–414.

Samson R, Mani S, Boddey R et al. (2005) The potential of C4 perennial grasses for developing a global BIOHEAT industry. Critical Reviews in Plant Sciences, 24, 461–495.

Sanderson MA (2008) Upland switchgrass yield, nutritive value, and soil carbon changes under grazing and clipping. Agronomy Journal, 100, 510.

Schmidt MWI, Torn MS, Abiven S et al. (2011) Persistence of soil organic matter as an ecosystem property. Nature, 478, 49–56.

Seltman HJ (2012) Experimental design and analysis. Available at: http://www.stat.cmu.edu/~hseltman/309/Book/Book.pdf (accessed 1 November 2012).

Silver W, Miya R (2001) Global patterns in root decomposition: comparisons of climate and litter quality effects. Oecologia, 129, 407–419.

Six J, Jastrow JD (2002) Organic matter turnover. In: Encyclopedia of Soil Science (ed. Lal R), pp. 936–942. Marcel Dekker, Boca Raton, FL.

Six J, Elliott ET, Paustian K (2000) Soil macroaggregate turnover and microaggregate formation: a mechanism for C sequestration under no-tillage agriculture. Soil Biology & Biochemistry, 32, 2099–2103.

Sobel ME (1982) Asymptotic confidence intervals for indirect effects in structural equation models. In: Sociological Methodology (ed. Leinhardt S), pp. 290–312. American Sociological Association, Washington, DC.

Stockmann U, Adams MA, Crawford JW et al. (2013) The knowns, known unknowns, and unknowns of sequestration of soil organic carbon. Agriculture, Ecosystems, and Environment, 164, 80–99.

Sumiyoshi Y (2011) Belowground impact of napier and Guinea grasses grown for biofuel feedstock production. Master's Thesis submitted to the University of Hawaii Manoa.

Tabachnick BG, Fidell LS (2007) Using Multivariate Statistics. Allyn and Bacon, Boston.

Tiemann LK, Grandy AS (2015) Mechanisms of soil carbon accrual and storage in bioenergy cropping systems. GCB Bioenergy, 7, 161–174.

Tilman D, Socolow R, Foley JA et al. (2009) Beneficial biofuels – the food, energy, and environment trilemma. Science, 325, 270–271.

Tran N, Illukpitiya P, Yanagida JF, Ogoshi R (2011) Optimizing biofuel production: an economic analysis for selected biofuel feedstock production in Hawaii. Journal of Biomass and Bioenergy, 35, 1756–1764.

Tufekcioglu A, Raich JW, Isenhart TM, Schultz RC (2001) Soil respiration within riparian buffers and adjacent crop fields. Plant and Soil, 229, 117–124.

Van Soest PJ (1963) Use of detergents in the analysis of fibrous feeds. II. A rapid method for the determination of fiber and lignin. Journal of the Association of Official Agricultural Chemists, 46, 829–835.

Van Soest PJ, Wine RH (1968) The determination of lignin and cellulose in acid detergent fiber with permanganate. Journal of the Association of Official Agricultural Chemists, 51, 780–787.

Virto I, Barré P, Burlot A, Chenu C (2012) Carbon input differences as the main factor explaining the variability in soil organic C storage in no-tilled compared to inversion tilled agrosystems. Biogeochemistry, 108, 17–26.

Vivanco L, Austin A (2006) Intrinsic effects of species on leaf litter and root decomposition: a comparison of temperate grasses from North and South America. *Oecologia*, **150**, 97–107.

Werling BP, Dickson TL, Isaacs R *et al.* (2014) Perennial grasslands enhance biodiversity and multiple ecosystem services in bioenergy landscapes. *Proceedings of the National Academy of Sciences*, **111**, 1652–1657.

Wider RK, Lang GE (1982) A critique of the analytical methods used in examining decomposition data obtained from litter bags. *Ecology*, **63**, 1636–1642.

Zhang D, Hui D, Luo Y, Zhou G (2008) Rates of litter decomposition in terrestrial ecosystems: global patterns and controlling factors. *Journal of Plant Ecology*, **1**, 85–93.

7

Characterization of 60 types of Chinese biomass waste and resultant biochars in terms of their candidacy for soil application

XIAOYIN SUN[1,*] (iD), RUIFENG SHAN[1], XUHUI LI[2], JIHUA PAN[1], XING LIU[1], RUONAN DENG[1] and JUNYAO SONG[1]

[1]Key Laboratory of Nansi Lake Westland Ecological Conservation & Environmental Protection (Shandong Province), College of Geography and Tourism, Qufu Normal University, Rizhao 276826, China, [2]College of Environment and Planning, Henan University, Kaifeng 475004, China

Abstract

The composition and pyrolysis characteristics of 60 types of biomass waste from the following six source categories were compared: agricultural residues, woody pruning waste from gardens and lawns, aquatic plant material from eutrophic water bodies, nutshells and fruit peels, livestock manure and residual sludge from municipal wastewater treatment. The yield and physicochemical characteristics of the biochar produced from these feedstocks at 350 °C, 500 °C and 650 °C were also examined. Results of correlation and canonical correspondence analysis between feedstock composition and biochar properties showed that feedstock type played an important role in controlling yield and properties of biochars. The yields of biochar dry ash-free (daf.) basis were positively correlated with cellulose, lignin and lignin/cellulose content of feedstock; and ash content hampered the biochar production. Furthermore, the intensity of correlation between biochar yield and its feedstock composition was improved with pyrolysis temperature and degree of feedstock decomposition. The fixed carbon content in biochar was also negatively influenced by ash content of feedstock, and it increased with increasing pyrolysis temperature when the ash content was below 34.57% in feedstock and decreased when the ash content exceeded. The fixed carbon production in biochar per unit ash-free mass (af.) was positively related to cellulose, lignin and lignin/cellulose content in feedstock, which were same with the yield of biochar (daf.). But on the contrary, the volatiles content in biochar (af.) had negative correlation with these organic constituents. For most feedstocks, the differences in the biochar characteristics among the biomass categories were greater than within any individual category. C/N, H/C and O/C atomic ratio and bulk density of biochar from different types of biomass were also compared. The results will provide guidance for the reutilization of biomass wastes and production of biochar with specified properties for soil amendment applications.

Keywords: agricultural residues, aquatic plant, biochar, canonical correspondence analysis, livestock manure, nutshell and fruit peel, physicochemical properties, pruning waste, residual sludge, yield

Introduction

Biochar is the carbon-rich product made from the pyrolysis of biomass under oxygen-limited conditions and relatively low temperatures (Lehmann & Joseph, 2015). As a soil amendment, biochar has received increasing attention due to its potential for soil carbon sequestration and the remediation of contaminated soil (Tripathi et al., 2016). Therefore, waste biomass material pyrolysis for biochar is considered to offer an attractive alternative to solid waste recovery and utilization. A wide range of biomass waste is available for producing biochar, including woody, agricultural, aquatic, human

and animal and industrial waste biomass (Vassilev et al., 2012). As an agricultural country, China produces approximately 800 million tons of agricultural straw residue (Jiang et al., 2012) and 223.5 million tons of livestock manure annually (Geng et al., 2013), although approximately 20% of the agricultural straw is burned in the field. In addition, there are about 14.4 million tons of garden waste biomass production annually in China (Shi et al., 2013), which included woody pruning waste and leaf litter from gardens and lawns. And the production of garden waste biomass is increasing quickly with the expansion of urban and greenspace areas in China, which is higher than the total annual harvest from national forests in the United States (Bratkovich et al., 2008). Part of garden waste is being recovered for composting, but a large proportion is simply discarded as

*Correspondence: Xiaoyin Sun
e-mail: xiaoyinsky@gmail.com

municipal solid waste. Therefore, biochar produced from these biomass wastes and applied to the soil could be a win-win approach, which not only reduces pollution and carbon emissions but also contributes to agricultural productivity, resource-use efficiency and soil bioremediation goals (Li & Wang, 2013).

In recent years, studies on biomass pyrolysis have revealed that the characteristics of the resulting biochar can vary significantly depending on feedstock types and processing temperatures (Lehmann & Joseph, 2015). The biomass composition and physicochemical properties vary greatly among different botanical species and even within a species depending upon the plant parts, growing conditions and harvest times (Lehmann & Joseph, 2015; Suliman et al., 2016). The properties of the biochar produced from these biomass materials are correspondingly diverse. Several studies have compared the yields and properties of biochar produced from different biomass feedstock; these studies suggest that higher biochar yields are generally obtained from feedstocks with high lignin and mineral contents (Demirbas, 2004, 2006; Lv et al., 2010; Nanda et al., 2013). According to Demirbas, hazelnut shell containing higher lignin as compared to oak wood and wheat straw has higher biochar yield (Demirbas, 2006). Similar findings were reported by Nanda et al. (2013) that pinewood containing most lignin as compared to other biomass has the highest biochar yield and wheat straw with the lowest lignin content produced the lowest biochar yield. Biomass waste can be classified into five categories by the source from where it is obtained, including woody biomass, agricultural residues, aquatic plant, human and animal waste and industrial waste biomass (Vassilev et al., 2012). Industrial biomass waste is the waste produced from the industry-related biomass raw materials, such as sugarcane residue from sugar refinery, waste from food processing factory and others. The constituents of biomass waste from industry are diverse and complex, but they are similar to the raw material which we used and some can be classified into the category of raw material based on their constituents. Biochar from feedstocks within the same category of source might show similar properties as those made from parent material of different types. For example, woody and agricultural biomass, which have high carbon and oxygen contents, typically produce more carbon-rich biochar than biochar from sewage sludge (Hossain et al., 2011) and livestock manure (Cao & Harris, 2010; Xu & Chen, 2013).

Many studies have investigated the impact of one or several sources of biomass feedstocks on the physicochemical properties of the resulting biochar (Demirbas, 2004, 2006; Lv et al., 2010; Nanda et al., 2015), which often fall into one or two biomass categories, and the results might seem less systematic. The properties of biochar determine its applications to soil, and a particular biochar may not be adapted to all types of soils. Therefore, optimizing biochar for a specific application may require a purposeful selection of feedstocks and production conditions to manufacture the biochar with desired characteristics. It is thus important to broadly and cohesively analyze the properties of different categories of biomass sources and their resulting biochar products under different production conditions.

To this end, this study collected 60 types of commonly available biomass waste, namely 23 types of agricultural residues that covered almost all of crop species in China, two types of livestock manure, 14 types of woody pruning waste from gardens and lawns, 12 types of nutshell and fruit peels from human food, eight types of aquatic plant waste from eutrophic water bodies and one type of industrial biomass, residual sludge from a municipal wastewater treatment plant. In China, agricultural residues are the most important source of biomass waste, followed by the forest and garden waste biomass, human and animal waste and industrial waste biomass. Aquatic plants selected in this study have been widely grown in eutrophic water bodies and constructed wetlands for water purification. The biomass of these plants must be properly harvested and disposed to protect the aquatic environment, and previous study demonstrated that these plants are valuable feedstocks for producing biochars (Cui et al., 2016). Therefore, the objectives of this study were to investigate the composition and pyrolysis characteristics of different types of biomass waste in China and to examine how feedstock sources and pyrolysis temperatures affect the yield and properties of biochar. The obtained results will provide guidance for the selection of biomass wastes and pyrolysis temperatures for the production of biochar with specified properties for soil amendment applications.

Materials and methods

Collection and characterization of biomass

Sixty types of biomass waste were collected in Shandong Province, China, from six categories: agricultural residues, woody pruning waste from gardens and lawns, nutshells and fruit peels from human food waste, aquatic plants from eutrophic water bodies, livestock manure and industrial waste (Table S1). The biomass studied covered almost all categories of biomass waste in China. Agricultural straw, stems and husks were collected, which included crop categories such as wheat, corn, peanut, soybean, rice, sorghum, cotton, sunflower, rape, batatas and other crops. All of the crop residues were obtained from same areas with the same climate and soil type in Shandong Province, China. Pruning waste was obtained from the campus of Qufu Normal University and included branches, leaves, and grass. Nutshells and fruit peels were obtained from campus food waste, and sugarcane waste was obtained from a sugar mill in Rizhao, Shandong Province. In this study, the sugarcane bagasse collected from the sugar refinery was classified into the nutshells and fruit peels category because of their similar composition. Aquatic plants included reed, lotus leaf, water hyacinth, Hydrocharis and Enteromorpha, the latter three of which are common plants in eutrophic waters. Reed, lotus leaf, water hyacinth and Hydrocharis were from Nansi Lake in Shandong Province, while Enteromorpha were from the Coast of Rizhao. Dairy manure and chicken manure were obtained from a dairy farm and chicken farm in Rizhao, respectively. Residual sludge was obtained from the second municipal wastewater treatment plant in Rizhao, which mainly treats domestic wastewater.

All of the feedstocks were oven-dried at 60 °C to reduce the moisture to less than 10% (w/w) and then milled to less than 2 mm for biochar production. Furthermore, the samples were passed through a 100-mesh sieve (0.154 mm) prior to feedstock characterization, including thermogravimetric, fabric, elemental, and proximate analyses. Thermogravimetric (TG) and derivative thermogravimetric (DTG) experiments were carried out on the powdered samples using a STA 449 thermal analyzer (Netzsch, Selb, Germany) under a nitrogen atmosphere which were heated from room temperature to 850 °C at the rate of 10 °C min^{-1}. Characteristic pyrolysis temperatures (including onset, midpoint, inflection and end temperatures), maximum rates of mass loss, mass changes and residual mass rates were calculated based on the TG curves using NETZSCH PROTEUS ANALYSIS software version 4.8.5 (Netzsch, Selb, Germany).

The cellulose and lignin contents in the biomass waste were determined using the Van Soest method (Van Soest *et al.*, 1991). Elemental (carbon, nitrogen, hydrogen and sulfur) analyses of the feedstocks were conducted on an elemental analyzer (Vario EL III, Elementar, Hanau, Germany), and the oxygen content was derived by subtraction of the C, N, H, S, and ash contents from the total mass of the sample. Proximate analysis methods were conducted using CNCA (China National Coal Association) Proximate Analysis Methods for Solid Biomass Fuel (GB/T 28731-2012) (CNCA, 2012). Briefly, the moisture content was determined from the weight loss of the biomass waste processed in a furnace preheated to 105 °C using a crucible with a cover for 2 h. To determine the volatile matter, a dedicated crucible with an inner-buckle cover (33 mm in diameter and 40 mm tall) was utilized to maintain the samples in air-free conditions. Each sample was weighed in a crucible and placed in the furnace preheated to 900 °C for 7 min; the weight loss of the sample was recorded as the volatile matter content. To determine the ash content, a dedicated crucible with a rectangular cuboid shape (length × width × height, 45 mm × 2 2 mm × 14 mm) was utilized. Samples weighed in a crucible were placed in the furnace, and the temperature was increased to 250 °C for 1 h and then increased to 550 °C for 2 h. The samples turned off-white in color after heating, and heating was continued if the color did not change into off-white completely. The weight of the residual was recorded as the ash content. The fixed carbon content was derived by subtracting the

ash, volatile matter and moisture values. Elemental analyses were conducted in duplicate, and proximate analyses were conducted in triplicate.

Biochar production and physicochemical properties

Biochars were produced from the above feedstocks at various temperatures under oxygen-limited conditions in a muffle. The inert gas used for pyrolysis was nitrogen with a flow rate of 10 L min^{-1}. The selected peak charring temperatures included 350 °C, 500 °C and 650 °C, and the heating rate was approximately 20 °C min^{-1} for each sample. The predried biomass was placed in a ceramic crucible, covered with a fitting lid and pyrolyzed under a N_2 atmosphere in a muffle for 2 h at a given temperature. After the temperature decreased to room temperature, the biochar products were manually removed and weighed to calculate the biochar product yield. All of the biochar samples were ground with a mortar and pestle and sieved through a 100-mesh sieve for the analyses.

Elemental analyses of the biochars were conducted on an elemental analyzer (Vario EL III, Elementar). Proximate analysis methods were used to determine the biomass. These results were then used to calculate the C/N, atomic H/C and O/C ratios, which are indicative of the bonding arrangement and polarity. The bulk densities of the biochar were analyzed according to the apparent density determination method of the granular activated carbon, which was determined by measuring the volume packed by a free fall from a vibrating feeder into an appropriately sized graduated cylinder and determining the mass of the known volume. For measurements of pH values, biochars were weighed to 1 ± 0.01 g and placed in 100-mL conical flasks, and then, 20 mL of deionized water was added to each flask. The flasks were agitated on an orbital shaker table for 2 h, and then, the pH of each equilibrated solution was determined. The filtered solution was used to determine the electrical conductivity. Elemental analyses were performed in duplicate, and the other analyses were performed in triplicate.

Data analysis

All of the experiments were conducted in duplicate or triplicate, and the average values were reported. The correlation between the biomass characteristics and the resulting biochar was made using SPSS for Windows version 20.0 (SPSS Inc., Chicago, IL, USA). Canonical correspondence analysis of the biochar yield and biomass properties was conducted with CANOCO 5.0 (Ter Braak & Šmilauer, 2012).

Results

Chemical composition of the biomass wastes

The results from the proximate, elemental and fiber analyses for the different categories of biomass wastes are presented in Table 1; the different biomass categories exhibited large variations. The data clearly show that the residual sludge and livestock manure have

Table 1 Compositions of the six categories of biomass waste. The results from the proximate, elemental and fiber analyses for the different categories of biomass wastes exhibited large variations

Biomass categories	Fiber analysis (%)		Proximate analysis (%)			Ultimate analysis (%)				
	Cellulose	Lignin	Volatiles	Fixed carbon	Ash	Carbon	Hydrogen	Nitrogen	Oxygen*	Sulfur
Agricultural residues	44.16 ± 10.89	14.23 ± 6.21	76.13 ± 3.20	16.99 ± 1.91	6.26 ± 3.23	44.39 ± 7.19	5.49 ± 0.78	0.80 ± 0.49	39.48 ± 6.93	–
Pruning waste	44.12 ± 11.96	14.81 ± 7.07	79.81 ± 1.87	15.63 ± 2.44	4.84 ± 1.74	53.28 ± 11.46	5.84 ± 0.42	1.16 ± 0.51	32.38 ± 12.62	–
Aquatic waste	40.71 ± 20.76	13.29 ± 4.33	67.98 ± 5.03	12.91 ± 5.93	18.04 ± 10.38	38.66 ± 4.82	9.65 ± 5.94	2.04 ± 0.92	26.50 ± 6.88	0.36 ± 0.28
Nutshell and fruit peel	41.44 ± 13.43	18.80 ± 8.18	76.87 ± 4.75	19.72 ± 4.45	3.40 ± 4.15	45.96 ± 7.62	6.90 ± 4.49	0.66 ± 0.50	39.84 ± 10.32	–
Livestock manure	17.81 ± 16.64	9.43 ± 1.67	49.47 ± 25.72	7.61 ± 2.59	38.98 ± 33.88	26.61 ± 12.38	3.75 ± 2.00	3.41 ± 3.75	23.30 ± 16.18	–
Residual sludge	–	–	52.80 ± 0.32	3.70 ± 0.26	41.19 ± 0.18	29.49 ± 0.35	4.62 ± 0.26	4.19 ± 0.31	20.43 ± 0.56	0.08 ± 0.34

–, Not detected.

*O was measured by the difference of C, H, N, S and ash from 100.

higher ash contents and lower organic constituents compared to the plant biomass. Moreover, the ash content of the aquatic biomass waste was the highest among all of the plant biomass waste types, although the volatile and fixed carbon contents were lower. The volatile and carbon contents of the woody pruning wastes ranked first in the plant biomass type. Aquatic biomass from eutrophic water bodies possessed the highest amount of hydrogen, nitrogen and sulfur. Sulfur and nitrogen contents were very low in the pruning and agricultural biomass, although they were high in aquatic plants and municipal residual sludge.

Among the category of agricultural residue, sunflower stem, sunflower heads, rice straw and rice husk contained higher ash contents at 11.5%, 11.2%, 11.9% and 10.08%, respectively, than other agricultural residues. The average content of other 19 agricultural residue samples was only 5.06%. The tall fescue collected from the lawns, as herbaceous plant, had a high ash content compared with other pruning wastes from the garden.

Different plant parts exhibited different compositions. In the agricultural biomass category, the constituents in straw and shells from the same crop showed many differences. For the same crop, the volatile, fixed carbon, carbon and lignin contents in straw were higher than in the shells mostly; the ash and cellulose contents in straw were also generally lower than those in the shells (Table S2). In the woody biomass category from pruning wastes, the compositions of leaves and branches from the same tree also varied greatly. The volatile, fixed carbon, carbon, lignin and cellulose contents in the branches were higher than in the leaves, while the ash content in the branches was lower than in the leaves (Table S3).

Thermal properties of the biomass wastes

Figure 1 shows the results of the thermogravimetric analysis (TGA) and differential thermograph (DTG) for the biomass waste samples to predict the pyrolysis behaviors of each biomass waste tested; the pyrolysis characteristics analyzed from TG curves are listed in Table S4. Pyrolysis characteristics of biomass varied with the biomass waste constituents, and the correlation between pyrolysis characteristics and composition of the biomass was significant (Table S5). As displayed in Table S5, the biomass maximum rate of mass loss and the corresponding temperature (inflection temperature) during pyrolysis were positively correlated with the content of volatiles, fixed carbon, oxygen, cellulose and lignin, while those were negatively correlated with the contents of ash, nitrogen and hydrogen. The characteristic temperatures of pyrolysis, including onset, midpoint,

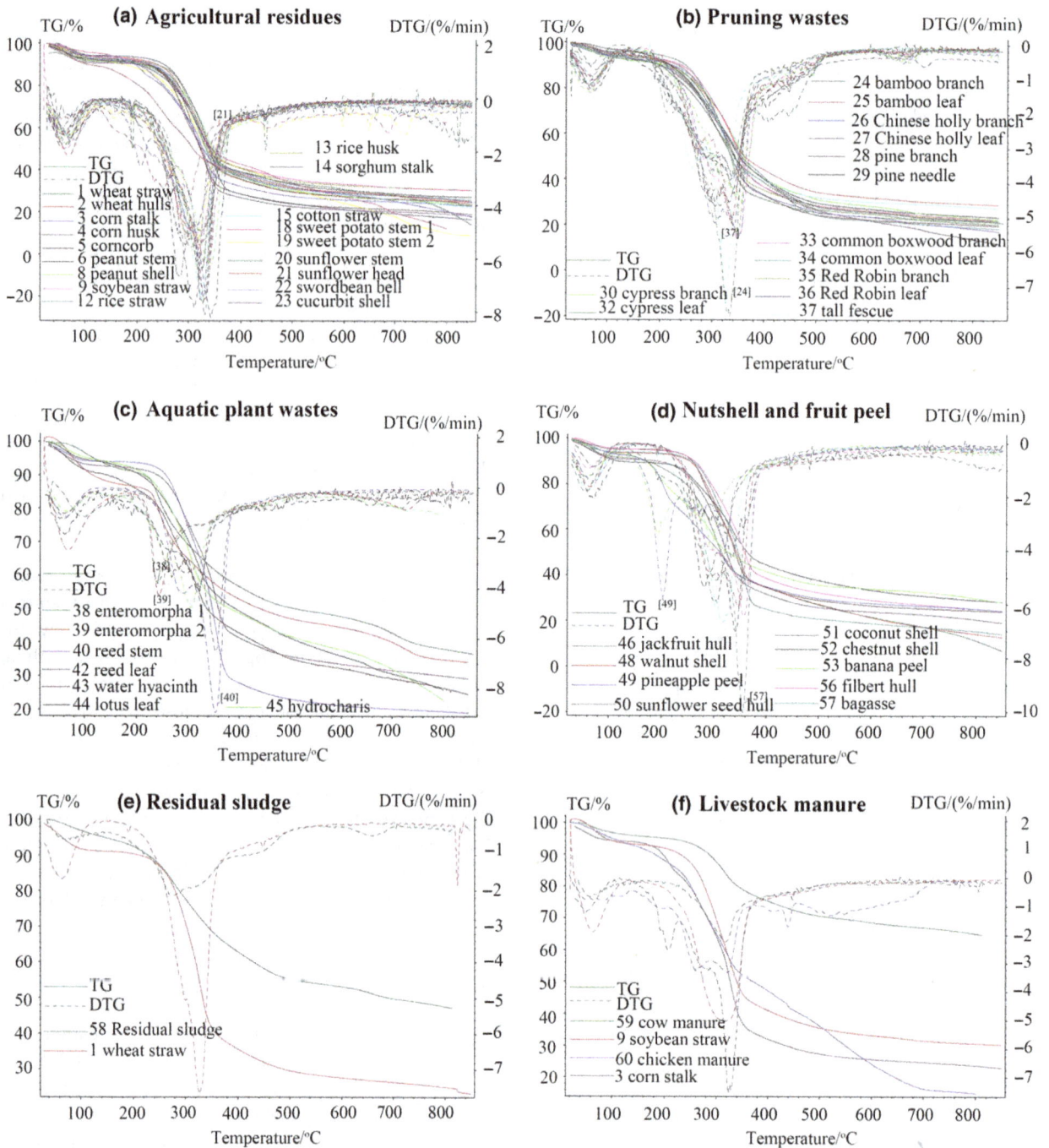

Fig. 1 Thermogravimetric-DTG curves to six categories of biomass wastes. The overall changes in TG and DTG profiles were roughly similar for the same categories of biomass waste. And the major maximal weight loss peak in DTG curves suggested the decomposition of fabric contents of biomass.

inflection and end temperature, were correlated with the cellulose content of biomass significantly, while the lignin content was correlated closely only with the first three characteristic temperatures.

Biomass waste samples exhibited different pyrolysis curves, although the overall changes were roughly similar for the same categories of biomass waste according

to TG and DTG curves, except for the individual types (Fig. 1). The maximal weight loss peaks and decomposition strength of hemicellulose, cellulose and lignin in different biomasses were sample dependent. Following the evaporation of the moisture at between 80 °C and 120 °C, the main mass loss actively occurred in the range from 120 °C to 600 °C due to the progressive

(a) Biochar mass yield

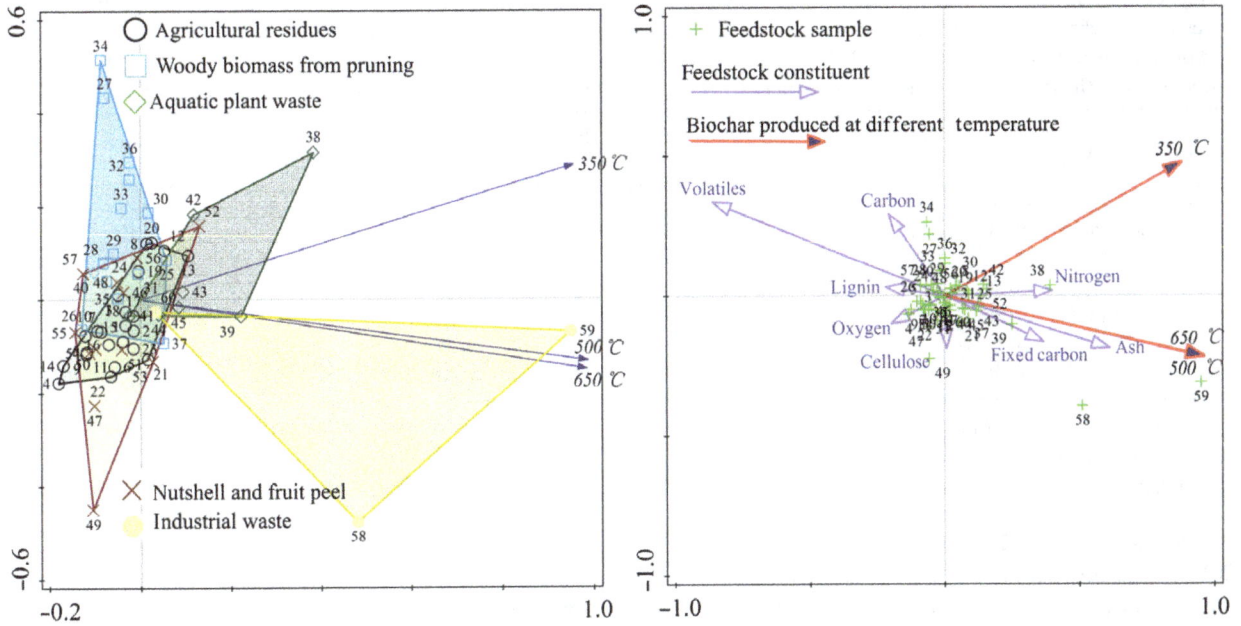

(b) Biochar yield(daf.)

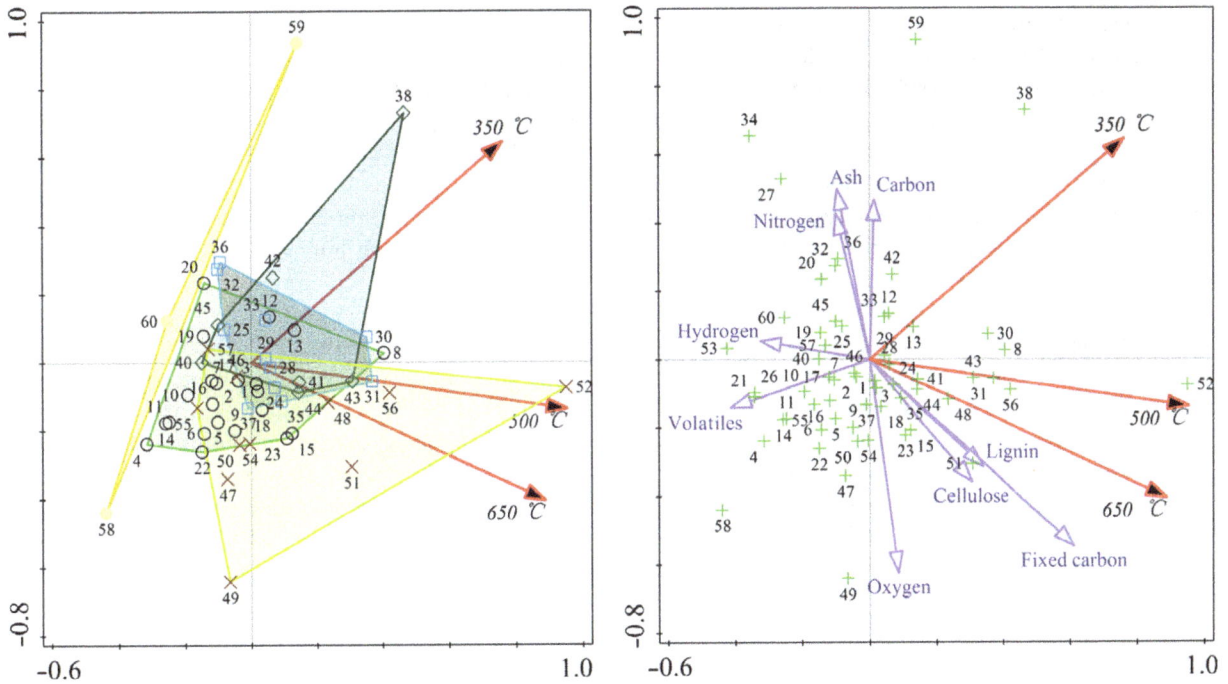

Fig. 2 Canonical correspondence analysis (CCA) biplot of the biochar yield and feedstock characteristics. The supplementary variables in CCA account for 85.8% and 74.7%, respectively, and adjusted explained variation is 81.3% and 70.9%, respectively. (a) Biochar yield was dependent on the ash content of feedstock mainly. (b) When the biochar yield is expressed on a dry ash-free (daf.) basis, the ash content of feedstock was negatively correlated with biochar yield (daf.), whereas the cellulose and lignin contents of feedstock were positively correlated with biochar yield.

hemicellulose, cellulose and lignin pyrolysis, even though the derivative weight loss peaked at different temperatures. Previous studies have assigned the major

maximal weight loss peak in DTG curves as per the degradation of hemicellulose (200–300 °C) and the shoulders with the degradation of cellulose (250–

Table 2 Correlation between the feedstock composition and the biochar product yield under different pyrolysis temperatures. The results of correlation for biochar yield and biochar yield expressed on a dry ash-free (daf.) basis had significant difference. On the whole, the biochar yields were negatively correlated with the feedstock organic contents, including cellulose, lignin, volatiles, fixed carbon, carbon and oxygen contents, while the ash and nitrogen contents were positively correlated with the biochar yield. The correlation between the feedstock and biochar yield (daf.) displays opposite results to that of total mass yield of the biochar

Biochar product yield		Fiber analysis			Proximate analysis			Ultimate analysis		
		Cellulose	Lignin	Lignin/Cellulose	Volatiles	Fixed carbon	Ash	Carbon	Nitrogen	Oxygen
Biochar mass yield	350 °C	−0.30*	–	–	−0.85**	−0.47**	0.85**	–	0.40**	−0.55**
	500 °C	−0.28	–	–	−0.96**	−0.45**	0.93**	−0.49**	0.40**	−0.38*
	650 °C	−0.25	0.28		−0.95**	−0.44**	0.93**	−0.49**	0.47**	−0.36*
Biochar yield (daf)	350 °C	–	–		–	–	–	–	–	–
	500 °C	0.39**	0.33*	0.24	–	0.57**	–	–	–	–
	650 °C	0.53**	0.47**	0.37*	0.30*	0.84**	−0.56**	–	−0.29	0.34*

−, Correlation is not significant.
*Correlation is significant at the 0.05 level (two-tailed).
**Correlation is significant at the 0.01 level (two-tailed).

350 °C) and lignin (200–500 °C) (Lehmann & Joseph, 2015). The intensities of peaks and decomposition strength of hemicellulose, cellulose and lignin in different biomasses were dependent on the content of these organic components in biomass. As shown in the DTG curves (Fig. 1), the average value for the maximum weight loss rate for the agriculture residue, pruning waste, aquatic plant, nutshell and fruit peel, livestock manure and municipal residual sludge were 6.9%, 5.0%, 4.9%, 5.6%, 2.8% and 2.1% min^{-1}, respectively (Table S4).

In the agricultural residues category, sunflower heads (sample 21) displayed different pyrolysis characteristics, as revealed by the TG and DTG curves. The DTG curve shape for crop stalk was similar to the corresponding husk curve because of the homogeneous composition. In the pruning wastes category, the maximum weight loss rates for bamboo branch (sample 24) and tall fescue (sample 37) were higher because of the higher cellulose contents. As green algae, the thermal properties for Enteromorpha (samples 38 and 39) showed significant differences from the aquatic and terrestrial plants. TG and DTG curves of nutshells and fruit peels varied greatly because there was great difference in the composition of these samples (Fig. 1d). As suggested by Fig. 1f, the overall changes in DTG curves for livestock manure were similar to those of the food source for livestock (samples 59 and 9 and samples 60 and 3), but the intensities of maximal weight loss peaks were different because of their fiber contents.

Biochar yields

As shown in previous studies (Lehmann & Joseph, 2015), the biochar yield decreased as the pyrolysis temperature increased. Between 500 °C and 650 °C, the

biochar yield did not change substantially, indicating that most of the volatile fractions were removed at the lower temperatures. The biochar product yields from the 60 types of biomass waste are displayed in Fig. 2a. The municipal residual sludge (sample 58) and dairy manure (sample 59) yielded much higher amounts of biochar than plant biomass. Biochar yields for the green algae Enteromorpha (samples 38 and 39) were ranked second. Yields for other aquatic plants, which included reed leaves (sample 42), water hyacinth (sample 43), lotus leaf (sample 44) and Hydrocharis (sample 45), and for rice straw and husk (samples 12 and 13, respectively) were ranked third. The biochar yields from other biomasses, including nutshells and fruit peels, agricultural residues and pruning wastes, were generally lower than those from aquatic plants.

The effects of the feedstock characteristics and biochar yield are illustrated in Table 2 and Fig. 2. On the whole, the biochar yields were negatively correlated with the feedstock organic contents, including cellulose, lignin, volatiles, fixed carbon, carbon and oxygen contents, while the ash and nitrogen contents were positively correlated with the biochar yield. There was no distinct correlation between the hydrogen content and the biochar yield in this study. With respect to the feedstock pyrolysis characteristics, the pyrolysis parameters generated by the TG for feedstock were significantly correlated with the biochar yield, with the exception of the onset temperature (Table S6).

Canonical correspondence analysis (CCA) of the biochar yields and the proximate, elemental and fiber analyses of the feedstock were performed. As shown in Fig. 2, arrows point in the direction of the steepest increase in the variable values, while the angle between the arrows indicates the correlation between the individual variables; the correlation increased with a

Table 3 Bulk density of the biochar and its feedstock. The results for the bulk density suggested that there were significant variations in the biochars derived from the different feedstock categories

Biomass categories	Bulk density of feedstock (g cm^{-3})	Bulk density of biochar (g cm^{-3})		
		HTT at 350 °C	HTT at 500 °C	HTT at 650 °C
Agricultural residues	0.230 ± 0.079	0.136 ± 0.051	0.185 ± 0.071	0.137 ± 0.058
Pruning waste	0.354 ± 0.058	0.222 ± 0.035	0.347 ± 0.073	0.257 ± 0.039
Aquatic waste	0.189 ± 0.049	0.185 ± 0.058	0.237 ± 0.082	0.162 ± 0.082
Nutshell and fruit peel	0.387 ± 0.168	0.213 ± 0.132	0.271 ± 0.118	0.248 ± 0.116
Livestock manure	0.658 ± 0.194	0.560 ± 0.313	0.607 ± 0.331	0.570 ± 0.297
Residual sludge	0.690 ± 0.011	0.649 ± 0.012	0.652 ± 0.014	0.539 ± 0.018

decrease in the angle. The approximate correlation is positive when the angle is sharp and negative when the angle is larger than 90°. The length of an arrow suggests the correlation between the feedstock factors and the biochar yield. The distance between the symbols approximates the dissimilarity of the biochar yield as measured by its Euclidean distance. The CCA results agreed with correlations in Table 2 roughly.

To remove the effects of moisture and ash, the biochar yield is expressed on a dry ash-free (daf.) basis. The correlation between the feedstock and biochar yield (daf.) is listed in Table 2, which displays opposite results to those of total mass yield of the biochar. With respect to the different pyrolysis temperatures, the correlations between the feedstock characteristics and the biochar yield were not significant at 350 °C, and the intensities of their correlations increased with pyrolysis temperature. The ash content of feedstock had a negative correlation with biochar yield (daf.); the cellulose, lignin, volatiles, fixed carbon and oxygen contents of feedstock were positively correlated with biochar yield at 650 °C. It is worth noting that the ratio of lignin to cellulose content in feedstock was found to be positively correlated with the biochar yields (daf.).

Biochar bulk density

In this study, the bulk (or apparent) densities of biochars were determined. The correlations for the bulk density between the feedstocks and the resulting biochars at 350 °C, 500 °C and 650 °C were 0.790, 0.832 and 0.885, respectively, which were significant. The results for the bulk density suggested that there were significant variations in the biochars derived from different feedstock categories (Table 3). Generally, the biochar bulk density results exhibited the following order, from greatest to least: sludge, livestock manure, pruning waste, fruit peels and agricultural residues. The bulk densities of the biochar produced from manure and sludge were higher than those of plant biomass because of the feedstock densities. Among the different types of plant biomass wastes, biochars produced from pruning waste and nutshells possessed the highest bulk density.

Proximate analyses

Proximate analyses of the biochar show quantitative concentrations of the volatile, fixed carbon and ash contents. As shown in Fig. 3, the ash contents of the biochar increased, while the volatile contents decreased, with the increase in the pyrolysis temperature. However, the changes in the fixed carbon contents with the pyrolysis temperature were dependent on the feedstock characteristics. The biochar fixed carbon contents increased with increasing feedstock treatment temperatures when the ash content was below 34.57% and the volatile content exceeded 55.16% (Fig. 4). In contrast, the biochar fixed carbon content decreased with the pyrolysis temperature.

Figure 5 and Table 4 display the relationship between feedstock composition and the yield of the fixed carbon

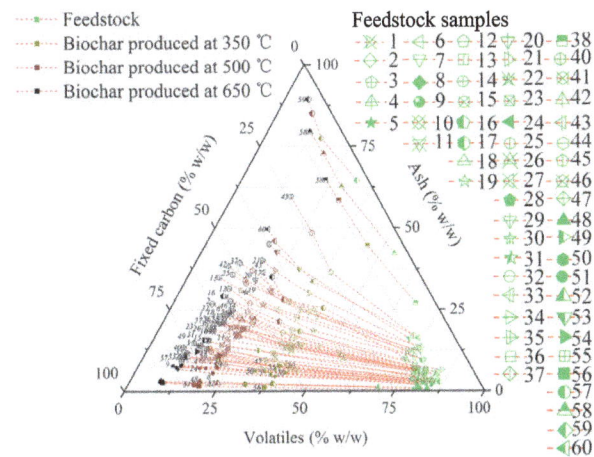

Fig. 3 Proximate analysis ternary diagram for biochars and their feedstocks. With the increase in the pyrolysis temperature, the ash contents of the biochar increased, while the volatile contents decreased. The changes in the fixed carbon contents with the pyrolysis temperature were dependent on the ash content of feedstock.

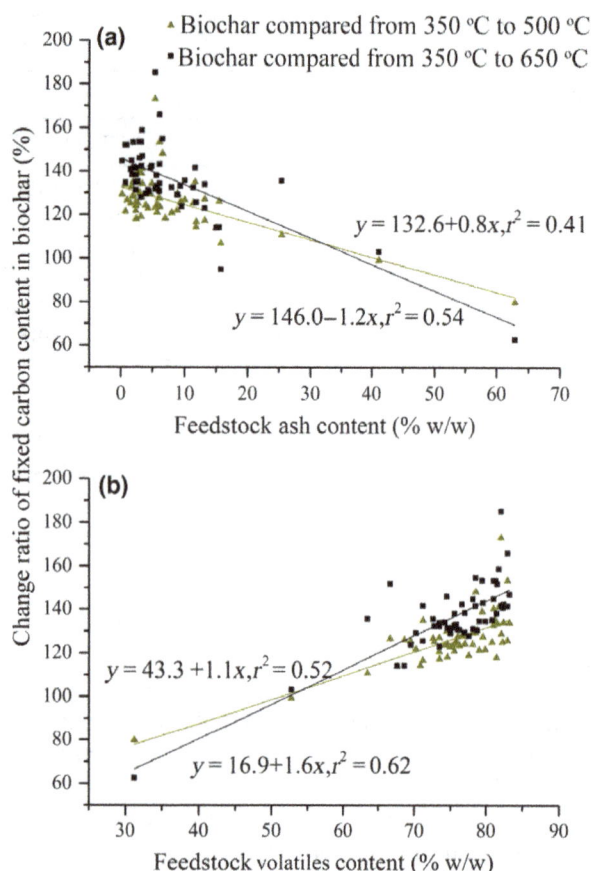

Fig. 4 Relationship between the increases in fixed carbon content in biochar and proximate analysis of feedstock. The biochar fixed carbon contents increased with increasing feedstock treatment temperatures when the ash content was below 34.57% and the volatile content exceeded 55.16%.

production per unit ash-free mass in biochar (af.). The ash content in feedstock had a negative correlation with the fixed carbon production in biochar (daf.); the contents of fixed carbon, volatiles, cellulose, lignin and lignin/cellulose in feedstock were positively related. Besides that, the nitrogen content was correlated negatively with the fixed carbon production in biochar (af.). In contrast, volatiles content in biochar (af.) had a negative correlation with the content of fixed carbon, volatiles, cellulose, lignin and lignin/cellulose in feedstock (Table 4 and Fig. 5).

Substantial differences were observed in the volatile, fixed carbon and ash contents in the biochar resulting from different biomass types (Fig. 3). As noted above, livestock manure, residual sludge and aquatic plants possessed large proportions of ash, and the biochar yields produced by these feedstocks were higher than those of other biomass types, which were inherited from feedstocks. The largest fixed carbon content and lowest volatile content were found in the nutshell biochars

(filbert hull, coconut shell, chestnut shell and walnut shell). Part of the agricultural residues and pruning wastes (peanut shell, bamboo branch, sunflower seed hulls, corncob, pine branch, soybean straw, sorghum stalk, etc.) were ranked second, followed by the other agricultural residues, pruning wastes and fruit peels. The fixed carbon content of the biochar produced by aquatic plant wastes was the lowest among the plant biomass waste types. Sludge and livestock manure biochar had the lowest fixed carbon contents among all of the biomass wastes.

Elemental analysis

The feedstock characteristics influenced the elemental composition of the biochar, and the correlation for the carbon contents between the feedstocks and biochars was significant and reasonable. The carbon content in the biochar at 500 °C ranged from 14.36% to 85.57%, and the average recovery of carbon from feedstock was 48.02%. Among all types of biochar produced by the biomass waste, nutshell biochars, such as coconut shell, walnut shell, filbert hull and chestnut shell, which had a carbon content exceeding 80%, contained higher carbon concentrations. In the case of agricultural residue category, the biochars produced from corn stalk and related residues (corncob and corn husk) possessed the highest carbon contents, at 80.04%, 79.02% and 77.80%, respectively. For the biochars from pruning waste, the carbon contents in the bamboo branch biochar were at the top of the list in this category, followed by pine branch. The biochars from cow manure and municipal residual sludge had the lowest carbon contents; for example, the carbon content of cow manure biochar was only 16.78% of the coconut shell biochar. The C/N ratios of biochar ranged from 7.03 to 192.38. The biochars from sludge, animal manure and aquatic plant had the lower C/N ratio, which was not above 30 mostly (Table S7). The average C/N ratios of fruit peel, agricultural residues and pruning waste were 88.75, 52.01 and 40.89, respectively. Previous studies have found that biochars with a C/N ratio above 30 could contribute to decreased emissions of N_2O from soil (Brassard et al., 2016).

The H/C and O/C atomic ratios of biochars were typically correlated with the degree of aromaticity and the polarity of the biochar; the H/C and O/C ratios are plotted against one another in a Van Krevelen diagram (Fig. 6). The H/C and O/C atomic ratios for the feedstocks ranged from 0.94 to 2.43 and 0.05 to 1.00, respectively, while those for the biochar at 500 °C ranged from 0.42 to 0.81 and 0 to 0.27, respectively. The average values of H/C atomic ratios for biochar from residual sludge, livestock manure and aquatic plants were 0.67,

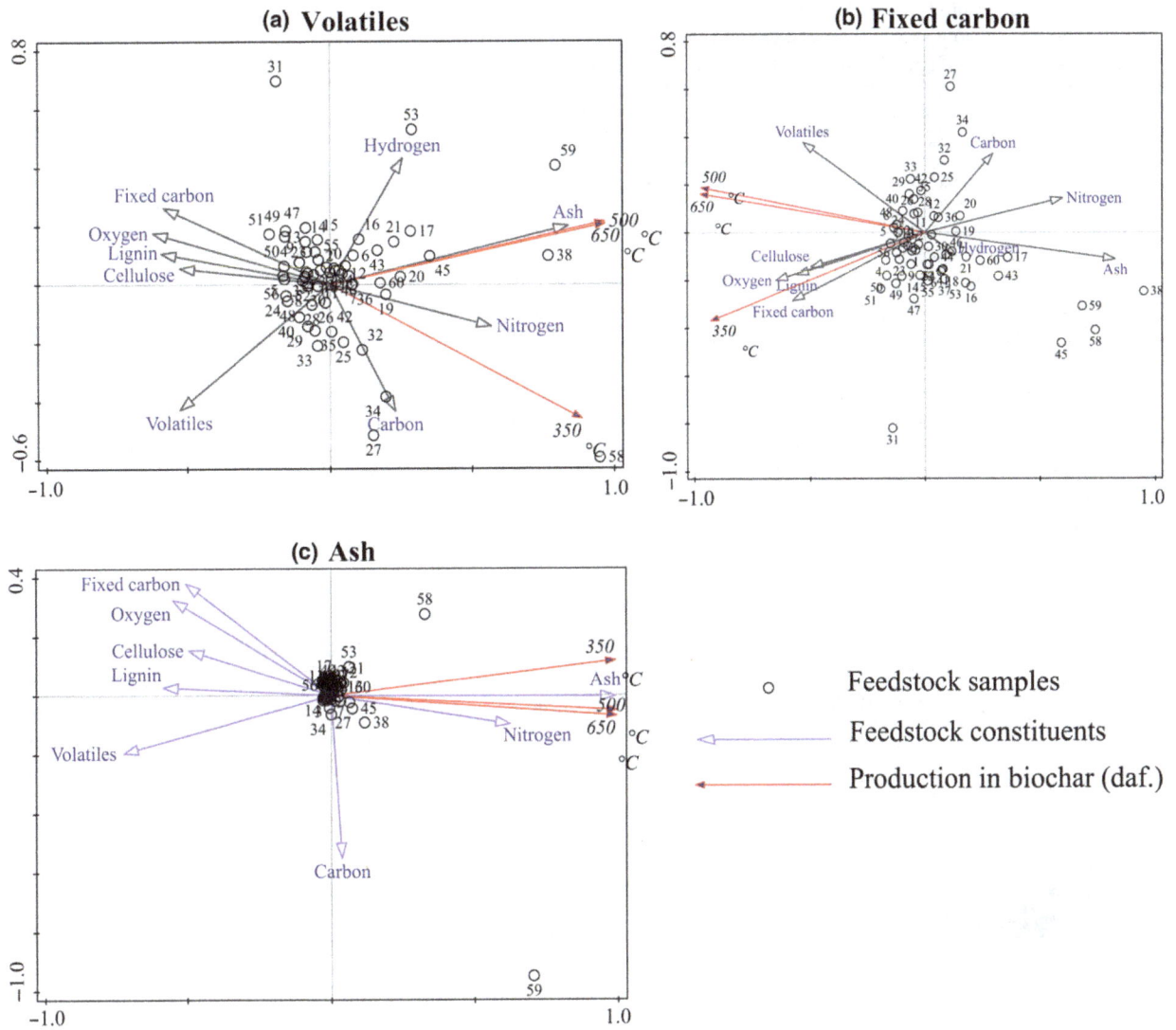

Fig. 5 Canonical correspondence analysis biplot of feedstock composition and the production of volatiles, fixed carbon and ash content in biochar (af.). The supplementary variables in CCA to volatiles (a), fixed carbon (b) and ash (c) accounted for 78.5%, 98.5% and 81.0%, respectively; adjusted explained variation was 71.5%, 74.9% and 98.1%, respectively.

0.66 and 0.57, respectively, and those of biochar from agricultural residues, pruning waste and fruit peel were 0.49, 0.48 and 0.47, respectively. In contrast, the average value of O/C atomic ratios for biochar decreased in the order: agricultural residues, pruning waste, fruit peel, aquatic plants, residual sludge and manure (Table S8).

EC and pH

Biochar pH and EC values increased with increases in the pyrolysis temperature (Fig. 7). Additionally, the correlation between the biochar pH and EC values was significant. The coefficients for the biochar at 350 °C, 500 °C and 650 °C were 0.723, 0.729 and 0.702, respectively. The pH values ranged from 7.14 for chestnut shell at 350 °C to 11.98 for banana peel at 650 °C, while

the EC values ranged from 99 μs cm^{-1} for chestnut shell at 350 °C to 23,500 μs cm^{-1} for banana peel. Figure 6 shows that the biochar pH and EC values were impacted by the ash content. The coefficients between the biochar ash contents and pH values at 350 °C, 500 °C and 650 °C were 0.37, 0.38 and 0.60, respectively, while the coefficients between the biochar ash contents and the EC values at 350 °C, 500 °C and 650 °C were 0.42, 0.36 and 0.37, respectively.

The average biochar pH and EC values for the six feedstock categories are listed in Table S9. As shown in Table S9, the biochar pH and EC values from pruning waste from garden and lawn were lowest of all categories. The biochar pH values from residual sludge were higher than those of other feedstock categories, although the biochar EC value, which was 1564 μs cm^{-1}

Table 4 The relationship between feedstock composition and the content of ash, fixed carbon and volatiles in biochar (af.). The ash content in feedstock had a negative correlation with the fixed carbon production in biochar (daf.); the contents of fixed carbon, volatiles, cellulose, lignin and lignin/cellulose in feedstock were positively related

Biochar (daf.)		Fiber analysis			Proximate analysis			Ultimate analysis			
		Cellulose	Lignin	Lignin/ Cellulose	Volatiles	Fixed carbon	Ash	Carbon	Nitrogen	Oxygen	Hydrogen
Ash content	350 °C	−0.36*	−0.57**	−0.36*	−0.91**	−0.57**	0.98**	−0.48**	0.62**	−0.52**	–
	500 °C	−0.38*	−0.58**	−0.37*	−0.90**	−0.56**	0.97**	−0.45**	0.66**	−0.55**	–
	650 °C	−0.38*	−0.56**	−0.35*	−0.89**	−0.53**	0.95**	−0.45**	0.62**	−0.57**	–
Fixed carbon	350 °C	0.42**	0.58**	0.36*	0.56**	0.67**	−0.71**	–	−0.52**	0.67**	–
	500 °C	0.28	0.48**	0.33*	0.75**	0.60**	−0.80**	–	−0.50**	0.60**	–
	650 °C	0.29	0.52**	0.36*	0.72**	0.60**	−0.76**	–	−0.58**	0.57**	–
Volatiles	350 °C	−0.43**	−0.57**	−0.36*	−0.57**	−0.68**	0.74**	–	0.47**	−0.61**	–
	500 °C	−0.34*	−0.52**	−0.34*	−0.77**	−0.62**	0.85**	−0.30*	0.42**	−0.55**	0.35*
	650 °C	−0.44**	−0.57**	−0.35*	−0.81**	−0.66**	0.90**	−0.31*	0.52**	−0.55**	0.31*

–, Correlation is not significant.

*Correlation is significant at the 0.05 level (two-tailed).

**Correlation is significant at the 0.01 level (two-tailed).

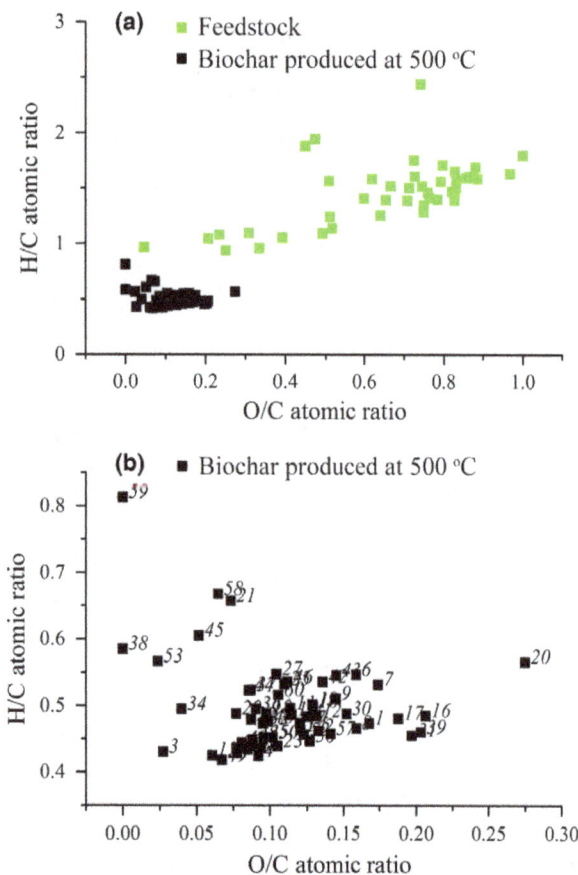

Fig. 6 The H/C and O/C atomic ratios for the feedstock and biochar. The H/C and O/C ratios of biochar were found to be decreased for feedstock. The comparison of feedstock and biochar was displayed on picture (a), and the comparison of biochars were displayed on picture (b).

at 650 °C, was low. On the whole, the biochar EC values from aquatic plant wastes were higher than those of other feedstocks. With respect to the agricultural residues, the rice straw biochar at 650 °C had the highest pH value of 11.49. Among pruning wastes, the biochar pH value for lawn waste (*Festuca arundinacea*) (pH = 11.16 at 650 °C) was higher than that for garden pruning waste (pH < 10.50).

Discussion

Biochar yield and characteristics are influenced by biomass feedstock and pyrolysis conditions (e.g., treatment temperature), so it is important to perform biochar manufacture with different feedstock types at different pyrolysis conditions, and to determine the properties of the resulting biochar. This work characterized a wide range of biochars from numerous feedstocks in China to identify the suitable biochar for agronomy application. In our study, chemical composition and properties varied with feedstock and treatment temperature; especially for biochar fixed carbon yield, the key indicator of potential ability of carbon sequestration, it increased with increasing treatment temperatures when the ash content was below 34.57% in feedstock and decreased when the ash content exceeded 34.57%. While expressed by per unit ash-free mass in biochar, the fixed carbon production decreased with increasing biochar manufacture temperatures consistently. The reason was that the feedstock ash content hampered the fixed carbon production in biochar (af.),supported by their negative correlations (Table 4). Similar findings were reported by Enders *et al.* (2012) that the fixed carbon content of

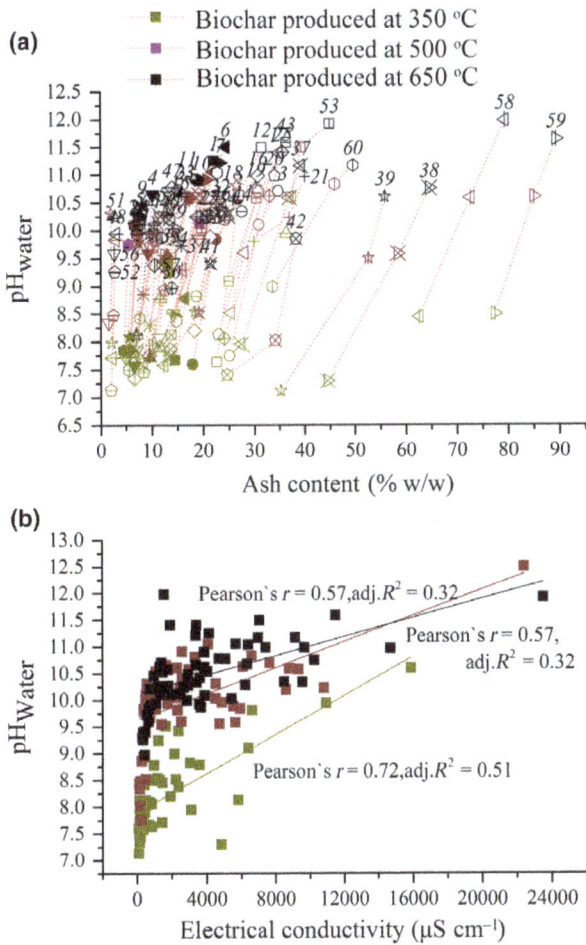

Fig. 7 Relationship between the ash contents, pH and EC values of biochars. Biochar pH values increased with increases in the pyrolysis temperature. The EC value of biochar correlated with pH significantly. (a) displayed the relations between pH and ash content of biochars, and (b) displayed the relations between pH and electrical conductivity of biochars.

biochar increased with increasing treatment temperature for low-ash feedstocks and decreased for high-ash feedstocks. The ash content, which was alkali and alkaline earth metallic species mainly, had a strong catalytic effect to pyrolysis and hampered the generation of fixed carbon (Bridgwater, 2007). The contents of cellulose, lignin and lignin/cellulose in feedstock were positively related to the fixed carbon production in biochar (af.). The correlations suggested that feedstock cellulose and lignin promoted the production of fixed carbon in biochar but fixed carbon generation was higher in the feedstock which had more lignin content as compared to cellulose. The lignin is more thermally stable than cellulose during pyrolysis; thus, the production of fixed carbon in biochar increased with the increase in lignin/cellulose content in biomass. It was noted that

nitrogen content hampered the production of fixed carbon in biochar, supported by their negative correlations. The biomass wastes possessed high nitrogen content, such as sludge, manure and aquatic plant from eutrophied water body, and had less efficiency to fixed carbon production as biochar manufacture.

The biochar product yield is an important factor for realizing economic benefit in biomass waste reutilization. The biochar yields of municipal residual sludge and dairy manure top the list, followed by aquatic plants and then the other biomass wastes. However, when we removed the effects of moisture and ash, the factors that impact biochar yield (daf.) showed contrasting correlations (Table 2), which suggested that the cellulose and lignin contents in feedstock enhanced biochar formation; the ash content of feedstock also had a negative effect on the biochar yield (daf.). Furthermore, the relationship between the biochar yield (daf.) and the lignin/cellulose content ratio was positive (Table 2). Some sporadic comparisons have reported similar findings that biomass samples having higher lignin contents produced higher amounts of char (Demirbas, 2004; Lv et al., 2010). The reason for this difference is that the cellulose component in the biomass is liable to produce volatile products, while the lignin content is important for the biochar yield (Tripathi et al., 2016).

To identify suitable temperatures for biochar manufacturing, the effect of treatment temperature on product yield and biochar properties should be examined. The extent of the correlations between the feedstock and biochar varied at different pyrolysis temperatures; the correlation increased with increasing pyrolysis temperatures because the cellulose and lignin contents of most biomass categories completely decomposed at 650 °C and the coefficients of feedstock composition and biochar yield were significant (Fig. 1). At lower pyrolysis temperatures, the degree of decomposition varied with the type of biomass and the intensity of the correlation between the feedstock and biochar decreased.

The characteristics of biochars from different biomass waste categories varied greatly and their performance in soil application also had large difference. Brassard et al. reported that biochars with a lower N content (C/N ratio >30) were more suitable for mitigation of N_2O emissions from soils. In this sense, biochars from residual sludge, animal manure and aquatic plant from artificial wetland or eutrophic water body were not appropriate for soil amendment; C/N ratios in biochar produced at 500 °C were under average of 30. On the other hand, they could improve soil fertility because of their higher nitrogen content and other mineral elements. In addition, among six categories of biomass feedstock, biochars from sludge and livestock manure had the lowest fixed carbon contents; the fixed carbon content of aquatic plant biochar was the lowest among all plant biomass waste types. Because of high ash content, biochar from sludge, livestock manure and aquatic plant possessed high pH values, which were suitable to acidic soil amendment.

Among the plant biomasses, the results suggest that high carbon and oxygen contents are characteristic of woody pruning wastes, agricultural residues, nutshell and fruit peel biomasses, which lead to increased biochar formation. Therefore, these waste categories are most suitable for biochar production. The ash, volatile and fixed carbon contents of biochar from all of the categories were mapped in a triangle plot, which displayed the private feedstock content and resulting biochar yields and can be useful as a tool for guiding biochar applications and selection for specific soils.

Acknowledgements

This work was supported by the Natural Science Foundation of Shandong Province (Project No. BS2013NY009), the Scientific and Technological Projects of the College in Shandong Province (Project No. J13LF02), the National Natural Science Foundation of China (Project No. 41471389 and 41501542) and the Science Foundation of Qufu Normal University.

References

Brassard P, Godbout S, Raghavan V (2016) Soil biochar amendment as a climate change mitigation tool: key parameters and mechanisms involved. *Journal of Environmental Management*, **181**, 484–497.

Bratkovich S, Bowyer J, Fernholz K, Lindburg A (2008) *Urban Tree Utilization and why it Matters*. Dovetail Partners, Inc. Available at: http://www.dovetailinc.org/files/DovetailUrban0108ig.pdf

Bridgwater AV (2007) The production of biofuels and renewable chemicals by fast pyrolysis of biomass. *International Journal of Global Energy Issues*, **27**, 160–203.

Cao X, Harris W (2010) Properties of dairy-manure-derived biochar pertinent to its potential use in remediation. *Bioresource Technology*, **101**, 5222–5228.

CNCA (2012) *Proximate Analysis Methods of Solid Biomass Fuel*. National Coal Association, Beijing, China.

Cui X, Hao H, He Z, Stoffella P, Yang X (2016) Pyrolysis of wetland biomass waste: potential for carbon sequestration and water remediation. *Journal of Environmental Management*, **173**, 95–104.

Demirbas A (2004) Effects of temperature and particle size on bio-char yield from pyrolysis of agricultural residues. *Journal of Analytical and Applied Pyrolysis*, **72**, 243–248.

Demirbas A (2006) Biodiesel production via non-catalytic SCF method and biodiesel fuel characteristics. *Energy Conversion and Management*, **47**, 2271–2282.

Enders A, Hanley K, Whitman T, Joseph S, Lehmann J (2012) Characterization of biochars to evaluate recalcitrance and agronomic performance. *Bioresource Technology*, **114**, 644–653.

Geng W, Hu L, Cui J, Bu M, Zhang B (2013) Biogas energy potential for livestock manure and gross control of animal feeding in region level of China. *Transactions of the Chinese Society of Agricultural Engineering*, **29**, 171–179.

Hossain MK, Strezov V, Chan KY, Ziolkowski A, Nelson PF (2011) Influence of pyrolysis temperature on production and nutrient properties of wastewater sludge biochar. *Journal of Environmental Management*, **92**, 223–228.

Jiang D, Zhuang D, Fu J, Huang Y, Wen K (2012) Bioenergy potential from crop residues in China: availability and distribution. *Renewable and Sustainable Energy Reviews*, **16**, 1377–1382.

Lehmann J, Joseph S (2015) *Biochar for Environmental Management: Science, Technology and Implementation*. Routledge, London and New York.

Li F, Wang J (2013) Estimation of carbon emission from burning and carbon sequestration from biochar producing using crop straw in China. *Transactions of the Chinese Society of Agricultural Engineering*, **29**, 1–7.

Lv D, Xu M, Liu X, Zhan Z, Li Z, Yao H (2010) Effect of cellulose, lignin, alkali and alkaline earth metallic species on biomass pyrolysis and gasification. *Fuel Processing Technology*, **91**, 903–909.

Nanda S, Mohanty P, Pant K, Naik S, Kozinski J, Dalai A (2013) Characterization of North American lignocellulosic biomass and biochars in terms of their candidacy for alternate renewable fuels. *BioEnergy Research*, **6**, 663–677.

Nanda S, Dalai AK, Berruti F, Kozinski JA (2015) Biochar as an exceptional bioresource for energy, agronomy, carbon sequestration, activated carbon and specialty materials. *Waste and Biomass Valorization*, **7**, 201–235.

Shi Y, Ge Y, Chang J, Shao H, Tang Y (2013) Garden waste biomass for renewable and sustainable energy production in China: potential, challenges and development. *Renewable and Sustainable Energy Reviews*, **22**, 432–437.

Suliman W, Harsh JB, Abu-Lail NI, Fortuna AM, Dallmeyer I, Garcia-Perez M (2016) Influence of feedstock source and pyrolysis temperature on biochar bulk and surface properties. *Biomass & Bioenergy*, **84**, 37–48.

Ter Braak C, Šmilauer P (2012) *Canoco 5, Windows Release (5.00). [Software for Canonical Community Ordination]*. Microcomputer Power, Ithaca.

Tripathi M, Sahu JN, Ganesan P (2016) Effect of process parameters on production of biochar from biomass waste through pyrolysis: a review. *Renewable & Sustainable Energy Reviews*, **55**, 467–481.

Van Soest PJ, Robertson JB, Lewis BA (1991) Methods for dietary fiber, neutral detergent fiber, and nonstarch polysaccharides in relation to animal nutrition. *Journal of Dairy Science*, **74**, 3583–3597.

Vassilev SV, Baxter D, Andersen LK, Vassileva CG, Morgan TJ (2012) An overview of the organic and inorganic phase composition of biomass. *Fuel*, **94**, 1–33.

Xu YL, Chen BL (2013) Investigation of thermodynamic parameters in the pyrolysis conversion of biomass and manure to biochars using thermogravimetric analysis. *Bioresource Technology*, **146**, 485–493.

8

Projected gains and losses of wildlife habitat from bioenergy-induced landscape change

NATHAN M. TARR[1], MATTHEW J. RUBINO[1], JENNIFER K. COSTANZA[2], ALEXA J. MCKERROW[3], JAIME A. COLLAZO[4] and ROBERT C. ABT[5]

[1]North Carolina Cooperative Fish and Wildlife Research Unit, Department of Applied Ecology, North Carolina State University, Campus Box 7617, Raleigh, NC 27695, USA, [2]Department of Forestry and Environmental Resources, North Carolina State University, 3041 Cornwallis Road, Research Triangle Park, NC 27709, USA, [3]U.S. Geological Survey, Core Science Analytics, Synthesis, and Libraries, Campus Box 7617, Raleigh, NC 27695, USA, [4]U.S. Geological Survey, North Carolina Cooperative Fish and Wildlife Research Unit, Department of Applied Ecology, North Carolina State University, Campus Box 7617, Raleigh, NC 27695, USA, [5]Department of Forestry and Environmental Resources, North Carolina State University, Campus Box 8008, Raleigh, NC 27695, USA

Abstract

Domestic and foreign renewable energy targets and financial incentives have increased demand for woody biomass and bioenergy in the southeastern United States. This demand is expected to be met through purpose-grown agricultural bioenergy crops, short-rotation tree plantations, thinning and harvest of planted and natural forests, and forest harvest residues. With results from a forest economics model, spatially explicit state-and-transition simulation models, and species–habitat models, we projected change in habitat amount for 16 wildlife species caused by meeting a renewable fuel target and expected demand for wood pellets in North Carolina, USA. We projected changes over 40 years under a baseline 'business-as-usual' scenario without bioenergy production and five scenarios with unique feedstock portfolios. Bioenergy demand had potential to influence trends in habitat availability for some species in our study area. We found variation in impacts among species, and no scenario was the 'best' or 'worst' across all species. Our models projected that shrub-associated species would gain habitat under some scenarios because of increases in the amount of regenerating forests on the landscape, while species restricted to mature forests would lose habitat. Some forest species could also lose habitat from the conversion of forests on marginal soils to purpose-grown feedstocks. The conversion of agricultural lands on marginal soils to purpose-grown feedstocks increased habitat losses for one species with strong associations with pasture, which is being lost to urbanization in our study region. Our results indicate that landscape-scale impacts on wildlife habitat will vary among species and depend upon the bioenergy feedstock portfolio. Therefore, decisions about bioenergy and wildlife will likely involve trade-offs among wildlife species, and the choice of focal species is likely to affect the results of landscape-scale assessments. We offer general principals to consider when crafting lists of focal species for bioenergy impact assessments at the landscape scale.

Keywords: biodiversity, bioenergy target, biofuel, habitat, landscape change modeling, renewable energy, southeastern United States, wildlife, wood pellets

Introduction

In the United States and Europe, government policies encourage the use of bioenergy as an alternative to fossil fuels. In the United States, the Energy Policy Act of 2005 and the Energy Independence and Security Act of 2007 set nationwide targets for a renewable fuel standard whose goals include replacing some dependence on fossil fuels through bioenergy production and reducing greenhouse gas (GHG) emissions (Energy Policy Act, 2005; Energy Independence and Security Act,

2007). The European Union set renewable energy targets that are expected to be met, in part, by harvesting forests in the southeastern United States to produce wood pellets (Abt *et al.*, 2014; Wang *et al.*, 2015; Galik & Abt, 2015). Besides alleviating fossil fuel usage and GHG emissions, bioenergy production can benefit rural, agricultural economies by diversifying markets, increasing demand, and supplementing profits for traditional crops (Dale *et al.*, 2010) or so-called purpose-grown bioenergy crops, such as switchgrass, sorghum, and short-rotation woody crops (SRWCs) composed of pine (*Pinus* spp.) or poplar (*Populus* spp.) (Coleman & Stanturf, 2006).

Correspondence: Nathan M. Tarr
e-mail: nathan_tarr@ncsu.edu

As renewable fuel targets and incentives increase bioenergy production, landscapes and the wildlife that inhabit them will be affected through the intensification of agriculture and forest management, influences on land use change (direct and indirect), including conversion of land to forest and agriculture, and the prevention of urbanization (Groom et al., 2008; Robertson et al., 2008; Eggers et al., 2009; McDonald et al., 2009; Dauber et al., 2010; Fargione, 2010; Wang et al., 2015). However, the specific effects of this increase will likely vary across spatial and temporal scales, geographic regions, the wildlife species considered, feedstock types, feedstock management practices, and land use histories (Fletcher et al., 2010; Wiens et al., 2011; Rupp et al., 2012; Immerzeel et al., 2014).

The conversion of land from natural ecosystems (e.g., forests, woodlands, grasslands) to ones dominated by crop monocultures could be one of the most important ecological consequences of the increase in demand for bioenergy because such changes are associated with losses of wildlife habitat and biodiversity (Tilman et al., 2002; Firbank, 2008; Dornburg et al., 2010; Fletcher et al., 2010; Wiens et al., 2011; Rupp et al., 2012; Immerzeel et al., 2014). Although some crops grown for bioenergy such as Miscanthus, sweet sorghum, switchgrass, and short-rotation woody crops (SRWCs) provide food and/or cover for some species, the crops differ in their value for wildlife and if grown in monocultures will likely have little value to some species because of their lack of heterogeneity and structure (Fletcher et al., 2010; Wiens et al., 2011; Bonin & Lal, 2012). In addition, crop management practices, such as the timing of harvest relative to animals' annual and breeding cycles, could also limit their value for wildlife, as well as reduce survival and reproduction (Rupp et al., 2012).

The degree to which land conversions to bioenergy crops negatively affect wildlife depends in part upon the initial land use or land cover. Whereas the conversion of many natural ecosystems to bioenergy crops would represent a loss of habitat value, there may be cases where it improves habitat (Meehan et al., 2010; Wiens et al., 2011; Rupp et al., 2012). For example, if existing monocultures were replaced with structurally heterogeneous grasslands composed of diverse perennial plants, they could support more diverse wildlife communities and increase habitat availability for some grassland species (Rupp et al., 2012; Uden et al., 2015). Such benefits to wildlife would likely be lower in fields with less heterogeneity in plant species composition and structure and lower than in native grasslands, depending upon crop harvest and management practices (Hartman et al., 2011; Wiens et al., 2011; Robertson et al., 2012). SRWC monocultures are associated with lower biodiversity than natural forest ecosystems but may have higher diversity

than agricultural croplands (Rupp et al., 2012). Little is known about herpetofaunal diversity on SRWCs (Rupp et al., 2012), but bird and mammal species richness and abundance in midwestern SRWC Populus plantations were lower than in forests and shrublands and higher than in nonhay croplands (Christian et al., 1998).

In the southeastern United States, biomass could also be derived from forests through elevated rates of harvest and thinning and removal of forest harvest residues. The specific characteristics of these harvests will determine their impact on species, although an experimental study failed to detect relationships between the amount of retained woody biomass in clear-cut pine plantations and herpetofauna diversity, evenness, or richness (Fritts et al., 2016) or shrew abundance (Fritts et al., 2015). Decisions regarding which forest types to extract biomass from (e.g., forested uplands vs. wetlands, natural vs. planted stands) will influence the magnitude of effects on individual species (Evans et al., 2013). Thinning forests can increase wildlife diversity by increasing structural complexity, but specific impacts vary across species, thinning technique, and thinning intensity (Rupp et al., 2012).

Land conversions, bioenergy crop cultivation, and forest biomass harvests at individual sites may scale up to biodiversity impacts that are measurable at the landscape level. Once again, the degree and nature of such changes depend upon the initial landscape and bioenergy sources utilized. Heterogeneous landscapes support greater biodiversity (Wiens et al., 2011), so the conversion of abandoned and marginal lands to crop agriculture, along with the concentration of biomass crops around processing facilities, could decrease heterogeneity on some landscapes and, therefore, negatively influence biodiversity (Fletcher et al., 2010; Wiens et al., 2011). Alternatively, planting perennial grasslands for bioenergy feedstocks could be an opportunity to increase the heterogeneity of landscapes that are currently dominated by agricultural crops (Wiens et al., 2011). Effects of the establishment of SRWCs on landscape- and regional-scale biodiversity will vary depending on the degree to which landscapes are already forested and the configuration of SRWC stands, but even relatively small amounts of SRWC plantations in forested landscapes with little existing shrub-dominated land would enhance landscape-scale biodiversity. This effect would not necessarily be the same for landscapes dominated by grasslands or managed forests (Rupp et al., 2012). Rupp et al. (2012) argued that forest thinning for biomass will likely increase the forest structural diversity in landscapes and consequently increase landscape-level species diversity and abundance.

The influence of biomass cultivation and harvest on wildlife habitat at large spatial and temporal scales has rarely been demonstrated (Immerzeel *et al.*, 2014). Few long-term or landscape-scale studies of bioenergy, effects on wildlife in the southeastern United States exist other than Evans *et al.* (2013), which examined the concurrence of wildlife habitat and forest biomass harvest risk under scenarios that differed in factors such as the types of forests harvested, the spatially explicit probability of pine plantation establishment, and the intensity of biomass extraction at harvest sites. In our study, we examined the potential impacts of meeting realistic bioenergy demands on habitat availability across a landscape over a 40-year period. While previous large-scale studies examined single sources of biomass for bioenergy, we assessed impacts of bioenergy production under different portfolios of biomass sources that included land conversions to bioenergy crops and harvests of forest biomass. We accomplished this with deductive habitat modeling and results from a timber supply model that was integrated with a spatially explicit state-and-transition simulation model (Costanza *et al.*, 2016). Our results account for complex economic processes that link demand for forest biomass to landscape change, as well as other landscape change drivers, such as urbanization and vegetation dynamics. Furthermore, comparing across different scenarios allowed us to isolate some relationships between specific biomass sources or land conversions and individual species to gain insights into patterns that may emerge when bioenergy demand is applied to landscapes. Our objectives were to (i) determine whether realistic levels of bioenergy demand are large enough to translate into gains or losses in the amount of habitat for individual species at the landscape scale, (ii) assess the degree of similarity in habitat effects among diverse bioenergy portfolios, and (iii) find insightful patterns and connections between specific sources of biomass and individual species. Exploring these topics will inform future efforts to understand how bioenergy demand will impact wildlife and help policymakers craft bioenergy targets and feedstock portfolios that support sustainable wildlife populations.

Materials and methods

Study area

North Carolina (NC) is located in the southeastern United States and includes three distinct physiographic regions, each encompassing different types of land use and vegetation: the Mountains, Piedmont, and Coastal Plain. The majority of publicly owned forested lands lie in the Mountains, which are heavily forested with hardwood and mixed pine and hardwood forests. Most of the urban areas occur within the Piedmont where pine, hardwood, and mixed pine–hardwood forests form a matrix with agricultural lands. Much privately

owned agricultural use is spread over the eastern third of the state in the Coastal Plain where diverse plant communities such as bottomland hardwood forests, pine woodlands, and freshwater wetlands also occur. Landscape change due to attempts to meet future timber and agricultural demands, including demands for bioenergy, will likely occur within the Coastal Plain and Piedmont regions. Traditional silviculture that is dominated by loblolly pine (*Pinus taeda*) plantation growth and management is also common in the Coastal Plain, and much of the future urbanization is predicted to occur in the Piedmont (Terando *et al.*, 2014; Costanza *et al.*, 2015). A state goal of replacing 10% of petroleum sold in NC with locally grown and produced liquid biofuels through 2017 was chosen in 2007 (Burke *et al.*, 2007). Additionally, existing or announced wood pellet plants could consume 3.45 million green tonnes (3.8 million green short tons) of forest biomass annually (Forisk Consulting LLC, 2014). We used deductive species–habitat modeling to translate future projected landscapes into potential trends in habitat availability within NC between 2010 and 2050 for 16 wildlife species under five alternative bioenergy production scenarios.

Bioenergy scenarios and land use/land cover

In the southeastern United States, bioenergy will likely be produced from agricultural crop residues, bioenergy crops (e.g., sorghum and switchgrass), SRWC, the harvest or thinning of forests, and forest harvest residues (Rupp *et al.*, 2012; Fritts *et al.*, 2014; Wang *et al.*, 2015). We could not predict the exact portfolio of sources in NC's future, so we explored Costanza *et al.*'s (2016) five bioenergy production scenarios along with their baseline that did not include increased bioenergy production (Table 1). The bioenergy scenarios differed in how much they utilized three sources of biomass: (i) feedstock sourced from the harvest and thinning of planted pine, natural pine, hardwood, mixed pine–hardwood, upland, and bottomland forests (hereafter, 'conventional forestry'); (ii) the conversion of agricultural lands on poor soils to purpose-grown feedstocks and subsequent harvests (i.e., switchgrass, sorghum, SRWC; hereafter 'conversion of Ag lands'); and (iii) the conversion of forests on poor soils to purpose-grown feedstocks and subsequent harvests (hereafter, 'conversion of forests').

One of the bioenergy scenarios ('Conventional') assumed that 3.63 million green tonnes (4.0 million green short tons) of forest biomass would be harvested, and thus met the anticipated demand for biomass for wood pellet production (Table 1). Two other scenarios ('Conventional-Ag', and 'Conventional-Ag-Forest') also included the harvest of 3.63 million green tonnes of forest biomass for bioenergy, as well as enough purpose-grown feedstocks to meet or exceed the 10% liquid if the forest biomass was used for biofuels instead of wood pellets (Costanza *et al.*, 2016). Those two scenarios differed in the types of lands converted to purpose-grown feedstocks. Conventional-Ag assumed that only marginal agricultural lands would be converted to purpose-grown crops, whereas Conventional-Ag-Forest assumed that marginal agricultural and forest lands, in equal proportions, would be converted to purpose-grown crops. The remaining two bioenergy scenarios ('Ag' and

Table 1 Six bioenergy production scenarios modeled by Costanza *et al.* (2016) and assessed here for their impacts on species' habitats over 40 years in North Carolina, USA

Scenario name	Description	Conv. forest* (%)	Marg. Ag.† (%)	Marg. forest‡ (%)	NCRFS§ (%)	Forest biomass¶ (M gt)	Harvest residue** (%)	Ag. converted†† (ha)	Forest converted‡‡ (ha)
Baseline	Business-as-usual production of conventional forest products; no bioenergy production	0	0	0	0	0	n/a	0	0
Conventional	Increased conventional forestry for bioenergy	100	0	0	0	3.63	40	0	0
Conventional-Ag	Increased conventional forestry and conversion of agricultural lands for bioenergy	15	85	0	100–148	3.63	40	425 000	0
Conventional-Ag-Forest	Increased conventional forestry and conversion of agricultural lands and forests for bioenergy	15	42.5	42.5	100–148	3.63	40	212 500	212 500
Ag	Conversion of agricultural lands for bioenergy	0	100	0	100–140	0	n/a	504 000	0
Ag-Forest	Conversion of agricultural lands and forests for bioenergy	0	50	50	100–140	0	n/a	252 000	252 000

*Percentage of scenario derived from planted and natural forests.
†Percentage of scenario derived from conversion of agricultural lands on marginal soils to purpose-grown feedstocks.
‡Percentage of scenario derived from conversion of forests on marginal soils to purpose-grown feedstocks.
§Potential percentage of North Carolina's 10% biofuel target met (see text for description of 10% target). Ranges are presented instead of point estimates because the amount of energy that can be derived from feedstocks varies with refinement process (Costanza *et al.*, 2016).
¶Biomass from conventional forestry (in million green tonnes).
**Percentage of forest harvest residues removed.
††Amount of agricultural land converted to switchgrass, sorghum, or SRWC.
‡‡Amount of forest land converted to switchgrass, sorghum, or SRWC.

'Ag-Forest') would meet the 10% biofuel target without increased conventional forestry. Ag assumed that marginal agricultural lands would be the only lands converted to purpose-grown crops, whereas Ag-Forest incorporated conversion of equal proportions of marginal agricultural and forest lands. All five bioenergy scenarios and the baseline scenario included urbanization, natural disturbance such as fire, ecological succession, and conversion between forest and agricultural land use according to recent change trends (Wear & Greis, 2013; Costanza *et al.*, 2016).

Costanza *et al.* (2015, 2016) translated demand for bioenergy into potential changes in NC's land cover using the Sub-regional Timber Supply Model (SRTS) (Abt *et al.*, 2009) and state-and-transition simulation model (STSM) software ST-Sim (ApexRMS, 2014). The SRTS model produced aspatial time series of areal amounts of forest thinning and harvest, as well as conversion between agricultural and forest land uses. For the three scenarios without biomass from conventional forests (including Baseline), the SRTS model incorporated demand for broad forest types and age classes according to recent trends in USDA Forest Service Forest Inventory and Analysis (FIA) data. For the three scenarios that included conventional forest biomass, the SRTS model projected forest harvest and thinning amounts that would result in 3.63 million green tonnes of

biomass extracted from forests, while accounting for demand displacement of other forest products (Costanza *et al.*, 2015).

The spatial STSM incorporated time series outputs of forest thinning, harvest, and land use conversion from SRTS and aspatial state-and-transition model pathways for each of 52 vegetation and land use classes. It also included several spatial inputs: maps of initial vegetation, land use, successional stages, and vegetation structure; an existing time series of future urbanization probability (Terando *et al.*, 2014); marginal agricultural and forest lands to be converted to purpose-grown crops for bioenergy; and lands under conservation ownership to be excluded from biomass production (Costanza *et al.*, 2016). To ensure that adjacent pixels were more likely to undergo disturbance, management, and land conversion transitions together, rules were set in the STSM that governed the sizes of these events so that no transitions were smaller than 2.0 ha (approximately six pixels), which was approximately the minimum patch size in the initial vegetation and land use raster for NC (Costanza *et al.*, 2016).

Initial vegetation and land use maps and most of the aspatial model pathways are detailed in Costanza *et al.* (2015). Costanza *et al.* (2016) added model pathways corresponding to each of the three purpose-grown crops. Switchgrass and sorghum pathways each consisted of a single state class, whereas the

SRWC pathway contained three state classes: a recently harvested class, an early-succession class, and a mid-succession class. When lands were converted to SRWC, they transitioned to the early-succession class. The planted pine pathway was divided into nine state classes: one recently harvested class, three recently thinned classes, and five classes that represented various age classes of closed (not recently thinned) forest (Table S1). The accuracy of the initial vegetation and land use map has not been assessed, but some accuracy metrics have been measured for the datasets from which it was created (Costanza et al., 2016). The current version of the GAP Land Cover Map has not been formally assessed, but a similar land cover mapping effort for the North Carolina Gap Analysis had per-class accuracies between 74% and 95% for general land cover classes and overall accuracy of 58% for the detailed classes based on conditional probabilities (McKerrow et al., 2006). Validation of the NLCD canopy cover model with canopy cover values from field data by Coulston et al. (2012) showed an R^2 value of 0.8 in the southeastern United States. LANDFIRE s-class data are based on LANDFIRE vegetation height data, which have a spatial bias of 3.8%, and LANDFIRE cover data had an overall agreement of 74% when compared with field data across the U.S. (Toney et al., 2012).

Costanza et al. (2016) identified marginal agricultural and forest lands as those that were row crop, pasture/hay, or forest in NLCD 2006, and were in nonirrigated soil capability classes 3 or 4 according to gridded Soil Survey Geographic Database (gSSURGO). They used GAP PAD-US data (U.S. Geological Survey National Gap Analysis Program, 2013) to identify lands under conservation ownership that would be ineligible for biomass production and conversion to urban or agricultural land. The STSM simulated changes in each of the mapped vegetation and land use classes due to forest management and land conversions specific to bioenergy production described above, as well as natural disturbance processes, ecological succession, urbanization, forest management, and background conversion between forest and agricultural lands.

The spatial outputs resulting from the STSMs had 60 m × 60 m cells. Each cell was assigned a land use or vegetation class. Vegetation classes were also assigned a successional stage (early, mid, or late succession) and a structure class (open or closed canopy, thinned, or recently harvested) (Table S2). Hereafter, we refer to these unique combinations of vegetation class, successional stage, and structure class as 'land cover classes'. Whereas SRTS and the STSMs were run on an annual time step, spatial outputs were produced every 10 time steps. In order to match the resolution of other model inputs, we resampled these spatial vegetation and land use outputs to 30 m × 30 m cell resolution using a nearest neighbor technique that assigned the values of larger cells to the each of the smaller, nested cells and used these finer resolution raster layers in the habitat models.

Focal species

While modeling all terrestrial vertebrate species within NC would have been ideal for our study, doing so was impractical given the effort and information required to build a habitat model. Therefore, we carefully selected 16 species that, collectively, allowed us to meet our objectives, focusing on birds and amphibians because numerous published studies on the habitat associations of these taxa existed (Table 2). To meet our objective of determining whether bioenergy demand could translate into changes in habitat amount, given the economic and ecological processes at play on the landscape, we sought a list that included species that were, collectively, associated with each of the major systems that will be affected by bioenergy demand (e.g., forests, woodlands, and agricultural lands) and could be sensitive to the changes in successional stage and vegetative structure likely resulting from biomass harvests, such as decreases in percent canopy cover or the development of dense shrub layers. To meet our objective of exploring the diversity of responses, we sought a list of species with diverse habitat preferences. Therefore, we chose species that contributed a unique suite of land cover class associations with the list and that select habitat in part on the basis of vegetative composition or structure at the level of detail concomitant with the land cover classification and changes in forest successional stage and vegetative structure in the STSM. We presumed species which do not select habitat based on vegetative structure would be unlikely to experience changes in habitat amount from forest biomass harvests and did not consider them for inclusion. Similarly, landscape change would be less likely to manifest in changes in habitat availability for habitat generalists than for habitat specialists, so we avoided adding individual species associated with the majority of the land cover classes in the land cover map. Our list included 10 species of greatest conservation need, three species of special concern, and two species listed as threatened by NC (NCWRC, 2015; Table S4). In addition, seven are on the Partners in Flight Watch List and four are on the IUCN Red List (Rich et al., 2004; IUCN, 2015; Table S4).

Habitat models

We built a deductive habitat model for each species that categorized STSM land cover classes as either inhabitable or uninhabitable in order to estimate future trends in species' habitat availability under each scenario (Table S2). Our deductive approach (habitat models) followed that of the United States Geological Survey's National Gap Analysis Program (GAP; Scott et al., 1993; Corsi et al., 2000; Overmars et al., 2007). GAP models are based on species–habitat requirement information gleaned from scientific literature and expert opinion. They associate individual species with land cover/land use classes (Gergely & McKerrow, 2013) and incorporate ancillary parameters (e.g., elevation, distance from forest edge, and distance from water) and range data to map the distribution of individual species' habitat at a landscape scale (1 : 100 000 map scale). Whereas national GAP models are based upon the GAP National Land Cover, ours corresponded with the land cover classification used in STSM, which has a finer thematic resolution than GAP's classification due to the specification of multiple states and stages (Table S2). For several species, we added constraints on model output related to maximum elevations and distances to water features as specified by the species' GAP model (Table S5) (http://gapanalysis.usgs.gov/). We

Table 2 We built deductive habitat models for 16 focal species to explore bioenergy demand effects on wildlife habitat in North Carolina, USA

Common name	Scientific Name	Regions	Forest	Open Woodland	Open Grass	Dense Shrub	Urban	% Habitat
Eastern Tiger Salamander	*Ambystoma tigrinum*	P, CP	X	X				4
Gopher Frog	*Lithobates capito*	CP		X				<1
Mole Salamander	*Ambystoma talpoideum*	M, P	X	X				1
Oak Toad	*Anaxyrus quercicus*	CP		X				1
Brown-headed Nuthatch	*Sitta pusilla*	M, P, CP	X	X			X	33
Cerulean Warbler	*Dendroica cerulean*	M, CP	X					14
Field Sparrow	*Spizella pusilla*	M, P, CP				X		20
Grasshopper Sparrow	*Ammodramus savannarum*	M, P, CP			X			13
Hairy Woodpecker	*Picoides villosus*	M, P, CP	X					54
Kentucky Warbler	*Oporornis formosus*	M, P, CP	X			X		27
Loggerhead Shrike	*Lanius ludovicianus*	P, CP		X	X			17
Prothonotary Warbler	*Protonotaria citrea*	CP	X					3
Red-headed Woodpecker	*Melanerpes erythrocephalus*	M, P, CP		X			X	12
Swainson's Warbler	*Limnothlypis swainsonii*	M, CP	X			X		19
Wood Thrush	*Hylocichla mustelina*	M, P, CP	X					41
Yellow-breasted Chat	*Icteria virens*	M, P, CP				X		16

The regions field indicates which ecoregions the species occurs in within NC: 'M' for Mountains, 'P' for Piedmont, and 'CP' for Coastal Plain. X's indicate that the species is associated with a land cover or vegetative structure type, but we caution that the associations listed here are highly generalized from the associations in the habitat models, which are described in Tables S2 and S3 and include specification of suitable successional stages and vegetative structure classes. Percent habitat refers to the proportion of the land area of North Carolina that our species–habitat models predicted as suitable for the species in 2010.

excluded some GAP model parameters because they are dependent upon habitat patch characteristics that had a high degree of uncertainty in the STSM output (e.g., patch size, whether a cell is forest edge or forest interior). We used ArcGIS 10.1 (ESRI, 2012) for all geoprocessing.

Model assumptions

Our modeling methods necessarily involved a set of assumptions about bioenergy feedstocks and habitat use patterns because some feedstocks are not yet widely present in NC or their characteristics and suitability for wildlife have not been documented. First, we assumed that forest thinning, whether under Baseline or in the bioenergy scenarios, removed 50% of biomass and therefore 50% of canopy cover. In the SRTS model, forest thinning was largely used for nonbioenergy products including sawtimber, so larger diameter trees would be included in the thinnings with and without bioenergy demand. Second, we assumed forest harvest under bioenergy scenarios removed 40% of residual biomass while retaining some snags (Costanza *et al.*, 2015). Third, Conventional-Ag, Conventional-Ag-Forest, Ag, and Ag-Forest included the cultivation of sorghum, switchgrass, and SRWC, but the extent to which wildlife will use these crops in NC has not been thoroughly studied. We assumed that none of our species will breed in sweet sorghum because they rarely breed in traditional row crops in NC; switchgrass crops will be tall, dense monocultures that may only be suitable for Yellow-breasted Chats; and our focal species would inhabit SRWC at rates similar to those in young planted pine stands (Dhondt *et al.*, 2007).

Another set of assumptions are related to species-specific complexities in habitat suitability. First, Cerulean Warblers are sensitive to the proportion of forest in landscapes where they occur but can benefit from commercial thinning within heavily forested landscapes (Wood *et al.*, 2013). We assumed thinned forests in the Mountains, which is heavily forested, would remain suitable for Cerulean Warblers but thinned forests in the Coastal Plain, which is largely agricultural, would render them unsuitable. However, we explored the sensitivity of our model to this assumption and determined that this component made little difference in the results in terms of total amount of habitat for the species. Second, Brown-headed Nuthatches and Red-headed Woodpeckers are known to inhabit some urbanized areas, but only where large trees and suitable nest cavities typically found in snags occur (Slater *et al.*, 2013). Therefore, we assumed that sites that transitioned from treeless land cover types to 'urban' remained unsuitable for 20 years and removed such sites from models. Third, Grasshopper Sparrows reportedly inhabit traditional row crops and switchgrass fields in Nebraska (Uden *et al.*, 2015). However, we did not attribute them to these classes because we assumed switchgrass fields will be too dense and homogenous for them in NC; we found no indication that they regularly breed in crops in NC; and the species exhibits regional variation in habitat preferences (Vickery, 1996).

Analysis and summary of habitat model outputs

The six scenarios (Baseline and five bioenergy portfolios) and five time steps (10-year increments from 2010 to 2050) yielded

30 separate model outputs for each of 16 focal species. We quantified the influence of each bioenergy feedstock portfolio on habitat availability for each species by calculating the differences in the amounts of habitat available between each bioenergy scenario and Baseline. To do this, we converted the number of cells in the habitat model output from each species-scenario-time step combination into hectares (Table S6) and then calculated the areal differences between individual bioenergy scenarios and baseline values at each time step for each species. We also summarized changes in habitat amount under Baseline in order to provide background references for both the amount of habitat initially available and projected trends given drivers of landscape change other than bioenergy production. However, we limit our presentation of baseline trends because they were somewhat difficult to interpret for some species due to modeled increases in new thinned and harvested land cover classes on the landscape during the first ten annual time steps of the STSM (Costanza *et al.*, 2015). Those large modeled landscape changes had the capacity to affect trends for species that use intact, thinned, or harvested forests during the same time period. We discuss this issue further in the Discussion.

Results

Our habitat models estimated that in 2010, the proportion of NC that was suitable ranged from <1% to 54% among our focal species (Table 2). For most species, between 11% and 20% of the state was suitable, but only three to four percent was suitable for the Prothonotary Warbler and Eastern Tiger Salamander, one percent for the Oak Toad and Mole Salamander, and less than one percent for the Gopher Frog. Among the species with the most habitat were Kentucky Warbler (27%), Brown-headed Nuthatch (33%), Wood Thrush (41%), and Hairy Woodpecker (54%).

Under Baseline, trends in the projected amount of habitat varied by species. Habitat was lost with each time step for the Eastern Tiger Salamander, Mole Salamander, Cerulean Warbler, Grasshopper Sparrow, Hairy Woodpecker, Swainson's Warbler, and Wood Thrush (Fig. 1). In addition, a net loss of habitat occurred by 2050 for the Kentucky Warbler, Loggerhead Shrike, and Prothonotary Warbler, but these species experienced gains and losses during the 40-year period. The Gopher Frog also experienced gains and losses during the 40-year period, but 2050 habitat amounts were similar to initial amounts. Mole and Eastern Tiger Salamanders experienced habitat losses in both upland and bottomland hardwood and conifer forests. Habitat was gained by 2050 for the Red-headed Woodpecker, Brown-headed Nuthatch, Oak Toad, Yellow-breasted Chat, and Field Sparrow (Fig. 1).

For several species, trends in habitat amount remained similar to Baseline under scenarios with bioenergy production; that is, the amount of habitat available did not deviate more than five percent from the amount

available under Baseline at any time step under any scenario (Fig. 2). Nevertheless, some patterns emerged for those species. Conventional, Conventional-Ag, and Ag were similar to Baseline for the Mole Salamander, Brown-headed Nuthatch, Eastern Tiger Salamander, Hairy Woodpecker, and Wood Thrush, but Conventional-Ag-Forest and Ag-Forest, which included the conversion of marginal forests, offered less habitat than Baseline for those species. Cerulean Warbler habitat amounts were most similar to Baseline under scenarios without increased conventional forestry (Ag and Ag-Forest) and declined over the 40-year period in scenarios with increased conventional forestry. However, those declines were never more than 4% below Baseline. Kentucky Warbler habitat amounts were most similar to Baseline under the scenario without conversion of agricultural lands or forests (Conventional) but alternated gains and losses of up to 5% from Baseline in scenarios with conversion of agricultural lands or marginal forests. Habitat for Swainson's Warbler also varied among time steps under each bioenergy scenario, but never deviated more than 4% from Baseline.

Among species that experienced >5% deviation from Baseline under a bioenergy scenario, the trends and magnitudes of differences varied among species, scenarios, and time steps (Fig. 2). The Gopher Frog, Grasshopper Sparrow, Loggerhead Shrike, and Yellow-breasted Chat experienced the greatest differences from Baseline with Gopher Frog habitat falling 20–25% below Baseline by 2050 under scenarios that included increases in conventional forestry. Grasshopper Sparrow habitat area was almost 20% below Baseline and Loggerhead Shrike habitat area was 12% below Baseline under Ag, in which all bioenergy was derived from the conversion of agricultural lands to purpose-grown feedstocks. The Yellow-breasted Chat gained 14% more habitat by 2040 under scenarios that included added conventional forestry and conversion of agricultural lands to bioenergy feedstocks.

Several noteworthy relationships between specific bioenergy feedstocks and species' habitat were identifiable. First, for a few species, the trends in habitat availability under scenarios with increased conventional forestry differed from trends under the other scenarios. For the Oak Toad, those scenarios were better than all other scenarios before 2040, although in 2050, scenarios without added conventional forestry (Ag and Ag-Forest) were also better than Baseline (4% and <1%, respectively). For the Gopher Frog, which has similar habitat associations with the Oak Toad except that it is not associated with hardwood forests (Table 2), increasing conventional forestry *decreased* the habitat amount to as much as 25% below Baseline in 2050 (25% under

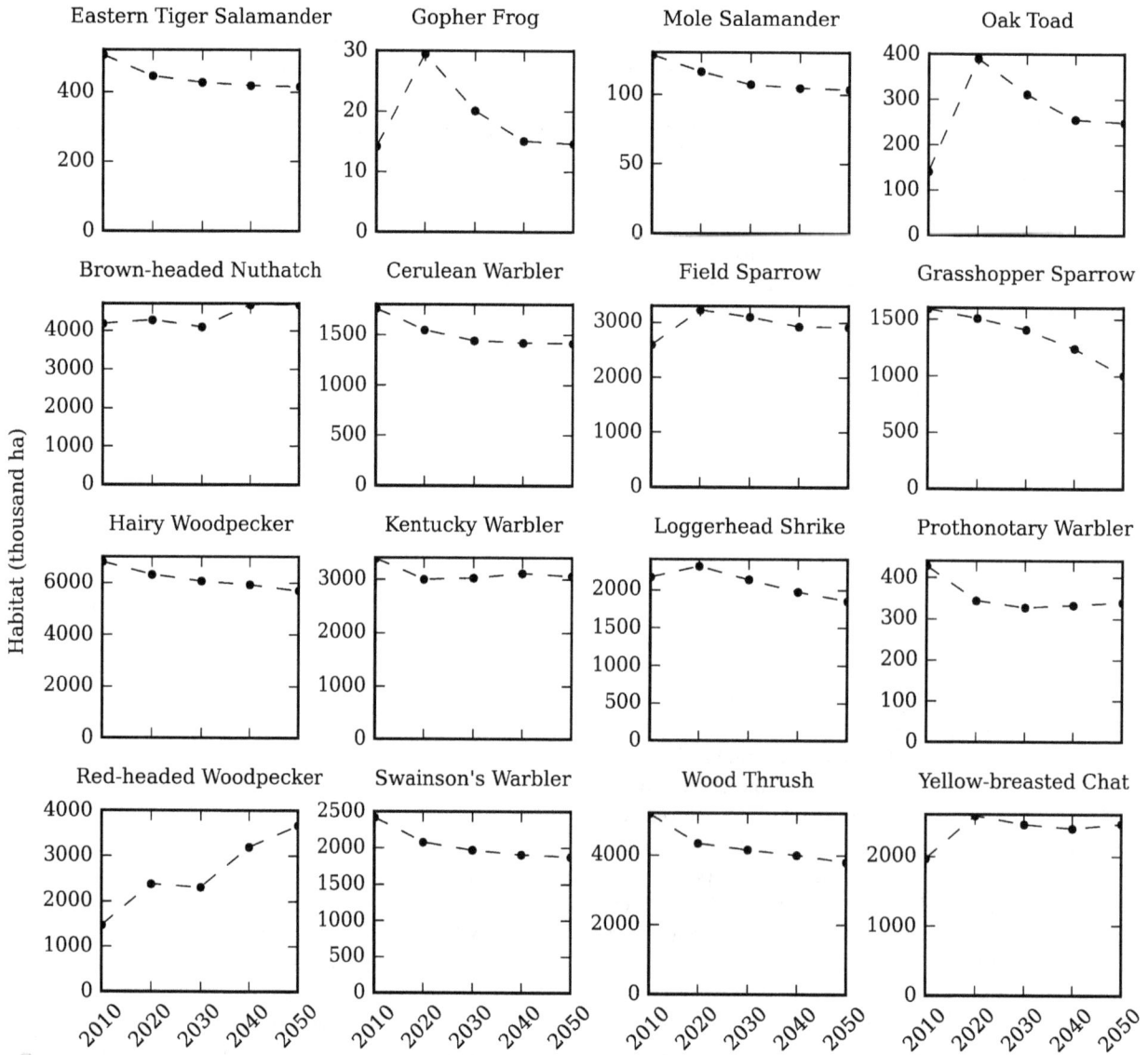

Fig. 1 Projected area of habitat in North Carolina, USA, under a 'business-as-usual' scenario (Baseline) that included urbanization, forestry, and vegetation dynamics (Costanza *et al.*, 2016).

Conventional-Ag-Forest; 23% under Conventional; 20% under Conventional-Ag). For the Prothonotary Warbler and the Red-headed Woodpecker, scenarios that included increased conventional forestry caused greater deviation from Baseline while remaining similar to each other, whereas habitat amounts under the other scenarios (Ag and Ag-Forest) were very similar to Baseline. However, Prothonotary Warbler habitat decreased to almost 10% below Baseline, whereas Red-headed Woodpecker habitat *increased* to around 5% above Baseline by 2030 and then decreased to 2–3% above Baseline by 2050 under Conventional, Conventional-Ag, and Conventional-Ag-Forest. Scenarios that included increased conventional forestry were better for the Field Sparrow than scenarios that did not, and converting marginal

forests to bioenergy feedstocks also increased the amount of its habitat. Conventional provided a steady gain in habitat relative to Baseline, but habitat amounts under the other scenarios varied among time steps; Conventional-Ag-Forest was consistently better than Baseline with up to 8% more habitat, whereas Ag was consistently worse with 8% less habitat in 2030.

Second, two species clearly responded negatively to the conversion of Ag lands to purpose-grown feedstocks: the Grasshopper Sparrow and the Loggerhead Shrike. For the Grasshopper Sparrow, sourcing biomass for wood pellets (Conventional) did not cause added habitat loss, but the conversion of agricultural lands did; ranking scenarios based on the proportion of bioenergy derived from agricultural land conversion (i.e.,

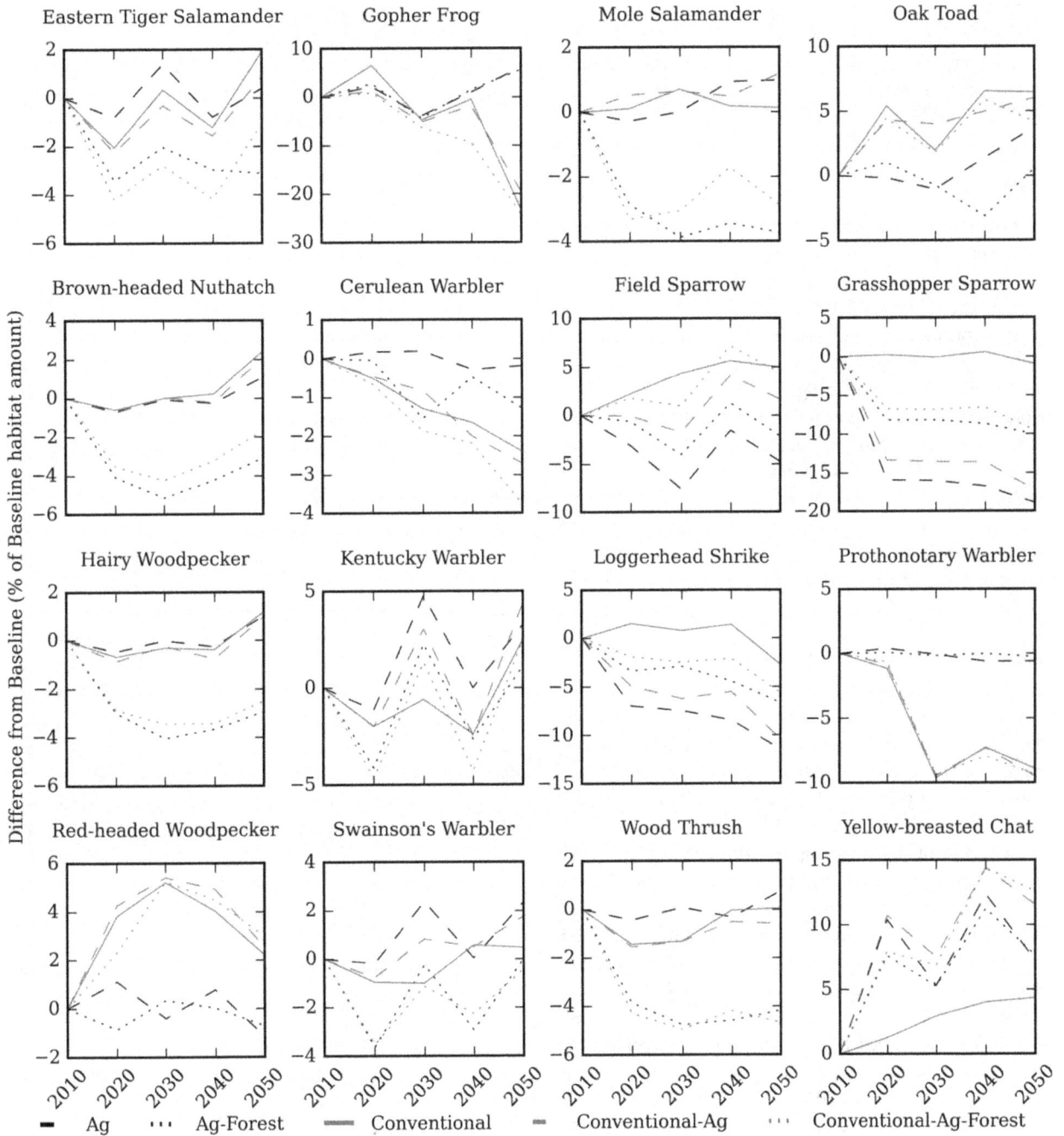

Fig. 2 Trends in habitat amount relative to Baseline for 16 species and five bioenergy production scenarios. Habitat amounts are presented as percent differences from Baseline amounts, calculated at each time step. For example, the models projected that in 2030 there would be almost ten percent less habitat for the Prothonotary Warbler under Conventional than under Baseline.

Conventional, Conventional-Ag-Forest, Ag-Forest, Conventional-Ag, Ag; Table 1) corresponded to their ranking in terms of habitat loss (Fig. 1; Conventional-Ag-Forest and Ag-Forest between 5% and 10%; Ag and Conventional-Ag between 13% and 19%). Patterns in the differences among scenarios for the Loggerhead Shrike were similar to the Grasshopper Sparrow, but

muted; the worst scenario, Ag, was 12% below Baseline by 2050.

Finally, all bioenergy scenarios provided more habitat for the Yellow-breasted Chat than Baseline. This species experienced a steady increase relative to Baseline over the 40-year period with increased conventional forestry to meet wood pellet demand and was

4% greater than the Baseline in 2050. Under scenarios capable of meeting the biofuel target (Conventional-Ag, Conventional-Ag-Forest, Ag, and Ag-Forest), habitat amounts were between 7% and 13% greater than Baseline in 2050.

Discussion

We projected future changes in habitat availability for 16 species under alternative scenarios of bioenergy production and found that realistic levels of demand for biofuels and wood pellets in NC would be capable of causing considerable changes in the amount of habitat available for some species. These changes could be positive or negative, depending upon a species' habitat associations and the sources of biomass utilized to meet demand. Several patterns emerged that will inform future research and policy regarding bioenergy and wildlife.

Species that inhabit dense, shrubby vegetation gained habitat in some bioenergy scenarios because forest biomass harvests (clear-cutting and thinning) encouraged development of shrubby vegetative structure (shrubs and saplings), and conversions of forests on marginal land to purpose-grown crops converted unsuitable forests to dense bioenergy crops that we assumed would be suitable for shrubland species. For example, the Field Sparrow experienced habitat gains when conventional forestry was increased, and the Yellow-breasted Chat gained habitat by 2050 under each of the bioenergy scenarios. Chats are associated with open, shrubby areas characteristic of regenerating and thinned forests (Eckerle & Thompson, 2001), and they were associated with SRWC and switchgrass in our models. Furthermore, it was unclear whether future chat habitat would be gained or lost compared with 2010 under Baseline because of their sensitivity to modeled landscape changes described below in *uncertainty and limitations*, but the four scenarios that included the conversion of marginal agricultural lands compensated for baseline losses between 2020 and 2040. This suggests that bioenergy feedstock harvest and cultivation could compensate for habitat losses for species that inhabit regenerating forests.

Clear-cutting and thinning natural and planted forests for bioenergy, such as to meet the foreign demand for wood pellets, would result in gains for some species and losses for others. Our results suggest that species associated with clear-cut, thinned, or regenerating forests could benefit from increases in forest biomass harvest within NC. Red-headed Woodpeckers are associated with woodlands and open areas containing snags, which we assumed would exist in harvested forests, so they benefited from increased thinning and clear-cuts. Likewise, the Oak Toad benefited from thinning closed canopy forests. Interestingly, Gopher Frog habitat, which consists of mostly open longleaf pine communities, initially increased relative to Baseline with additional forest thinning, but decreased when suitable forest types transitioned from an open to closed canopy state during the final time step. This highlights the fact that forest biomass harvests result in somewhat temporary changes because of vegetative successional processes, and, therefore, repeated harvests may be necessary to sustain forests in a state beneficial to species preferring open canopies and regenerating forests.

Increased conventional forestry resulted in habitat losses for some mature forest species including Prothonotary, Swainson's, and Cerulean Warblers. These species are associated with bottomland hardwood forests, which are important habitats because they harbor high levels of biodiversity (Mitchell *et al.*, 2009). Prothonotary Warbler habitat decreased under scenarios that included the harvest and thinning of these forest types (increased conventional forestry). Similarly, Evans *et al.* (2013) found that large amounts of forested wetlands are likely to be harvested in NC's Coastal Plain and among their indicator species, the Prothonotary Warbler had the greatest relative risk of habitat changes due to biomass harvest for wood pellets. There was also a large amount of overlap between Swainson's Warbler habitat and areas of high harvest risk when forested wetland harvests were allowed in their models, but we found little difference from Baseline for the Swainson's Warbler under any bioenergy scenario. Our models indicated a response similar to Prothonotary Warbler's for the Cerulean Warbler, but the magnitude of changes relative to Baseline was smaller. However, this muted response may have been a result of the Cerulean Warbler's distribution within NC. The species mostly occurs in the Mountains, with a small, isolated population inhabiting forests along the Roanoke River in the Coastal Plain (Fig. S1). Any changes for the coastal population were likely diluted by habitat dynamics occurring in the Mountains. Had our area of interest for this study been limited to the Coastal Plain, results for the Cerulean Warbler may have been different. The case of these bottomland hardwood-associated species suggests a need for further research and monitoring of how species depend upon them and underscores the need to identify important vegetative communities and understand their value to species when crafting policy that could result in changes to the landscapes where they occur.

Converting pasture and croplands on poor soils to purpose-grown bioenergy feedstocks exacerbated habitat loss for two pasture-associated species, the Grasshopper Sparrow and Loggerhead Shrike. These results support concerns about turning to marginal agricultural lands for bioenergy feedstock cultivation (Meehan *et al.*, 2010; Wiens *et al.*, 2011). Grasshopper Sparrow and Loggerhead Shrike habitats, which are characterized by open, grassy conditions (Vickery, 1996; Yosef, 1996), were lost under all scenarios that included the conversion of marginal agricultural lands to purpose-grown feedstocks. That is, the greater the proportion of bioenergy that came from conversion of marginal agricultural lands, the worse the scenario was for these species. This result was specifically associated with the conversion of pasture located on marginal soils. Our analysis indicates a clear relationship where increasing the cultivation of marginal lands for bioenergy would decrease the amount of habitat for grassland-associated species. Such a decrease is troubling because, in NC, pasture is already expected to be lost with urbanization, a process which was evident in loss of habitat under our 'business-as-usual' scenario. Furthermore, grassland birds have undergone the steepest national population decline of any group of birds in the last 40 years (Wiens *et al.*, 2011; Rupp *et al.*, 2012), and species associated with open vegetation may be most vulnerable to future land use change (Martinuzzi *et al.*, 2015). The importance of two drivers of landscape change for the Grasshopper Sparrow in our study indicates that considering bioenergy impacts within the context of other drivers of landscape change, rather than alone, will improve the accuracy of assessments of their impacts on biodiversity.

The conversion of forests located on poor soils to purpose-grown bioenergy feedstocks, including SRWC, was beneficial to the Yellow-breasted Chat, Field Sparrow, and Kentucky Warbler which will presumably inhabit dense, shrubby growth, and/or low canopy cover that will likely exist in SRWC stands. Conversely, converting forests resulted in habitat loss for forest species, such as the Eastern Tiger Salamander, Mole Salamander, Brown-headed Nuthatch, Hairy Woodpecker, Wood Thrush, and Swainson's Warbler. The diverse responses of species to land conversions in our study echo a point that others have made that impacts of land conversions on wildlife depend upon the initial types of land cover involved in the conversions (Meehan *et al.*, 2010; Wiens *et al.*, 2011; Rupp *et al.*, 2012). Our results suggest that the choice of which species to examine deserve to be added to the list of factors that influence whether land conversions for bioenergy will be deemed 'good' or 'bad'.

By virtue of numbers, species with small ranges, such as rare and range restricted species, are more sensitive to small changes in habitat acreage than common species with large ranges. We estimated that <1% of NC was suitable in 2010 for the Gopher Frog. This species

has undergone large range-wide population declines (Jensen & Richter, 2005) and is listed as near threatened by the IUCN and a species of special concern within NC (Hammerson & Jensen, 2004; North Carolina Wildlife Resources Commission, 2015). The overall losses that emerged suggest increased conventional forestry within its range would challenge conservation of this species. However, the initial increase in habitat over Baseline from additional conventional forestry suggests that forest biomass harvests for bioenergy could benefit the species if closed canopy forests were thinned in ways appropriate for it (Wilson, 1995). Little of the state was potentially suitable for the Prothonotary Warbler in 2010 (3%), and whereas its association with mature bottomland hardwood forests is very different than the Gopher Frog's association with open forests and woodlands on sandy soils, habitat amount for this species also declined (10%) under scenarios with increased conventional forestry. Other species for which habitat was less abundant in 2010 could fare better; the Oak Toad could gain habitat relative to baseline projections by 2050 under all bioenergy scenarios. Mole and Eastern Tiger Salamanders could gain habitat by 2050 under scenarios with increased conventional forestry. These patterns underscore that consideration of the spatial, temporal, and vegetative characteristics (i.e., successional stage and vegetative structure) of landscape changes due to bioenergy demand would benefit conservation assessments of populations of rare species because they have the capacity to emerge as important influences on habitat amounts. Our results suggest that bioenergy policy guidelines should consider the spatial coincidence of biomass production and species that are rare or have small ranges.

Uncertainty and limitations

Some aspects of future bioenergy demand, biomass source characteristics, and species' habitat use were uncertain in NC, even for the well-described species that we examined. We, therefore, had to make some assumptions in our modeling. The utility of novel bioenergy crops, thinned forests, and harvested forests to wildlife cannot be accurately modeled because their composition and structure will be dictated by cultivation and harvest practices that have not yet been fully developed and described. We, therefore, associated species with some land cover classes under assumptions of those classes' characteristics. If those assumptions prove inaccurate, then the accuracy of our projections will suffer. For example, if all snags are generally removed from clear-cuts, then Red-headed Woodpeckers will not gain as much habitat from increased conventional forestry as we indicated. Similarly, if switchgrass crops in

NC turn out to have vegetative composition or structure that is unsuitable for the Yellow-breasted Chats, then the species may not gain as much from bioenergy production as we projected. In light of these necessary assumptions, we caution against applying our results to questions other than those that we have addressed, or treating them as predictions. Future work to thoroughly describe and document the characteristics of bioenergy crops and the wildlife species that inhabit them will greatly benefit wildlife impact assessments.

Our study assessed general changes to the landscape, but the importance of finer-scale habitat factors should not be underestimated. For wildlife, the value of a site depends upon site, patch, and landscape characteristics; local management practices; and a myriad of other 'natural' influences on individual survival and reproduction not captured in our models. These factors could render patches of suitable land cover only marginally suitable or ecological traps (Schlaepfer et al., 2002). Similarly, creation of habitat via bioenergy crops and forest harvests will depend upon patch characteristics (e.g., field or forest patch size and shape), management practices, and landscape composition (Robertson et al., 2011, 2013; Blank et al., 2014). In forests, influences on survival and reproduction could include predation on adults, juveniles, and nests; nest parasitism; and prey densities (Wilcove, 1985; Streby et al., 2011; Zitske et al., 2011). In crops, management practices and field characteristics are still uncertain (Rupp et al., 2012; Blank et al., 2014; Immerzeel et al., 2014), but management practices including tillage; crop residue harvest intensities; and the intensity, frequency, and timing of harvest will be important (Rupp et al., 2012). Furthermore, without plant diversity and heterogeneity within fields, bioenergy crops could be of limited value to wildlife (Bonin & Lal, 2012). While we knew that patch and landscape characteristics influence whether sites are used by some of our focal species, we could not include these parameters in our models because predicting the sizes, shapes, and configurations of future habitat patches was not possible in our modeling framework. Understanding site-scale relationships between feedstocks and habitat quality better, as well as increasing the ability to predict future patch characteristics and landscape trends would improve our ability to understand the impacts of bioenergy on habitat (Immerzeel et al., 2014).

Our baseline projections of habitat availability changed more during the first time step than for subsequent time steps for some species (Fig. 1). For example, our model for the Gopher Frog projected large initial gains (120% by 2020) that were lost by 2050, with the amount of habitat in 2050 similar to 2010. We suspect that this was in part an amplification of landscape changes in the first few time steps of the STSM during which new thinned and harvested land cover classes increased on the landscape (Costanza et al., 2015) in combination with the species' distribution and the aging of recently harvested longleaf pine woodlands in the modeled landscape. These modeled landscape changes had the capacity to affect baseline trends for our species that use intact, thinned, or harvested forests and may have amplified early losses under Baseline for the Mole Salamander, Hairy Woodpecker, Eastern Tiger Salamander, Cerulean Warbler, Prothonotary Warbler, Swainson's Warbler, and Wood Thrush. There may have been a greater effect on gains and losses by 2050 for other species. For example, habitat was more abundant in 2050 than 2010 under Baseline for the Oak Toad, Gopher Frog, Field Sparrow, and Yellow-breasted Chat, but this overall gain in 2050 may have existed because of the large increase from 2010 to 2020 for those species (Fig. 1).

General insights for research, conservation, and policy

Biomass harvests and land conversions have the capacity to add or remove large amounts of habitat from a landscape under realistic levels of demand for bioenergy. Such changes will occur against a backdrop of other drivers of landscape change, and there may be cases where bioenergy will amplify or mediate changes in habitat amount from other change agents. Future increases in bioenergy production will likely have a variety of implications for wildlife because of the diversity of species' habitat requirements and biomass sources, along with the importance of how biomass harvests and cultivation will overlap with species' distributions. Each species that we examined experienced some variation among the bioenergy feedstock portfolios that we explored, but one species benefitted under all bioenergy scenarios. None of the scenarios that we explored were 'best' or 'worst' for all of our focal species, and the impacts of scenarios varied over time for some.

Our models indicated that there will be both 'winners' and 'losers' under a given bioenergy portfolio, and how a species fares may vary over time. Consequently, bioenergy policies will likely involve trade-offs for wildlife, and choices about which species to include in assessments of wildlife impacts will influence the assessments' conclusions. When evaluating the sustainability implications of bioenergy production, researchers and policymakers should therefore carefully consider whether to focus on all species, or subgroups of species occurring in an area of interest. If subgroups are examined, then the groups should be carefully defined to match the objectives. We present four general principles to consider when undertaking such a task, although there are likely others. One, species that inhabit regenerating forests may benefit from bioenergy demand. Two, species that are largely dependent upon the mature state of a single land cover type that could be harvested for biomass are at risk to be negatively impacted by bioenergy demand. Three, bioenergy demand could exacerbate habitat loss for species that primarily inhabit land cover types that are also being lost to other drivers of landscape change (e.g., urbanization). Four, species with small ranges can be more sensitive to landscape changes from bioenergy and deserve special consideration.

Acknowledgements

We thank Todd Earnhardt and Steve Williams for technical advice and support, as well as Dr. Sarah Fritts and three anonymous reviewers whose comments greatly improved this paper. Funding for this work was provided by the USGS Gap Analysis Program and the Biofuels Center of North Carolina. Any use of trade, firm, or product names is for descriptive purposes only and does not imply endorsement by the U.S. Government.

References

Abt RC, Cubbage FW, Abt KL (2009) Projecting southern timber supply for multiple products by sub region. *Forest Products Journal*, **59**, 7–16.

Abt KL, Abt RC, Galik CS, Skog KE (2014) *Effect of policies on pellet production and forests in the U.S. South: a technical document supporting the Forest Service update of the 2010 RPA Assessment*. Gen. Tech. Rep. SRS-202, U.S. Department of Agriculture Forest Service, Southern Research Station, Asheville, NC, 33 p.

Apex Resource Management Solutions (2014) ST-Sim state-and-transition simulation model software. Available at: http://www.apexrms.com/stsm (accessed 1 July 2014).

Blank PJ, Sample DW, Williams CL, Turner MG (2014) Bird communities and biomass yields in potential bioenergy grasslands. *PLoS ONE*, **9**, 1–10.

Bonin C, Lal R (2012) Agronomic and ecological implications of biofuels. In: *Advances in Agronomy*, Vol 117 (ed. Donald LS), pp. 1–50. Academic, New York, USA.

Burke S, Hall BR, Shahbazi G, Tolson EN, Wynne JC (2007) North Carolina's Strategic Plan for Biofuels Leadership. Raleigh, NC, USA, 20 p. Available at: https://www.ncbiotech.org/sites/default/files/biofuels_plan_0.pdf (accessed 11 December 2015).

Christian DP, Hoffman W, Hanowski JM, Niemi GJ, Beyea J (1998) Bird and mammal diversity on woody biomass plantations in North America. *Biomass and Bioenergy*, **14**, 395–402.

Coleman MD, Stanturf JA (2006) Biomass feedstock production systems: economic and environmental benefits. *Biomass and Bioenergy*, **30**, 693–695.

Corsi F, de Leeuw J, Skidmore A (2000) Modeling species distribution with GIS. In: *Research Techniques in Animal Ecology* (eds Boitani L, Fuller T), pp. 389–434. Columbia University Press, New York.

Costanza JK, Abt RC, McKerrow AJ, Collazo JA (2015) Linking state-and-transaction simulation and timber supply models for forest biomass production scenarios. *AIMS Environmental Science*, **2**, 180–202.

Costanza JK, Abt RC, McKerrow AJ, Collazo JA (2016) Bioenergy production and forest landscape change in the southeastern U.S. *GCB Bioenergy*, Accepted.

Coulston JW, Moisen GG, Wilson BT, Finco MV, Cohen WB, Brewer CK (2012) Modeling percent tree canopy cover: a pilot study. *Photogrammetric Engineering & Remote Sensing*, **78**, 715–727.

Dale VH, Kline KL, Wiens J, Fargione J (2010) *Biofuels: implications for land use and biodiversity*. Biofuels and Sustainability Reports: 1–13. Available at: http://www.esa.org/esa/wp-content/uploads/2014/11/ESA-Biofuels-Report-1.pdf (accessed 7 July 2016).

Dauber J, Jones MB, Stout JC (2010) The impact of biomass crop cultivation on temperate biodiversity. *Global Change Biology Bioenergy*, **2**, 289–309.

Dhondt AA, Wrege PH, Cerratani J, Sydenstricker KV (2007) Avian species richness and reproduction in short rotation coppice habitats in central and western New York. *Bird Study*, **54**, 12–22.

Dornburg V, van Vuuren D, van de Ven G et al. (2010) Bioenergy revisited, key factors in global potentials of bioenergy. *Energy and Environmental Science*, **3**, 258–267.

Eckerle KP, Thompson CF (2001) Yellow-breasted chat (*Icteria virens*). In: *The Birds of North America Online* (ed. Poole A), Cornell Lab of Ornithology, Ithaca. Available at: http://bna.birds.cornell.edu/bna/species/575 (accessed 11 December 2015).

Eggers J, Troltzsch K, Falcucci A et al. (2009) Is biofuel policy harming biodiversity in Europe? *Global Change Biology Bioenergy*, **1**, 18–34.

Energy Independence and Security Act of 2007 (2007) 42 U.S.C. § 17001.

Energy Policy Act of 2005 (2005) 42 U.S.C. § 15801.

ESRI (2012) *ArcGIS Desktop: Release 10.1*. Environmental Systems Research Institute, Redlands, CA.

Evans JM, Fletcher RJ Jr, Alavalapati JRR et al. (2013) *Forestry Bioenergy in the Southeast United States: Implications for Wildlife Habitat and Biodiversity*. National Wildlife Federation, Merrifield, VA.

Fargione J (2010) Is bioenergy for the birds? An evaluation of alternative future bioenergy landscapes. *PNAS*, **107**, 18745–18746.

Firbank LG (2008) Assessing the ecological impacts of bioenergy projects. *Bioenergy Research*, **1**, 12–19.

Fletcher RJ, Robertson BA, Evans J, Doran PJ, Alavalapati JRR, Schemske DW (2010) Biodiversity conservation in the era of biofuels: risks and opportunities. *Frontiers in Ecology and the Environment*, **9**, 161–168.

Forisk Consulting LLC (2014) Wood bioenergy US database 2013. Available by subscription.

Fritts SR, Moorman CE, Hazel DW, Jackson BD (2014) Biomass harvesting guidelines affect downed woody debris retention. *Biomass and Bioenergy*, **70**, 382–391.

Fritts SR, Moorman CE, Grodsky SM, Hazel DW, Homyack JA, Farrell CB (2015) Shrew response to variable woody debris retention: Implications for sustainable forest bioenergy. *Forest Ecology and Management*, **336**, 35–43.

Fritts SR, Moorman CE, Grodsky SM, Hazel DW, Homyack JA, Farrell CB, Castleberry SB (2016) Do biomass harvesting guidelines influence herpetofauna following harvests of logging residues for renewable energy? *Ecological Applications*, **26**, 926–939.

Galik CS, Abt RC (2015) Sustainability guidelines and forest market response: an assessment of European Union pellet demand in the southeastern United States. *Global Change Biology Bioenergy*, **8**, 658–669.

Gergely KJ, McKerrow A (2013) Land cover—National inventory of vegetation and land use: U.S. Geological Survey Fact Sheet 2013-3085, 1 p. Available at: http://pubs.usgs.gov/fs/2013/3085/ (accessed 11 December 2015).

Groom MJ, Gray EM, Townsend PA (2008) Biofuels and biodiversity: principles for creating better policies for biofuel production. *Conservation Biology*, **22**, 602–609.

Hammerson G, Jensen J (2004) Lithobates capito. The IUCN Red List of Threatened Species. Version 2015.2. Available at: www.iucnredlist.org (accessed 20 August 2015).

Hartman JC, Nippert JB, Orozco RA, Springer CJ (2011) Potential ecological impacts of switchgrass (*Panicum virgatum* L.) biofuel cultivation in the Central Great Plains, USA. *Biomass and Bioenergy*, **35**, 3415–3421.

Immerzeel D, Verweij P, Hilst F, Faaij AP (2014) Biodiversity impacts of bioenergy crop production: a state-of-the-art review. *Global Change Biology Bioenergy*, **6**, 183–209.

IUCN (2015) *The IUCN Red List of Threatened Species*. Version 2015-3. Available at: www.iucnredlist.org (accessed 4 November 2015).

Jensen JB, Richter SC (2005) Rana capito, Gopher Frogs. In: *Amphibian Declines: The Conservation Status of United States Species* (ed. Lannoo MJ), pp. 536–538. University of California Press, Berkeley, CA, USA.

Martinuzzi S, Pidgeon AM, Radeloff VC, et al. (2015) Impacts of future land-use changes on wildlife habitat: insights from Southeastern US. *Ecological Applications*, **25**, 160–171.

McDonald RI, Fargione J, Kiesecker J, Miller WM, Powell J (2009) Energy sprawl or energy efficiency: climate policy impacts on natural habitat for the United States of America. *PLoS ONE*, **4**, e6802.

McKerrow AJ, Williams SG, Collazo JA (2006) The North Carolina Gap Analysis Project: Final Report. North Carolina Cooperative Fish and Wildlife Research Unit. Submitted to: The National Gap Analysis Program. U.S. Geological Survey, Biological Resources Division. Available at: http://www.basic.ncsu.edu/ncgap/NCFinal%20Report.pdf (accessed 11 December 2015).

Meehan TD, Hurlbert AH, Gratton C (2010) Bird communities in future bioenergy landscapes of the Upper Midwest. *Proceedings of the National Academy of Sciences of the United States of America*, **107**, 18533–18538.

Mitchell R, Engstrom T, Sharitz R (2009) Old forests and endangered woodpeckers: old-growth in the southern coastal plain. *Natural Areas Journal*, **29**, 301–310.

North Carolina Wildlife Resources Commission (2015) *North Carolina Wildlife Action Plan*. 1328 p. Available at: http://www.ncwildlife.org/Portals/0/Conserving/

documents/2015WildlifeActionPlan/NC-WAP-2015-All-Documents.pdf (accessed 11 December 2015).

Overmars KP, de Groot WT, Huigen MGA (2007) Comparing inductive and deductive modeling of land use decisions: principles, a model and an illustration from the Philippines. *Human Ecology*, **35**, 439–452.

Rich TD, Beardmore CJ, Berlanga H *et al.* (2004) *Partners in Flight North American Landbird Conservation Plan*. Cornell Lab of Ornithology, Ithaca, NY.

Robertson GP, Dale VH, Doering OC *et al.* (2008) Sustainable biofuels redux. *Science*, **322**, 49–50.

Robertson BA, Doran PJ, Loomis LR, Robertson JR, Schemske DW (2011) Perennial biomass feedstocks enhance avian diversity. *Global Change Biology Bioenergy*, **3**, 235–246.

Robertson BA, Rice RA, Sillett TS *et al.* (2012) Are agrofuels a conservation threat or opportunity for grassland birds in the United States? *Condor*, **114**, 679–688.

Robertson BA, Landis DA, Sillett TS, Loomis ER, Rice RA (2013) Perennial agroenergy feedstocks as en route habitat for spring migratory birds. *Bioenergy Research*, **6**, 311–320.

Rupp SP, Bies L, Glaser A *et al.* (2012) Effects of bioenergy production on wildlife and wildlife habitat. Wildlife Society Technical Review 12-03. The Wildlife Society, Bethesda, Maryland, USA.

Schlaepfer MA, Runge MC, Sherman PW (2002) Ecological and evolutionary traps. *Trends in Ecology & Evolution*, **17**, 474–480.

Scott JM, Davis F, Csuti B *et al.* (1993) Gap analysis: a geographic approach to protection of biological diversity. *Wildlife Monographs*, **123**, 3–41.

Slater GL, Lloyd JD, Withgott JH, Smith KG (2013) Brown-headed Nuthatch (*Sitta pusilla*). In: *The Birds of North America Online* (ed. Poole A), Cornell Lab of Ornithology, Ithaca. Available at: http://bna.birds.cornell.edu/bna/species/349 (accessed 11 December 2015).

Streby HM, Peterson SM, Andersen DE (2011) Invertebrate availability and vegetation characteristics explain use of nonnesting cover types by mature-forest songbirds during the postfledging period. *Journal of Field Ornithology*, **82**, 406–414.

Terando AJ, Costanza J, Belyea C, Dunn RR, McKerrow A, Collazo JA (2014) The southern megalopolis: using the past to predict the future of urban sprawl in the Southeast U.S. *PLoS ONE*, **9**, e102261.

Tilman D, Cassman KG, Matson PA, Naylor R, Polasky S (2002) Agricultural sustainability and intensive production practices. *Nature*, **418**, 671–677.

Toney C, Peterson B, Long D, Parsons R, Cohn G (2012) Development and applications of the LANDFIRE forest structure layers. In: *Moving from status to trends: Forest Inventory and Analysis (FIA) symposium 2012 December 4–6; Baltimore MD* (eds Morin RS, Liknes GC), pp. 305–309. USDA Forest Service Northern Research Station GTR NRS-P-105, Newtown Square.

Uden DR, Allen CR, Mitchell RB, McCoy TD, Guan Q (2015) Predicted avian responses to bioenergy development scenarios in an intensive agricultural landscape. *Global Change Biology Bioenergy*, **7**, 717–726.

U.S. Geological Survey National Gap Analysis Program (2013) Protected Areas Database-US (PAD-US), Version 1.3. Available at: http://gapanalysis.usgs.gov/padus/ (accessed 1 July 2014)

Vickery PD (1996) Grasshopper sparrow (*Ammodramus savannarum*). In: *The Birds of North America Online* (ed. Poole A), Cornell Lab of Ornithology, Ithaca. Available at: http://bna.birds.cornell.edu/bna/species/239 (accessed 11 December 2015).

Wang W, Dwivedi P, Abt R, Khunna M (2015) Carbon savings with transatlantic trade in pellets: accounting for market-driven effects. *Environmental Research Letters*, **10**, 114019.

Wear DN, Greis JG (2013) *The Southern Forest Futures Project: technical report*. Gen. Tech. Rep. SRS-178. U.S. Department of Agriculture Forest Service, Southern Research Station, Asheville, NC, 542 p.

Wiens J, Fargione J, Hill J (2011) Biofuels and biodiversity. *Ecological Applications*, **21**, 1085–1095.

Wilcove DS (1985) Nest predation in forest tracts and the decline of migratory songbirds. *Ecology*, **66**, 1211–1214.

Wilson LA (1995) *The Land Manager's Guide to the Amphibians and Reptiles of the South*. The Nature Conservancy, Chapel Hill, NC.

Wood PB, Sheehan J, Keyser P *et al.* (2013) *Management Guidelines for Enhancing Cerulean Warbler Breeding Habitat in Appalachian Hardwood Forests*. American Bird Conservancy, The Plains, Virginia.

Yosef R (1996) Loggerhead Shrike (*Lanius ludovicianus*). In: *The Birds of North America Online* (ed. Poole A), Cornell Lab of Ornithology, Ithaca. Available at: http://bna.birds.cornell.edu/bna/species/231 (accessed 11 December 2015).

Zitske BP, Betts MG, Diamond AW (2011) Negative effects of habitat loss on survival of migrant warblers in a forest mosaic. *Conservation Biology*, **25**, 993–1001.

9

The role of bioenergy and biochemicals in CO_2 mitigation through the energy system – a scenario analysis for the Netherlands

IOANNIS TSIROPOULOS[1] (ID), RIC HOEFNAGELS[1], MACHTELD VAN DEN BROEK[1], MARTIN K. PATEL[2] and ANDRE P. C. FAAIJ[3]

[1]Copernicus Institute of Sustainable Development, Utrecht University, Heidelberglaan 2, 3584 CS Utrecht, The Netherlands, [2]Energy Group, Institute for Environmental Sciences and Forel Institute, University of Geneva, Boulevard Carl-Vogt 66, 1205 Geneva, Switzerland, [3]Energy and Sustainability Research Institute, University of Groningen, Nijenborg 4, 9747 AC Groningen, The Netherlands

Abstract

Bioenergy as well as bioenergy with carbon capture and storage are key options to embark on cost-efficient trajectories that realize climate targets. Most studies have not yet assessed the influence on these trajectories of emerging bioeconomy sectors such as biochemicals and renewable jet fuels (RJFs). To support a systems transition, there is also need to demonstrate the impact on the energy system of technology development, biomass and fossil fuel prices. We aim to close this gap by assessing least-cost pathways to 2030 for a number of scenarios applied to the energy system of the Netherlands, using a cost-minimization model. The type and magnitude of biomass deployment are highly influenced by technology development, fossil fuel prices and ambitions to mitigate climate change. Across all scenarios, biomass consumption ranges between 180 and 760 PJ and national emissions between 82 and 178 Mt CO_2. High technology development leads to additional 100–270 PJ of biomass consumption and 8–20 Mt CO_2 emission reduction compared to low technology development counterparts. In high technology development scenarios, additional emission reduction is primarily achieved by bioenergy and carbon capture and storage. Traditional sectors, namely industrial biomass heat and biofuels, supply 61–87% of bioenergy, while wind turbines are the main supplier of renewable electricity. Low technology pathways show lower biochemical output by 50–75%, do not supply RJFs and do not utilize additional biomass compared to high technology development. In most scenarios the emission reduction targets for the Netherlands are not met, as additional reduction of 10–45 Mt CO_2 is needed. Stronger climate policy is required, especially in view of fluctuating fossil fuel prices, which are shown to be a key determinant of bioeconomy development. Nonetheless, high technology development is a no-regrets option to realize deep emission reduction as it also ensures stable growth for the bioeconomy even under unfavourable conditions.

Keywords: bioeconomy, CO_2 mitigation, cost-minimization, emerging sectors, scenario analysis

Introduction

In line with long-term climate targets agreed upon at the 21st Conference of Parties in Paris (UNFCCC, 2015), the European Union (EU) set out to increase its renewable energy supply to 27% and to achieve 40% greenhouse gas (GHG) emission reduction by 2030 compared to 1990, towards a 80–95% reduction by 2050 (EC, 2015). Large-scale modern bioenergy deployment, carbon capture and storage (CCS), and their combination (bioenergy with carbon capture and storage; BECCS) are among the key energy supply and carbon mitigation options required to embark on cost-efficient trajectories

that pursue climate goals (IPCC, 2014; Rose *et al.*, 2014; Matthews *et al.*, 2015; Winchester & Reilly, 2015).

Within the EU, bioenergy supply is shown to be significant in sectors such as heat and road transport (Stralen *et al.*, 2013). Increasingly, there is evidence to suggest that emerging bioeconomy sectors such as aviation and chemicals, which have few or no other renewable alternatives than biomass, and CCS and BECCS will also be needed. Based on mid-term demand projections, biochemicals and bioplastics (frequently referred to as nonenergy uses of biomass) may consume 9–24% of global biomass demand by 2050 (Piotrowski *et al.*, 2015). Other studies show 15–17% of total biomass to be used for nonenergy applications (18–27 EJ yr^{-1}) and to supply approximately 7–11 EJ yr^{-1} of global nonenergy biomass products (Daioglou *et al.*, 2015). In other

Correspondence: Ioannis Tsiropoulos
e-mail: i.tsiropoulos@uu.nl

sectors, such as aviation, the EU has the ambition to reach 88 PJ (2 Mt, assuming 44 GJ t^{-1} heating value) renewable jet fuel (RJF) consumption, which is about 3.7% of its projected jet fuel demand by 2020 (EC, 2003, 2011). These new sectors are particularly relevant for countries with relatively large refining capacity and energy intensive industry such as the Netherlands. The Netherlands consumes about a quarter of its total final energy for nonenergy purposes (585 PJ in 2013, CBS, 2016) and within the EU, it has the largest petrochemical capacity next to Germany (OGJ, 2012). Regarding emission reduction, at a global level, BECCS would need to contribute between 2 and 10 Gt CO_2 yr^{-1} in 2050 in order to ensure compliance with the 2 °C target (4–22% of the 1990 baseline; Fuss *et al.*, 2014). Based on Rose *et al.* (2014), modern bioenergy supply may reach 37% (or up to about 250 PJ) over total primary energy supply by 2050 and is largely combined with BECCS. Despite these expectations, comprehensive assessments of extended bioeconomy sectors (i.e. aviation, chemicals) in energy system models, interactions with other renewable energy sources (RES; e.g. wind or solar) and mitigation technologies (i.e. CCS, BECCS) at a national or regional level, are scarce.

Such an analysis requires an integrated energy systems assessment framework that takes into account emerging bioeconomy sectors next to modern bioenergy and that addresses key factors of uncertainty with sufficient level of detail on the energy system's structure and on the complex flows of the petrochemical industry. To obtain the necessary detail, we focus on the energy system of the Netherlands, which requires a significant transformation for the country to meet its renewable energy and GHG mitigation goals, in line with the EU targets (Roelofsen *et al.*, 2016; Vuuren *et al.*, 2016). Albeit having an efficient agricultural sector, the Netherlands is dependent on biomass imports in order to support large-scale bioeconomy developments, similar to the EU (Stralen *et al.*, 2013). This is deemed possible due to its advanced logistics infrastructure. While modelling outcomes are pertinent to the Netherlands, they are useful to provide insights in the implications of large-scale bioeconomy developments also in the EU.

Our earlier study incorporated the chemicals and aviation sector in a national energy systems model of the Netherlands and demonstrated that biomass conversion technologies may be cost-competitive compared to other fossil and renewable alternatives by 2030 to achieve renewable energy goals (Tsiropoulos, 2016). With respect to biomass conversion, industrial heat from biomass, lignocellulosic sugar production, biochemicals from sugar fermentation and Fischer–Tropsch (FT) road transport fuels from solid biomass gasification were shown to be most promising options. These findings are in line with other research (Ren & Patel, 2009; Ren *et al.*, 2009; Saygin *et al.*, 2013, 2014; Gerssen-Gondelach *et al.*, 2014). However, our earlier study also showed that while the renewable energy technology portfolio was stimulated by renewable energy policies, emission reduction targets of 40% by 2030, compared to 1990, were not met. Therefore, additional insights are needed as to the required preconditions to pursue those targets. One limitation of the abovementioned study is that it only assessed the influence of technology development as a factor of future uncertainty, while other crucial parameters such as varying fossil fuel prices and availability of low-cost biomass in combination with technological progress may also affect bioeconomy developments and the pathways to emission reduction. These uncertainties need to be assessed under a technology-neutral setting, with climate policy such as a CO_2 tax being the only driver for the deployment of a cost-optimal technology portfolio.

Such an assessment is performed in the present study using a national cost-minimization linear programming model developed for the Netherlands (MARKet ALlocation MARKAL-NL-UU; Tsiropoulos, 2016) that apart from technology characterization of the fossil energy system also includes key biomass conversion technologies, other renewables and mitigation options (CCS, BECCS). Using scenario assessment for a combination of uncertainty factors on technology development, biomass cost-supply and fossil fuel prices, we estimate the achieved CO_2 emission reduction, the required technology portfolio, the demand for biomass and supply of bioenergy and biochemicals in each case.

Materials and methods

We focus on bioeconomy activities that relate with the energy system and the chemical industry (i.e. bioenergy, biochemicals) that have the potential to replace fossil fuels in the energy system. Other economic activities based on biomass, for example food, feed, traditional biomass uses (lumber products), are not included in the framework. We translate key parameters of future uncertainty of the bioeconomy development (technology development, biomass cost-supply, fossil fuel prices) to scenarios and then perform scenario analysis by comparing outputs derived from a cost-minimization linear programming energy system model developed for the Netherlands.

Model

The MARKAL-NL-UU applied in this study uses cost-minimization linear programming techniques to define the technology portfolio required to meet demand for energy (electricity, heat, fuels) and chemicals that lead to least total system costs. The model can be described by three core modules: energy supply, energy and chemicals conversion, and energy and chemicals demand.

The electricity sector and the CCS technology portfolio for the Netherlands are described in van den Broek *et al.* (2008, 2011). The model's extension to the road transport sector is included in van Vliet *et al.* (2011). Finally, emerging bioeconomy sectors have been included by Tsiropoulos (2016). The technology portfolio of MARKAL-NL-UU for electricity, heat, road transport and jet fuels, and chemicals is described in Tables S1–S3.

Energy supply

In the energy supply module, cost-supply trajectories of fossil, nuclear and biomass resources are included. For fossil fuels, the price develops according to the International Energy Agency World Energy Outlook 2015 (IEA-WEO) New Policies Scenario (OECD/IEA, 2015), unless stated otherwise. Fossil fuel price variation is a key aspect of future uncertainty, which is taken into account in scenario assessment (section 'Fossil fuel prices').

Biomass cost-supply curves are estimated based on the sourcing region (domestic, European, global) and are specified for different biomass types. Road-side costs and potentials for biomass are determined for 2010–2030, based on the Intelligent Energy Europe project Biomass Policies (Elbersen *et al.*, 2015). In this database, biomass represents the net available potential for bioenergy, thereby excluding competition with traditional sectors such as food, feed and fibres. Costs refer to market prices for already traded biomass types and to road-side costs for biomass markets that are not developed (Elbersen *et al.*, 2015). To these costs, we add transport costs to the Netherlands using a geographical explicit biomass intermodal transport model (BIT-UU; described in Hoefnagels *et al.*, 2014a,b). Biomass transport costs, calculated at Nomenclature of territorial units for statistics 2 level, are aggregated based on the weighted average for 4 EU regions as described in Tsiropoulos (2016). From the regional biomass supply potential, it is assumed that approximately 5% may be available for export to the Netherlands, based on the share of the Dutch total primary energy supply over the EU's to 2030. In OECD/IEA (2014), the EU demand is 61 EJ under the 450 ppm scenario in 2030. For comparison, the Dutch demand is 3.2 EJ and the assumed biomass in 2030 is about 430 PJ or 13% of the country's total primary energy supply.

These assumptions may lead to conservative biomass cost-supply estimates for two reasons. Firstly, transport costs are based on wood chip logistics, thereby ignoring cost-efficiency gains that can be achieved if biomass is densified at the sourcing region, for example, to wood pellets. Secondly, each country may supply larger potential than the 5% we allocated if markets are well-developed. These factors are addressed in scenarios (section 'Biomass cost-supply').

Next to biomass from EU sources, five commodities from extra-EU sources are included, namely raw sugar, wood pellets, first- and second-generation ethanol, vegetable oil and biodiesel. Ultimately, it depends on the total production system costs, which include feedstock and conversion, to indicate the cost-optimal use of intra-EU or extra-EU resources. A total of 400 PJ of solid biomass and 50 PJ of liquid biomass are assumed to be available for imports to the Netherlands. Such

potential is approximately 26 Mt in wood pellet equivalent, which is rather large considering that it corresponds to global wood pellet consumption in 2015 (about 25.5 Mt; AEBIOM (2015)). However, there is sufficient evidence that suggests that these volumes may be available (Chum *et al.*, 2011; Ganzevles, 2014; Smeets, 2014). The influence of extra-EU import is assessed in a separate scenario, which assumes that only domestic and intra-EU biomass is available (section 'Biomass cost-supply'). CO_2 emissions from biomass production in the Netherlands contribute to the national total CO_2 emissions. Indirect emissions from extraction and import of fossil resources to the Netherlands or biomass production outside the Netherlands do not contribute to the national total.

Technologies for energy and chemicals conversion

The model includes a large portfolio of fossil (natural gas, oil, coal), nuclear and renewable energy technologies (e.g. biomass conversion, wind turbines, photovoltaics) that convert primary resources to electricity, heat, fuels for the energy system, and feedstocks or end products for the fossil-based and bio-based chemical industry. Fossil, nuclear and renewable energy conversion technologies are characterized based on their cost structure at a specific year and scale and technical parameters (process energy input, process efficiency). Annual costs consist of capital investment costs (e.g. process components, buildings, contingency), fixed costs (operation and maintenance, administrative costs) and variable costs (e.g. feedstock, utilities, labour). CO_2 emissions from conversion of primary to secondary energy carriers represent the emissions of the energy system, including industrial process emissions. Non-CO_2 GHG emissions that are not associated with the energy system are not included in the boundaries of the model (e.g. methane emissions by activities in agriculture). These represent approximately 16% of national total GHG emissions in 2014 (Table S9).

Biorefineries (biochemical and thermochemical) are also included in the model. Conventional coal gasification and FT-synthesis to fuels is excluded as an option. Similar to other multi-output processes such as combined heat and power plants, biorefineries deliver outputs to several sectors (e.g. to fuels and electricity) as opposed to, for example a wind turbine, which delivers only to the electricity sector. This enables access to different demand markets in direct competition with other technologies thus reducing total system costs.

An overview of technologies is presented in Tables S1–S3. The cost structures are described in Tables S4–S7, and the cross-sectoral flows are described in Tsiropoulos (2016). Technology development is rather uncertain and therefore assessed by scenarios in this study (section 'Rate of technology development and technology diffusion'; Figs S2 and S3).

Energy and chemicals demand

The final energy demand for electricity, heat, and the production volume of chemicals and aviation fuels is exogenously determined and specified for the Netherlands based on demand projections from EC (2003), Saygin *et al.* (2009), Chèze et al. (2011) and ECN (2015) as described in Tsiropoulos (2016). The final demand for road transport (liquid fuels, electricity) is endogenously calculated based on the assumed demand for vehicle-kilometres (van Vliet et al., 2011). While projections for key energy applications such as heat and electricity are relatively stable over time, for nonenergy uses future demand poses higher uncertainties. This in turn can determine to a large extent the deployment potential of biochemicals. In an additional scenario, we assume that the chemical sector follows a negative growth rate trajectory (section Other sensitivity scenarios). Demand projections are provided in Table S8 and Fig. S1.

Scenarios

Scenario analysis of 'if-then' propositions is shown to be useful to the extent that it provides insights that improve strategic management by better understanding uncertainties and robustness of decisions under a wide range of possible futures. These can be stirred by strategies but can also be influenced by uncontrolled variables (Schwartz, 1996; Moss et al., 2010).

Baseline scenarios (Base) in this study give a plausible indication on how the energy and nonenergy system may develop if no focus is placed on renewable energy and climate goals beyond 2020. We then deploy a set of scenarios that assess the effect of climate policies, namely a CO_2 tax that corresponds to meeting the 2 °C (OECD/IEA, 2015), in combination with bioeconomy strategies focused on the conversion and supply side.

Policy context of scenarios. Scenario parameters to 2020: To assess the cost-efficient contribution of biomass and other RES to CO_2 mitigation pathways, conversion technologies should compete on a level playing field. Scenarios that are technology-neutral avoid distortion caused by policies or support schemes (e.g. subsidies on specific technologies). However, up to 2020, binding policy goals at the EU level and national measures are already agreed and implemented. They include support to electricity, heat and road transport fuels up to 2020 and are assumed to be achieved in all scenarios. These include the following:

- the renewable energy share (14% for the Netherlands) and the biofuel target (10% including double-counting of biofuels from waste and residues, and contribution of renewable electricity in road transport by a factor 2.5; EU Renewable Energy Directive [RED]; EC, 2009);
- the retirement of old coal-fired power plants built before 1990 and wind deployment as part of national plans to meet the EU RED targets (SER, 2013);
- maximum of 25 PJ electricity produced by cofiring biomass in coal power stations (SER, 2013);

In addition, we assume an emission tax as part of the climate policies to 2020 (i.e. 15 € t CO_2^{-1} in 2020), based on the IEA-WEO 2015 New Policies Scenario. The tax is applied to emissions from all sectors (i.e. including transport and residential heat).

Scenario parameters from 2020 onwards. Beyond 2020, all sectors compete on a level playing field. Therefore, cost-competitiveness of secondary energy carriers and chemicals is the only determinant of technology deployment, biomass contribution to demand and achieved CO_2 emission reduction.

In line with the EU Intended Nationally Determined Contribution, mid-term emission reduction (2030) needs to reach 40% compared to 1990 (EC, 2015). We use CO_2 tax as the only policy instrument that stimulates emission reduction based on the IEA-WEO 2015 450 ppm scenario. Tax levels are 42 € t CO_2^{-1} in 2025 and 69 € t CO_2^{-1} in 2030 (OECD/IEA, 2015). The CO_2 tax applies to generated emissions in the Netherlands, as opposed to the carbon content of fossil feedstocks used. For biochemicals, this entails that only savings from energy use in industry and process emissions are affected by the tax and contribute to CO_2 emission reduction, as large part of the carbon in biomass feedstock remains embedded in final biochemical products. Emission savings from biochemicals may occur outside the geographical boundaries of the Netherlands, at their end-of-life in demand regions. However, such savings are not assessed in this study. Similarly, GHG emission savings may be achieved in sectors other those of the energy system (e.g. reduction in methane emissions in agriculture). However, these are excluded from the scope of this study due to the dominant role of CO_2 emissions in climate change in the long-term and their direct relationship with the energy system (Vuuren et al., 2016).

We assess fossil fuel prices and climate policy scenarios separately. The policy context beyond 2020 as described above is used in the *reference (Ref), biomass cost-supply (LowBio, HighBio)* and *fossil fuel price scenarios (LowFos, HighFos)* (section 'Scenario definitions'; Table 1).

Scenario definitions. The emission mitigation pathways are based on key strategies for development of biomass production systems across the supply chain, from feedstock to conversion.

Rate of technology development and technology diffusion (LowTech, HighTech). Technology development based on learning and subsequent cost reductions can considerably influence the competitiveness of biomass conversion technologies. Technology costs decline by a constant factor with each doubling of cumulative capacity (BCG, 1968). However, this occurs at a global level, which is outside of the regional scope of this study. Incremental improvements over time such as in efficiency may also affect conversion costs. These factors are not endogenized in MARKAL-NL-UU. Therefore, we capture the uncertainty of technical progress on cost reduction in biomass conversion technologies and the role of BECCS to 2030 using two technology pathways that follow *low (LowTech)* and *high technology (HighTech) development* progress (Figs S2 and S3). These pathways vary technology parameters and assume different learning rates for biomass conversion. More specifically, the two scenarios differ in technology portfolio, rate of incremental improvements, year of technology availability and scales as described in detail in Tsiropoulos (2016). *LowTech* assumes that little support is provided to conversion technologies by means of stimulating research and development (R&D),

Table 1 Overview of the scenarios assessed in this study. Baseline scenarios assume CO_2 tax up to 2020. All other variants assume CO_2 tax up to 2030

		Scenario variable: biomass cost-supply				
		Low-cost biomass	Reference-cost biomass	High-cost biomass		
Baseline (no CO_2 tax beyond 2020)	High technology development	n.a.*	HighTechBase	n.a.*	Reference fossil fuel price	Baseline (no CO_2 tax beyond 2020)
	Low technology development	n.a.*	LowTechBase	n.a.*	Reference fossil fuel price	
Scenario variable: technology development	High technology development (HighTech)	n.a.*	HighTech(RefBio_ HighFos)	n.a.*	High fossil fuel price	Scenario variable: fossil fuel price
		HighTech (LowBio_RefFos)	HighTechRef†‡	HighTech(HighBio_ RefFos)	Reference fossil fuel price	
		n.a.*	HighTech(RefBio_ LowFos)†	n.a.*	Low fossil fuel price	
	Low technology development (LowTech)	n.a.*	LowTech(RefBio_ HighFos)	n.a.*	High fossil fuel price	
		LowTech (LowBio_RefFos)	LowTechRef†‡	LowTech(HighBio_ RefFos)	Reference fossil fuel price	
		n.a.*	LowTech(RefBio_ LowFos)†	n.a.*	Low fossil fuel price	

*Combination of scenarios is not assessed in the present study.
†Scenario variables used to assess the sensitivity of the biochemical sector in low chemical demand and delayed decommissioning of steam crackers.
‡Scenario variables used to assess the impact of complete closure of coal-fired power plants on CO_2 emissions.

fast deployment of 1st-of-a-kind plants, support to technologies to go beyond the valley of death, rapid scale up and so forth. On the other hand, *HighTech* assumes that these conditions are met through coordinated action of business, industry and government.

To avoid supply of all demand in the transport sector by a single technology, we apply market constraints on second-generation technologies based on de Wit *et al.* (2010). Individual technologies can supply up to 10% of demand in 2030.

Biomass cost-supply (LowBio, HighBio). Low-cost biomass: The extra-EU and intra-EU cost-supply of biomass in baseline and reference scenarios are conservative for two reasons. Firstly, the price of extra-EU wood pellets is based on mill-gate costs of around 6 € GJ^{-1}. However, studies indicate that these can be as low as 3.9 € GJ^{-1} (Uslu *et al.*, 2008) or 2.3–3.1 € GJ^{-1} by 2030 (Batidzirai *et al.*, 2014) when low-cost biomass is used for pellets. These could be achieved, for example, using surplus or abandoned agricultural land for energy crops and intensification of agricultural productivity (de Wit & Faaij, 2009; Wicke, 2011). Secondly, the cost-price of intra-EU biomass delivered to the Netherlands, as assumed in this study, is conservative because transport costs are estimated based on wood chip logistics (section 'Energy supply'). The cost-competitiveness of biomass chains can improve if efforts focus on biomass

densification to reduce transport and handling costs (e.g. torrefaction and pelletization). Such efforts are assumed to take place in the *LowBio* scenario resulting to lower upstream cost-supply of solid biomass. The approach we used to estimate low-cost biomass supply curves for extra-EU and intra-EU biomass resources is presented in detail in S1.2 in Appendix S1.

High-cost biomass. As a consequence of worldwide increase in biomass demand, it is expected that global biomass trade will continue in the future, thereby allowing cost-efficient distribution of biomass from supply to demand regions. However, it is uncertain how trade and markets will develop. If the EU is the only region that supports bioeconomy developments, then EU demand regions like the Netherlands will have access only to intra-EU resources. The *HighBio* scenario assumes that extra-EU import of biomass is not possible, which decreases the total potential by 450 PJ compared to the reference. This effectively leads to increased costs of solid biomass as a large potential below 7.5 € GJ^{-1} for solid biomass becomes unavailable.

The cost-supply curves of solid biomass used across the different scenarios are presented in Fig. 1. The cost-supply curves exclude energy maize, solid waste, fuelwood, landscape wood and road-side grasses, as unlike the solid biomass feedstocks included in Fig. 1, they are linked with specific end-use

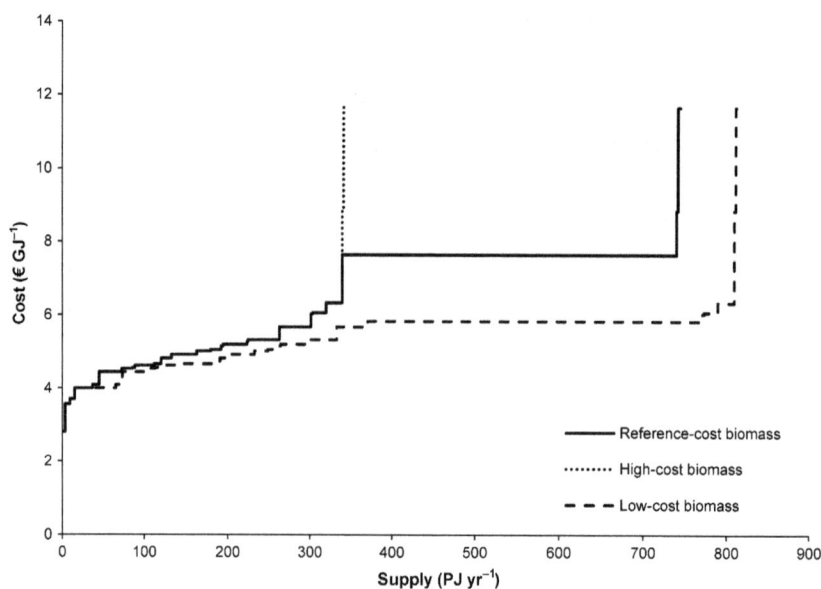

Fig. 1 Solid biomass cost-supply curves in the different scenarios of this study in 2030.

applications (e.g. energy maize with codigestion, solid waste with energy incineration and energy recovery and so forth).

Fossil fuel prices (LowFos, HighFos). Fossil fuel prices are uncertain, are subjected to change over time (OECD/IEA, 2015) and are key determining, but uncontrolled factor of the success or failure of bioeconomy development strategies and other RES. To capture the uncertainty that such variables may have on emission mitigation, we deploy the following scenarios:

- *LowFos*: To obtain insights into the magnitude of biomass and other RES development in an unfavourable environment, we use fossil fuel prices reported in IEA-WEO 2015 Low oil price scenario (OECD/IEA, 2015). Compared to the New Policies Scenario, these prices are lower approximately 35% for oil, 20% for natural gas and 6% for coal compared to the reference fossil fuel prices.

- *HighFos*: To obtain insights into the magnitude of biomass and other RES deployment under favourable conditions, a 50% higher fossil fuel prices is assumed compared to those of the New Policies Scenario reported at IEA-WEO 2015 (OECD/IEA, 2015).

The variation of fossil fuel prices is presented in Fig. 2.

Other sensitivity scenarios

Several drivers, such as contraction of the economy or competition from other regions (Broeren *et al.*, 2014), may saturate or even decrease the production demand for chemicals assumed for the Netherlands over time. Future reduction in the demand for chemicals in combination with no decommissioning of existing steam cracking capacity in the Netherlands is assessed as an additional sensitivity scenario. A 10% reduction in demand for chemicals in 2030 compared to 2010 is assumed based on the reduction in the size of the Dutch petrochemical industry according to van Meijl *et al.* (2016).

Furthermore, in the EU, several governments consider reducing support on, divesting in or even dismantling coal-fired power plants as this may compromise the diffusion of other RES and CO_2 emission reduction goals (Nicola & Andresen, 2015; Yeo, 2015; Pieters, 2016; Sterl *et al.*, 2016). We assess this possible future in a scenario, which assumes that electricity from coal cannot be produced in the Dutch energy system after 2020.

Other studies show that the role of biomass in the energy system varies, depending on the electricity mix. With exogenously determined electricity supply from other RES ranging between 17% and 80% and strong climate policy, biomass use for power generation in Europe ranges between 2.5% and 33% (0.4–2.1 EJ) of total fuel use in 2050 (Brouwer *et al.*, 2016), without, however, taking competition by other sectors into account. Furthermore, improvements on energy efficiency could reduce heat demand in the industry and residential sector. This suggests that a large number of additional scenarios can be defined to investigate the sensitivity of the system and competition for biomass, which, however are excluded from this study.

Indicators and overview of the modelling framework

For each scenario in Table 1 we assess the following:

- the final production output from RES per sector in 2030. For the energy sectors (electricity, heat, fuels), production output is expressed in final energy terms, while for the chemical sector it is estimated based on the lower heating value of biochemicals;

- the contribution of renewable energy on the total final energy produced by each sector;

- the renewable energy share (i.e. excluding the nonenergy use of the chemical sector);

- biomass demand that reflects total biomass consumption in primary energy terms, same as in Tsiropoulos (2016);

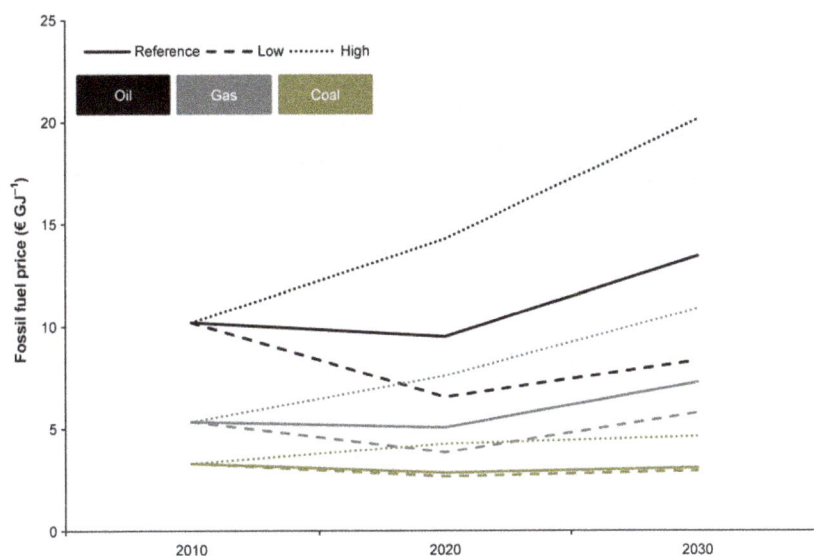

Fig. 2 Fossil fuel price in the reference scenarios and their fossil fuel price variants.

- the direct CO_2 emissions in the Netherlands related to the supply of energy services in all sectors and process emissions by industrial activities. Direct CO_2 emissions are those emitted in the Netherlands; they exclude emissions from production or extraction and transport of resources (biomass, fossil) to the Netherlands, consistent with IPCC (2006);

- total annual system costs in 2030, compared to *HighTechBase*.

Figure 3 presents an overview of the framework used in this study.

Results

Final production from biomass and other RES, and their contribution to each sector are shown for the reference scenario in combination with the two technology development variants by bars (*HighTechRef, LowTechRef*), while the range of outcomes based on the biomass cost-supply and fossil fuel price scenarios is indicated with whiskers (Fig. 4). Outcomes for the baseline situation in combination with the technology development scenarios are presented with markers (*HighTechBase, LowTechBase*). Results are presented for 2030.

For the indicators renewable energy share (Fig. 5), biomass consumption (Fig. 6) and CO_2 emissions (Fig. 7), we present the influence of technology development in reference conditions (i.e. CO_2 tax) in comparison with the baseline for the period 2010–2030. As the results for the two technology development variants do not differ significantly in 2020, we only show the 2010–2030 trajectory of the *HighTech* scenario for the biomass cost-supply and the fossil fuel price variants. For comparison, we include results for the *LowTech* scenario in

2030. Apart from the range due to the variation of scenario parameters, results also include the consumption under baseline and reference conditions. For all scenarios, the difference of their total annual system costs from *HighTechBase* is plotted against the corresponding difference in total direct CO_2 emission reduction in 2030. Results with sector-specific assumptions are presented in section 'Other scenarios' (Figs 10 and 11).

All results per scenario and sector are presented in Tables S10–S12.

Renewable energy

Final production from renewable resources was estimated to be between 460 and 510 PJ in 2030 and does not differ significantly between *HighTechRef* and *LowTechRef* (Fig. 4a), thereby indicating that under reference assumptions on CO_2 tax, biomass cost-supply and fossil fuel price, technology development is not the only driver for cost-efficient supply of bioenergy and other renewable energy. Other drivers include cost-supply of biomass and fossil fuel prices as indicated by the whiskers in Fig. 4 (range 230–745 PJ). The renewable energy output (Fig. 4b) corresponds to a 23–24% share on final energy consumption excluding chemicals or to a 18–20% share on final energy consumption including chemicals. More than two-thirds (73–79%) of renewable energy output is attributed to biomass (Fig. 4a, b), which is higher than the anticipated contribution of biomass in the energy system based on the EU RED targets for 2020 (Rijksoverheid, 2010; Stralen *et al.*, 2013). This is primarily due to the ambitious climate policy assumed in this study beyond 2020 (by means of high CO_2 tax). In addition, as in this study the CO_2 tax also applies to

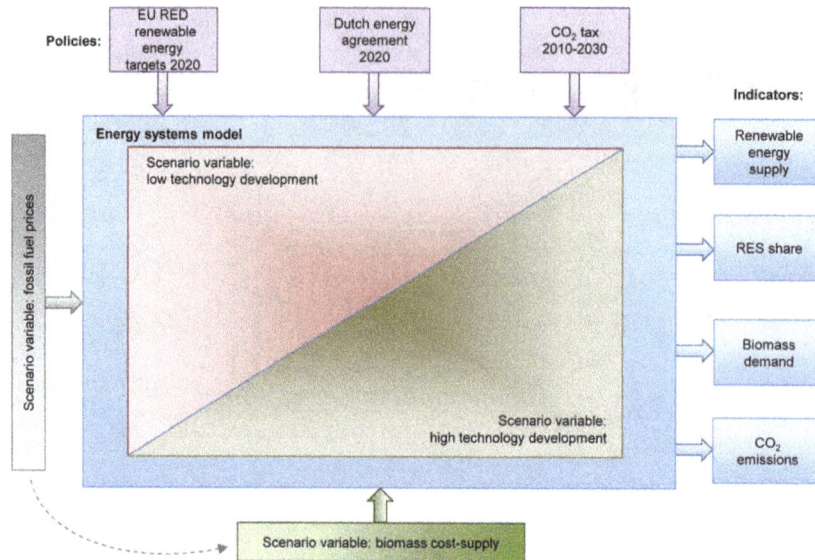

Fig. 3 Modelling framework used in the present study.

emissions by transport, it results to high biofuel output especially in *HighTechRef*. The remainder mainly represents renewable electricity by other renewable resources (wind and solar).

At a sector level and in absolute terms, technology development scenarios affect the supply of industrial biomass heat, biofuels and biochemicals (bars in Fig. 4a). These are also found to be the sectors with the largest bio-based output. Under *LowTech* assumptions, it is more cost-effective to supply solid biomass in industrial biomass boilers and produce heat as biofuel production technologies for road transport and RJF are not cost-competitive. Under *HighTech*, the reverse occurs as biomass conversion technologies to biofuels for road transport and aviation become cost-competitive. The trade-off between biomass heat and biofuel output is also observed in Tsiropoulos (2016) when using different policy assumptions beyond 2020. Electricity from biomass remains small (20–50 PJ; primarily from biorefineries, cofiring and municipal solid waste incineration), as wind becomes the key supplier of renewable electricity.

Biochemicals are produced even under *HighTechBase* and *LowTechBase* as a result of the retirement of steam-cracker capacity (20–50 PJ; 5–10% of the sector's output Fig. 4). While the CO_2 tax only affects the process emissions of the chemical sector, Fig. 4 shows that in *HighTechRef* the output of biochemicals almost doubles (about 100 PJ; 17%) compared to *LowTechRef*. This is a result of multi-output technologies that produce both chemicals and road transport fuels, with the latter being affected by the CO_2 tax.

The electricity sector is most sensitive to assuming high fossil fuel prices, which lead up to a factor 2.5

increase in electricity from other RES (primarily wind power) compared to reference scenarios. Regarding other scenario variants (e.g. low fossil fuel prices), electricity from other RES is not affected, as most of the wind capacity is installed by 2020 to confirm with Dutch EU RED targets.

Regarding electricity from biomass, at a sector level scenario variants, namely biomass cost-supply and fossil fuel prices, do not have significant influence, as ranges are found to be comparable with those of technology development scenarios (whiskers in Fig. 4a). Brouwer *et al.* (2016) suggest that low biomass prices could place electricity generation from biomass earlier in the merit order than electricity from natural gas; biomass could have a larger role in the electricity sector. Similarly, other assumptions, such as higher CO_2 emission taxes, or higher targets of RES for low-carbon power systems could lead to different outcomes regarding other RES (Brouwer *et al.*, 2016).

While in absolute terms, final energy supply from biomass per sector is comparable across all scenario variants (Fig. 4a), in relative terms, its contribution to the sector's final energy varies (Fig. 4b). Most notable is the contribution of biofuels to road transport fuels, which goes beyond 60% under *HighTech(RefBio_High-Fos)*. This occurs due to the increased biofuel supply (20% higher than in *HighTechRef*) and due to reduced fuel demand by the sector (roughly 1/3 or 130 PJ decrease compared to *HighTechRef*) as more efficient vehicles are deployed. These are primarily wheel motor hybrid vehicles with 76% higher efficiency compared to regular petrol cars found in reference scenarios (i.e. 927 compared to 526 km driven GJ^{-1}; van Vliet *et al.*, 2011). While electrification of transport is included in the

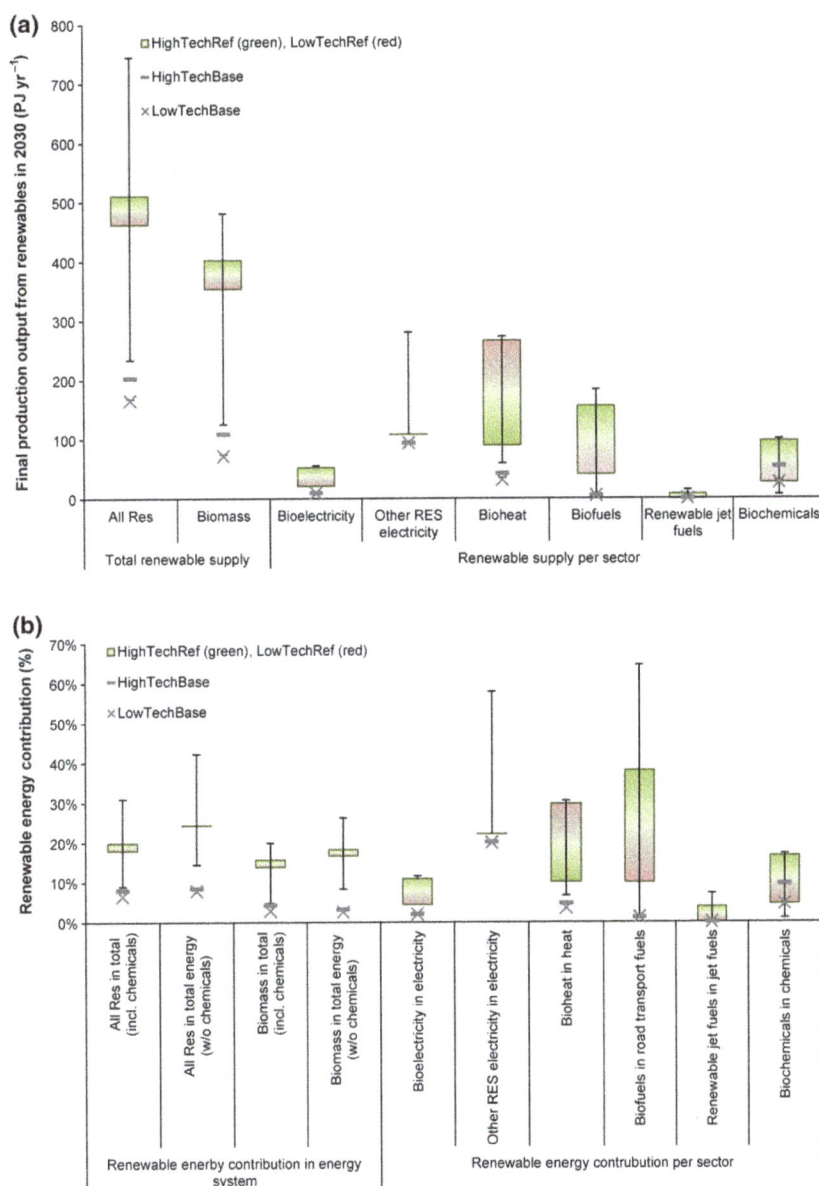

Fig. 4 (a) Final renewable energy and nonenergy supply renewable resources in the energy system and per sector and (b) contribution of renewable energy and nonenergy in the energy system and per sector (remainder is fossil fuels) in the Netherlands in 2030. Bars indicate ranges of reference scenarios, whiskers indicate range of biomass cost-supply and fossil fuel price scenarios (see scenario descriptions in Table 1).

technology portfolio, model outcomes show that for the energy system of the Netherlands by 2030, biofuel supply in *HighTech* scenarios and efficient hybrid vehicles (in *HighTech(RefBio_HighFos)*) is a more cost-effective option, partly owing to high costs of electric vehicles. This suggests that to increase electrification in transport, other support instruments are required (e.g. subsidies or tax exemptions).

The market constraint on individual second-generation biofuel technologies (section 'Rate of technology development and technology diffusion') limits the

production output of FT-fuels in *HighTechRef* and *HighTech(LowBio_RefFos)*. In *HighTech(RefBio_HighFos)*, it limits second-generation ethanol production due to the deployment of hybrid petrol engines for which ethanol is the substitute. Across scenarios, the share of first-generation biofuels over the total transport fuel supply is 1–13% (5–66 PJ) and 1–12% (5–55 PJ) in *LowTech* and *HighTech* scenario variants, respectively. In *LowTech* scenarios, all biofuels are supplied to road transport. In *HighTech* scenarios, 4–8 PJ of hydrotreated used cooking oil is supplied to aviation,

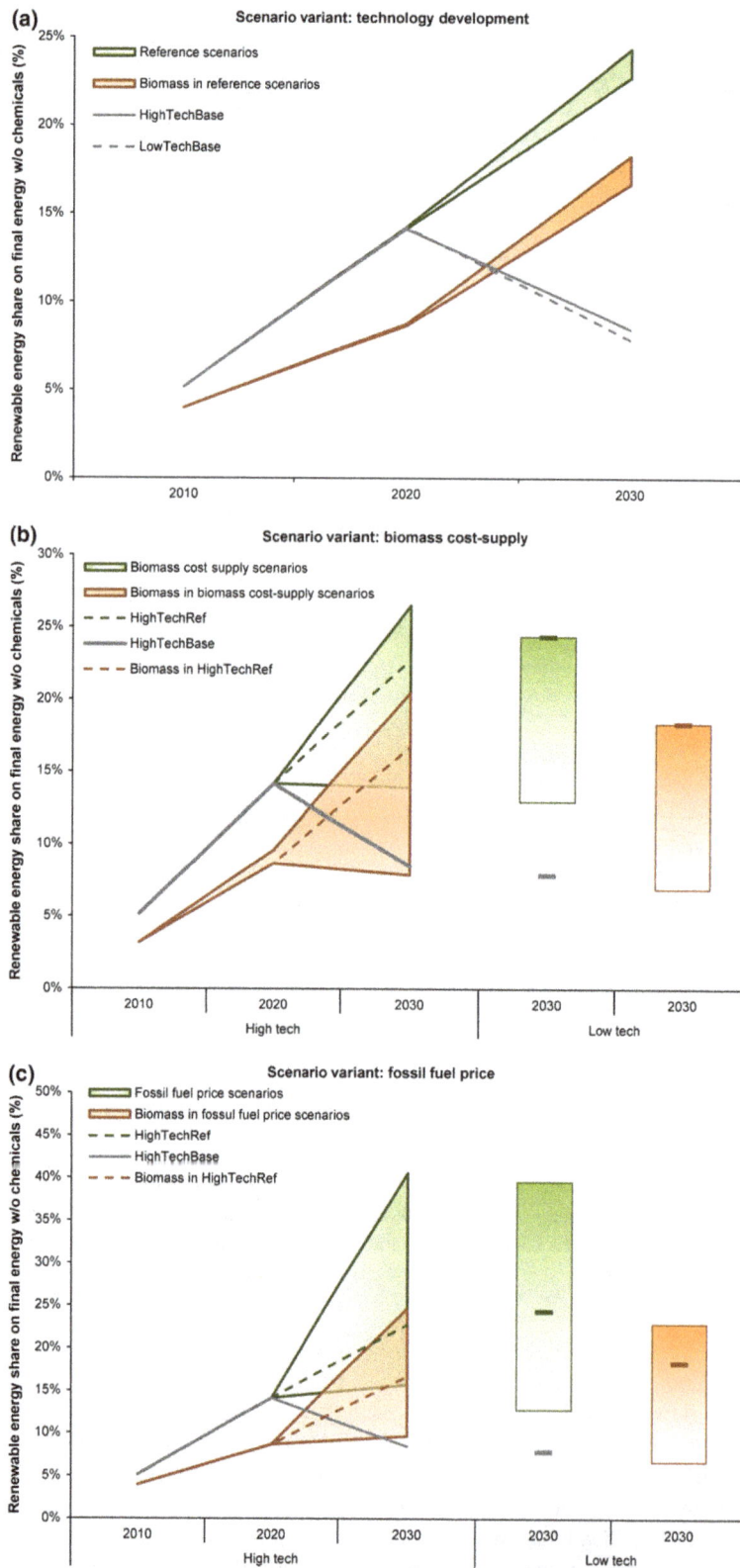

Fig. 5 Renewable energy share on final energy and biomass contribution in the Netherlands in 2010–2030 under high technology development compared to low technology development in 2030 for (a) technology development (b) biomass cost-supply and (c) fossil fuel price scenarios. In the Low tech variant, grey markers indicate the baseline and green markers the reference result (see scenario descriptions in Table 1).

while the remainder of first-generation biofuels are supplied to road transport. The share of second-generation biofuels over the total transport fuel demand is 0–3% (0–15 PJ) and 10–29% (60–136 PJ) in *LowTech* and *HighTech* scenario variants, respectively. In *HighTech* variants, 6–8 PJ of FT-RJF are supplied to aviation, while the remainder of second-generation biofuels are supplied to road transport.

In *HighTech* scenarios, while the output of RJF (7–14 PJ) and biochemicals (22–100 PJ) to total final energy supply is relatively small, their contribution to their sectors is up to 7% and 17%, respectively. This is quite significant considering that today's output is limited and that within a 15-year timeframe, such developments can obtain a large market share.

Figure 5 shows the renewable energy share and the contribution of biomass in more detail. Biomass contributes more than two-thirds to the share of renewable energy in reference scenarios (Fig. 5a) and biomass cost-supply scenarios (Fig. 5b). Access to low-cost biomass increases the renewable energy share and the contribution of biomass by approximately 17% but only under *HighTech(LowBio_RefFos)*. Under limited access to low-cost biomass, the contribution of renewable energy in 2030 is maintained similar to 2020 levels, that is 14% share from RES, where biomass supplies approximately 50% of renewable energy. Under *LowTech(LowBio_RefFos)* access to low-cost biomass does not increase the RES share or its contribution. Restricted access to low-cost supply coupled with low technology development can pose barriers to cost-efficient deployment of biomass in the Netherlands.

Fossil fuel price variation leads to wider ranges (Fig. 5c). First and foremost, under high fossil fuel prices the renewable energy share almost doubles compared to reference scenarios; biomass contribution does not follow the same relative growth due to the increase in electricity from wind turbines in the energy system. Under *HighTech(RefBio_LowFos)* and *LowTech(RefBio_LowFos)*, RES and biomass contribution remain in 2020 levels for, similar to *HighTech(LowBio_RefFos)* and *LowTech(LowBio_RefFos)*.

Biomass consumption

Early in the time horizon (2020), biomass consumption driven by technology development is relatively small, at approximately 200 PJ in both technology development scenarios and is comparable to baseline projections. Nevertheless a factor 2 growth is observed compared to 2010. By 2030, due to the CO_2 tax, biomass consumption is 330–460 PJ higher than the baseline (Fig. 6a).

Access to low-cost biomass shows that additional 100 PJ are used in the energy system (Fig. 6b), however, only under *HighTech(LowBio_RefFos)*. Total biomass consumption exceeds 700 PJ, which as seen in Fig. 5b also increases by roughly 4% the contribution of RES and biomass to the energy system. On the other hand, high biomass costs can reduce consumption levels significantly (to slightly above 300 PJ) even *HighTech(High-Bio_RefFos)*. The range found between *LowTech (HighBio_RefFos)* and *LowTech(LowBio_RefFos)* by 2030 is significantly smaller than the one found between *High-Tech(HighBio_RefFos)* and *HighTech(LowBio_RefFos)*, that is 240 PJ compared to 400 PJ. *HighTech* leads to growth in biomass consumption but *LowTech* leads to fairly constant consumption levels between 2020 and 2030. The above indicates that *LowTech* could impede long-term bioeconomy growth as indicated by biomass consumption.

Biomass consumption is also highly sensitive to the assumed level of fossil fuel prices to 2030 (Fig. 6c; reference fossil fuel prices: oil 13.4 € GJ^{-1}, natural gas 7.3 € GJ^{-1}, coal 3.1 € GJ^{-1} in 2030, see Fig. 2). For *Low-Tech* and *HighTech* scenarios, biomass consumption is more sensitive to low than high fossil fuel price assumptions. A 50% increase in fossil fuel prices leads to approximately 25% increase in biomass consumption in *HighTech(RefBio_HighFos)* and *LowTech(RefBio_High-Fos)* compared to *HighTechRef* and *LowTechRef*, respectively (indicated in Fig. 6c by the area above the dotted line and the upper marker in high and low technology development, respectively). Low fossil fuel prices lead to 50–60% reduction of biomass consumption found in reference scenarios. *HighTech(RefBio_HighFos)* and *High-Tech(RefBio_LowFos)* lead to similar consumption levels with *HighTech(LowBio_RefFos)* and *HighTech(HighBio_R-efFos)*, respectively (Fig. 6b, c). Therefore, even under unfavourable conditions induced by low fossil fuel prices, high technology development scenarios demonstrate small but stable growth in biomass consumption. On the contrary, *LowTech(RefBio_HighFos)* consumes maximum 560 PJ under most favourable conditions induced by high fossil fuel prices, which are comparable to *HighTechRef* (the upper range of the bar is comparable to the upper range of the dotted line in Fig. 6c).

CO_2 emissions

In Fig. 7a, it is shown that the CO_2 tax leads to emission reduction compared against projected baseline emissions in the range of 35–43% for *LowTechRef* and *HighTechRef*, respectively. Compared to *LowTechRef*, the additional emission reduction in *HighTechRef* is 15 Mt CO_2. The decreasing trend in emissions is steeper beyond 2020 as a result of higher CO_2 tax levels and

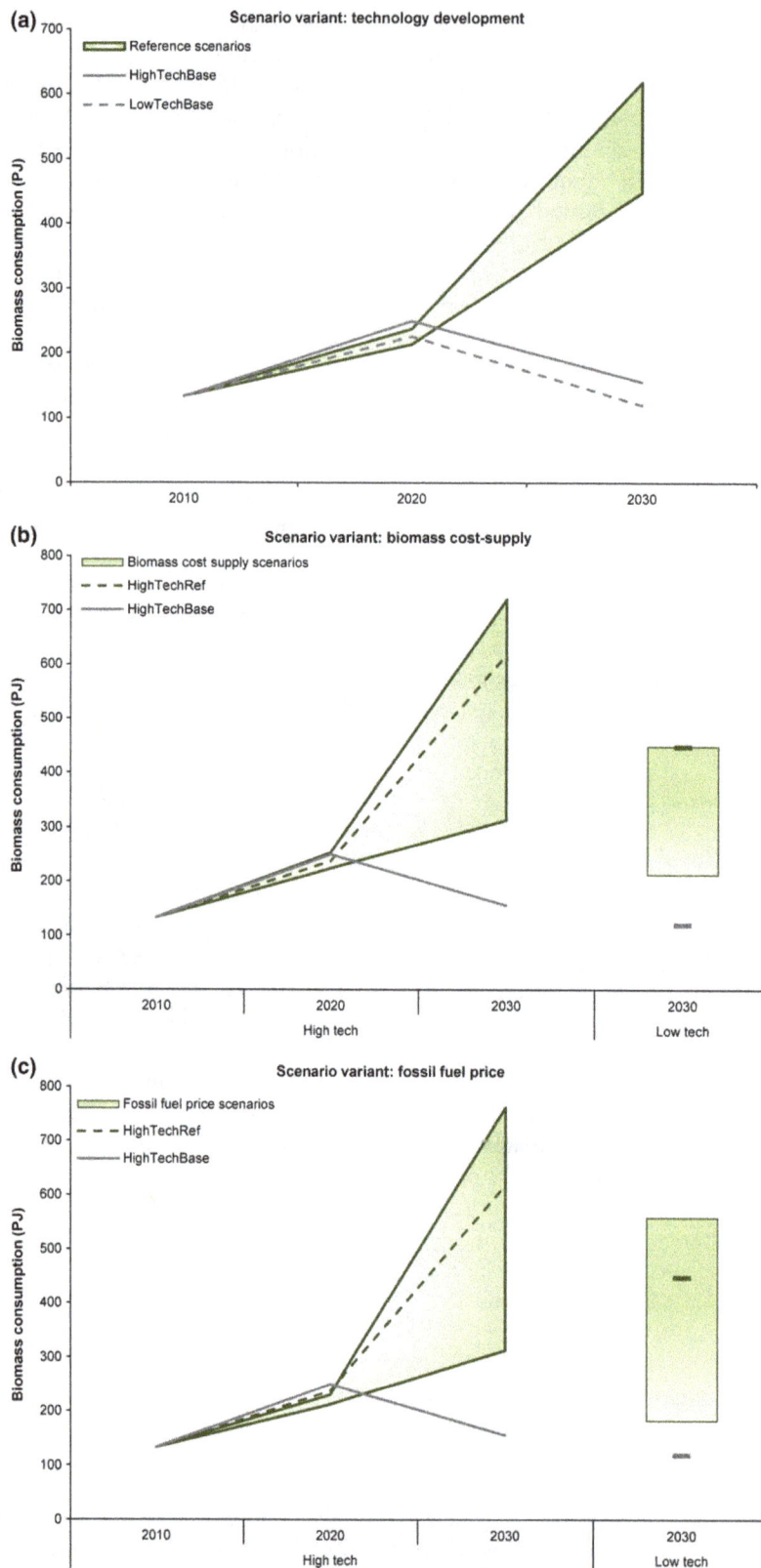

Fig. 6 Biomass consumption in the Dutch bioeconomy in 2010–2030 under high technology development compared to low technology development in 2030 for (a) technology development (b) biomass cost-supply and (c) fossil fuel price scenarios (see scenario descriptions in Table 1). In the Low tech variant, grey markers indicate the baseline and green markers the reference result.

additional technological options. Nevertheless, even under *HighTechRef* additional 20 Mt CO_2 emission reduction is required to reach the 40% emission reduction target compared to 1990. *HighTech(LowBio_RefFos)* leads to additional 5 Mt CO_2 reduction, partly bridging the gap with the target (Fig. 7b). As *LowTech(LowBio_RefFos)* does not utilize additional low-cost biomass compared to *LowTechRef* (Fig. 6b), no additional emission reduction is achieved (Fig. 7b). *HighTech(HighBio_RefFos)* leads to emission reduction levels comparable with those achieved in *LowTechRef* (i.e. approximately 35% compared to the baseline or direct CO_2 emissions in the range of 120 Mt CO_2). Similar reduction is observed in *HighTech(RefBio_LowFos)* (Fig. 7c). However, under such circumstances, the distance to the 40% emission reduction target is 30% (or 40 Mt CO_2). To remain in cost-efficient emission reduction trajectories, *HighTech* seems to be a no-regret solution even under unfavourable conditions shaped by high-cost biomass or low fossil fuel prices as they offer significant potential for deeper emission reduction. More specifically, results indicate that under *HighTech(RefBio_HighFos)*, the 40% emission reduction target is reached. In *LowTech(RefBio_HighFos)*, however, CO_2 mitigation is 12 Mt CO_2 behind the target (Fig. 7c). Note that these emissions exclude those that occur outside the geographical boundaries of the Netherlands from production and transport of biomass, land-use change, extraction and transport of fossil fuels and jet fuels (section 'Indicators and overview of the modelling framework').

Figure 8 shows the amount of carbon captured and stored by CCS and BECCS across the different scenarios (19–41 Mt CO_2). The contribution of CCS and BECCS in emission reduction is significant (42–60% compared to the baseline). The remainder of emission reduction is primarily achieved through biomass (20–40 Mt CO_2), as with the exception of high fossil fuel price scenarios, the capacity of wind and other RES does not increase significantly compared to the baseline.

Carbon capture and storage is stimulated by the high CO_2 tax while in baseline scenarios, no CCS is deployed. The key difference across *HighTechRef* and *LowTechRef* is the deployment of BECCS in gasification technologies that supply FT-fuels to the transport sector. These technologies are assumed not to be available in *LowTech* scenarios. Carbon capture by the power sector is primarily associated with retrofitted coal-based power plants. It represents more than 65% of the emissions captured and stored by the sector. The remainder is associated with gas-based capacity and is similar across the technology development scenarios. In *HighTech* scenarios, BECCS represent 16–50% of the emissions captured and stored.

In scenarios that assume high fossil fuel prices, CCS in *LowTech(RefBio_HighFos)* and in addition BECCS in *HighTech(RefBio_HighFos)* represent 10–20% of the emission reduction achieved compared to the baseline (10–20 Mt CO_2 is stored). In these scenarios, significant emission reduction is achieved through other RES, as the output of wind electricity increases by approximately a factor 3, compared to the baseline (emission reduction from bioenergy and other RES is 70–75 Mt CO_2). In addition, less coal capacity is projected to be used. Due to the decrease in demand for transport fuels by deployment of efficient vehicles the transport sector also contributes to emission reduction.

System costs

We compare total annual system costs and total direct CO_2 emissions in 2030 between all scenario variants and *HighTechBase* (Fig. 9). Total system costs in most scenarios increase from 0.6 to 13.1 billion € yr^{-1} compared to *HighTechBase*. An exception are *LowTech(RefBio_LowFos)* and *HighTech(RefBio_LowFos)* that show lower costs of about 6.5 billion € yr^{-1}. Annual system costs are most sensitive to fossil fuel price variation. A 35% decline in oil prices (section 'Fossil fuel prices'; Fig. 2) reduces annual system costs by about 9% and a 50% increase in oil prices increases annual system costs by approximately 18–19% in 2030, compared to *HighTechBase* (Table S12). *HighTech* scenarios consistently show lower system costs and CO_2 emissions in 2030 and cumulative system costs and CO_2 emissions over the period 2010–2030 (Table S11) when compared to their *LowTech* scenario counterparts. This illustrates that *HighTech* is a no-regrets solution also when costs are taken into account. Note that total system costs do not include technology development costs (e.g. R&D, 1st-of-a-kind plant, production at low capacity factors or high downtime) nor investment and dismantling lead-times and costs. As technology development occurs in larger regions and cannot be allocated to the Netherlands, these costs are excluded from both *LowTech* and *HighTech* scenarios and are not expected to affect the relative conclusions drawn in this study. Including such costs requires a modelling framework with a wider regional scope. In absolute terms, however, should investment and dismantling costs be included they would increase total system costs.

Other scenarios

Low demand for chemicals. Figure 10 shows a decline in demand for chemicals over time, which in combination with delayed decommissioning of old steam cracking capacity in the Netherlands beyond 2030 affects the production output of biochemicals. This is noticed early in

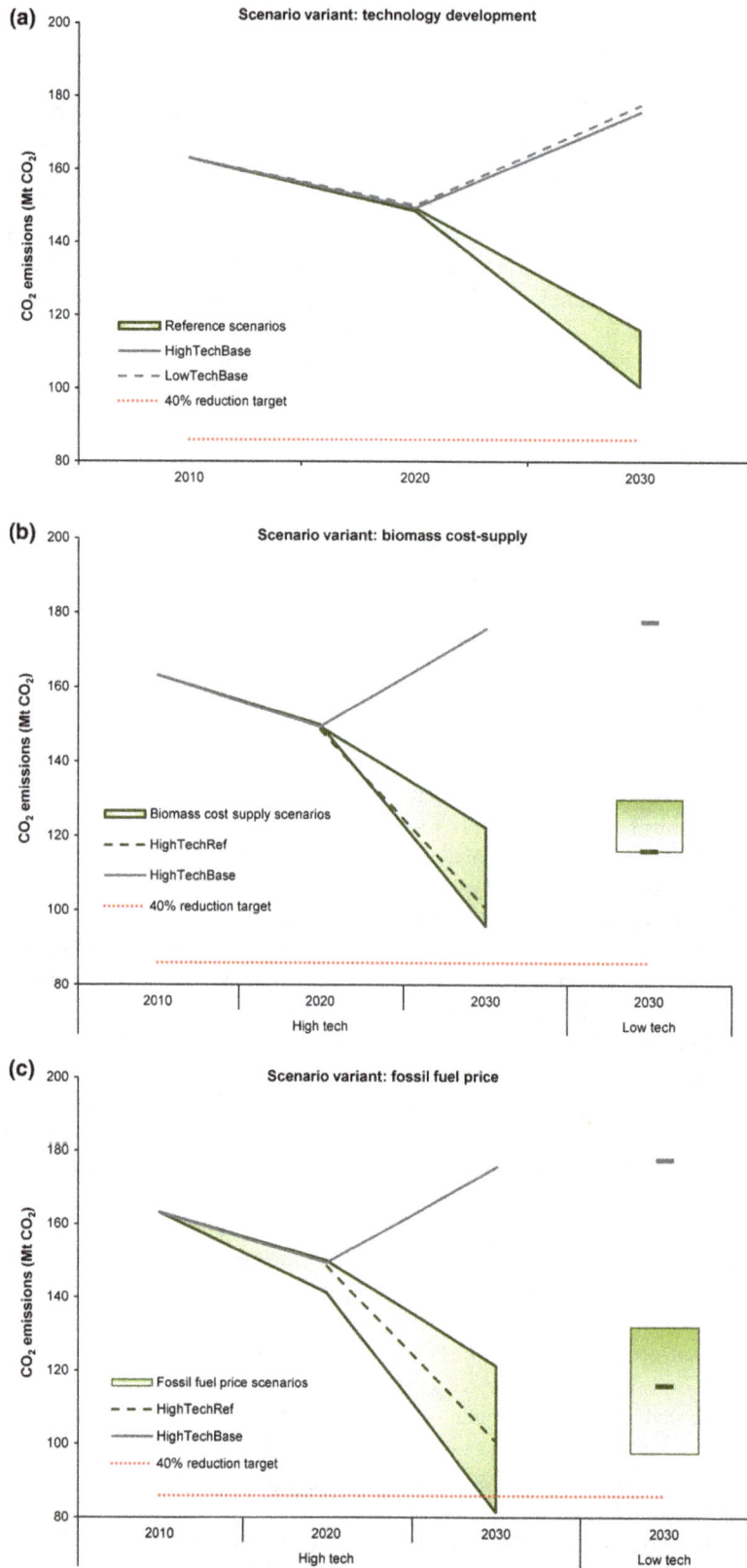

Fig. 7 CO$_2$ emissions in the Netherlands in 2010–2030 under high technology development compared to low technology development in 2030 for (a) technology development (b) biomass cost-supply and (c) fossil fuel price scenarios (see scenario descriptions in Table 1). In the Low tech variant, grey markers indicate the baseline and green markers the reference result.

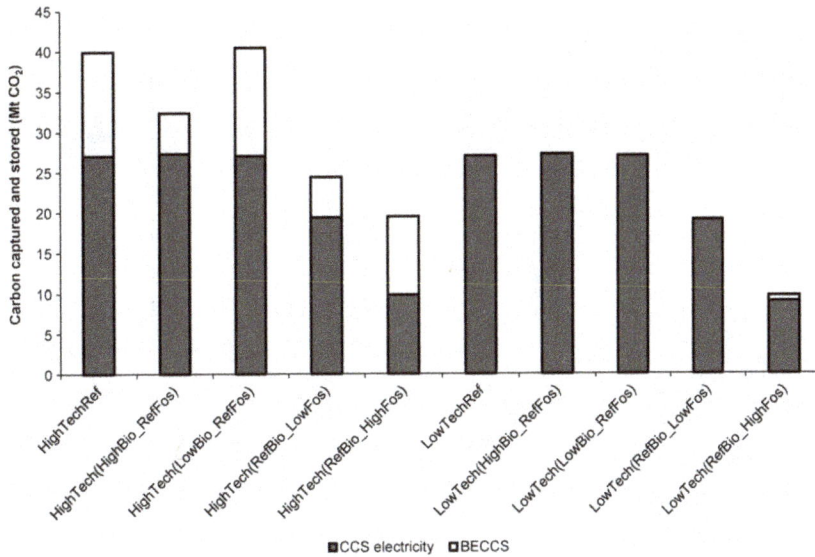

Fig. 8 Carbon captured and stored across different scenario variants in the Netherlands in 2030.

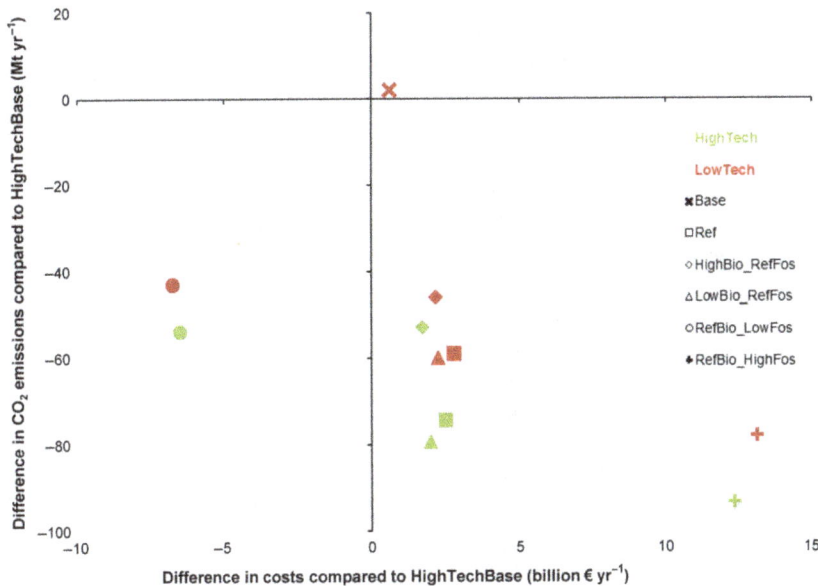

Fig. 9 Difference of total annual system costs and total direct emissions in 2030 between HighTechBase and the other scenario variants.

the time horizon (2020), when under *LowTech* assumptions, no production of biochemicals takes place, and under *HighTech* assumptions, the production output is reduced by 75% compared to *HighTechRef*. The difference in production output between scenarios becomes smaller by 2030, and lower demand for petrochemicals leads to a 16–37% reduction of output in *LowTech* and *HighTech* compared to their reference. However, assuming low fossil fuel prices creates an uncompetitive environment for biochemicals throughout the modelling period. This may be also an outcome of the limited

number biochemicals that are assumed in this study combined with the fact that the CO_2 tax does not affect nonenergy use (i.e. biochemicals do not receive any credit for their bio-based carbon content). As Fig. 11 shows, assuming lower demand for chemicals does not affect the direct CO_2 emissions of the Dutch energy system. Compared to their reference scenarios, the low chemical demand scenarios lead to 4–5% lower CO_2 emissions, primarily due to less process energy emissions (electricity, heat) as a result of decrease in industrial demand for energy.

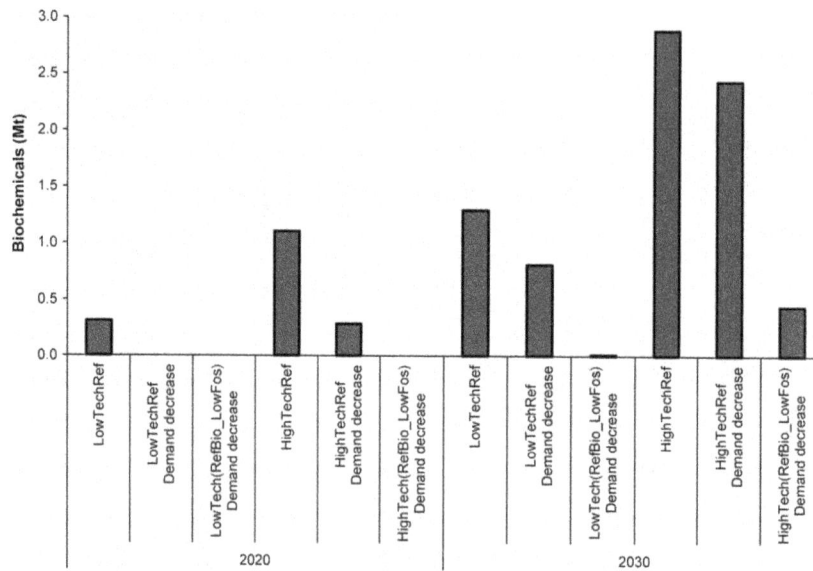

Fig. 10 Biochemical output in reference and low fossil fuel price scenarios assuming different growth rates for the chemical industry in the Netherlands in 2020–2030.

Decommissioning of coal-based power capacity. The results show that decommissioning coal-fired power stations in the Netherlands after 2020 increases wind electricity by 55–75% (48–65 PJ; offshore wind turbine capacity increase of 5.2 GW_e and 3.9 GW_e in low and high technology development, respectively, compared to *LowTechRef* and *HighTechRef*, respectively) and 13–18% (30–39 PJ) in natural gas-based electricity. By 2030, offshore wind turbines are expected to become more cost-efficient than other electricity production options leading to higher renewable energy share and contribution in the electricity sector. Deployment of onshore wind reaches constraint levels (8 GW_e) across all scenarios with high CO_2 tax by 2030. Despite the significant deployment of wind power, direct CO_2 emissions remain at levels comparable with *LowTechRef* and *High-TechRef* because CCS combined with coal power plants is no longer an available mitigation option (Table S12). Overall the total carbon removed and stored by CCS is lower by 15 Mt CO_2 and 12.5 Mt compared to *LowTechRef* and *HighTechRef*, respectively. Decommissioning coal power plants leads to additional system costs of 6.7–9 M€ yr^{-1} from 2020 onward and increases CO_2 mitigation costs by 12% and 14% in *HighTechRef* and *LowTechRef*, respectively (Tables S12 and S13).

Discussion

This study compared multiple scenario outcomes of an energy systems model to gain insights in CO_2 emission reduction that can be achieved by renewable energy, CCS and BECCS deployment in the energy system

when driven by cost competition with fossil fuel alternatives. We used the MARKAL-NL-UU model, which includes a representation of modern and emerging biomass conversion technologies, other renewable and fossil fuel conversion technologies. We did not incorporate any policy assumptions beyond 2020 to allow for free competition between all options. Using CO_2 tax as the only instrument for emission reduction, we assessed the achievement of or the distance to the EU's 40% emission target in 2030 compared to 1990.

We incorporated different biomass cost-supply curves to assess how deployment of biomass conversion technologies at a sectoral level and emission reduction at a systems level can be affected. We also assessed how dependent the national bioeconomy and the renewable energy system is on fossil fuel price variation. By combining biomass cost-supply and fossil fuel price scenarios with different assumptions on technology development, which vary in learning progress and technical parameters of technologies, we captured key uncertainties of mid-term bioeconomy development.

There are important considerations that should be taken when interpreting the outcomes.

Firstly, we used CO_2 emission pricing as the instrument to stimulate emission reduction as opposed to applying a cap on national emissions. In most scenarios, the 40% emission reduction target is not reached albeit significant reduction is realized (46–97 Mt CO_2 across scenarios compared to baseline; Fig. 7). The CO_2 price assumptions of IEA-WEO 2015 reflect the EU and not the required level for an individual country such as the

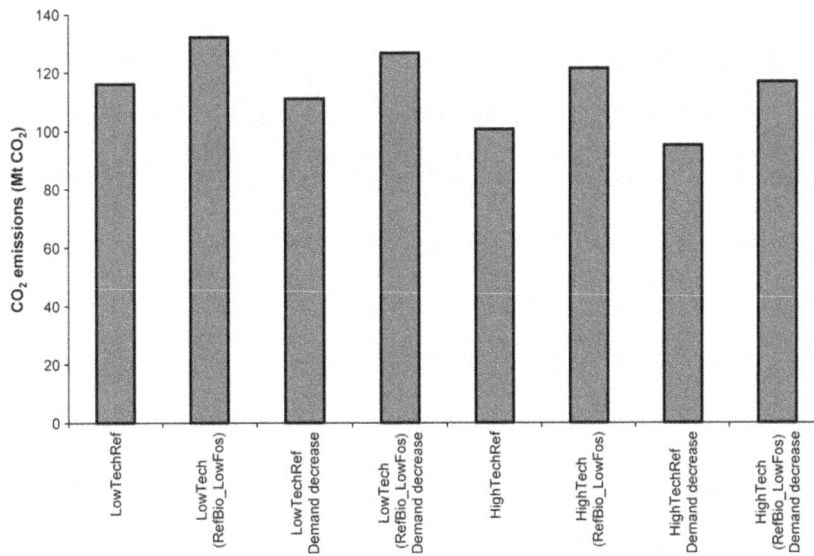

Fig. 11 Direct CO_2 emissions in reference and low fossil fuel price scenarios assuming different growth rates for the chemical industry in the Netherlands in 2020–2030.

Netherlands to achieve the target. Evidently, a higher CO_2 price would be required for the Netherlands. In addition, the assumed CO_2 price is an outcome of simulation where other policy measures and technologies such as energy efficiency are taken into account. Such measures are not included in our model. It could be argued that the assumed CO_2 price would be adequate to achieve the target in all scenarios had low-cost efficiency measures such as insulation of buildings been included. Then, the abatement achieved by biomass, other RES, CCS and BECCS could be lower. Related to the above, is that the CO_2 tax is assigned on sectoral emissions and not on the fossil carbon they consume. This is relevant for the chemical sector, which consumes large volumes of fossil carbon as feedstock that remains embedded in the products. Applying the tax on fossil carbon consumption similar to other studies (Daioglou et al. (2014)) could lead to different system dynamics because the benefit of avoiding CO_2 emissions from waste management would be taken into account. However, for a national model, this entails an improved representation of the end-of-life phase of products, where cascading uses and exports of chemicals are taken into account (Tsiropoulos 2016). This study finds that significant volumes of biochemicals could potentially be produced by 2030 (5–20% of total chemical output in final energy terms; Fig. 4b), even under baseline assumptions (5–10% of total chemical output in final energy terms; Fig. 4b). This entails that there may be a high potential for cascading uses of biomass from higher to lower value applications (Keegan et al., 2013). While this is not modelled in the study, it is important to point out that cascading uses would lead to increase in efficient

biomass use in the energy system and possibly to an increased output of biomass heat and electricity. Note that, based on Dammer et al. (2013), the production capacity of biochemicals in the Netherlands is comprised primarily of starch-blends and functional polymers (approximately 0.13 Mt). These applications are excluded from our study as we focus on large-scale production of biochemicals with significant potential for the substitution of fossil energy in the chemical industry. A combination of factors such as delayed retirement of steam cracking capacity, low fossil fuel price environment and decline in demand have an impact on the competitiveness of biochemicals as the output becomes negligible. Regarding the aviation sector, it was found that RJF may be supplied only under high technology development assumptions.

Furthermore, the outcomes represent only domestic emissions that occur in the Netherlands. The emissions related to production and transport of fossil fuels and biomass outside the Netherlands are not included. Consequently, neither are emissions from direct and indirect land-use change. Emissions from indirect land-use change are rather uncertain (Wicke et al., 2012). However, the present study demonstrates that large volumes of biomass may be consumed in the Netherlands in the mid-term and direct as well as indirect land-use change emissions may influence global CO_2 emission reduction efforts. In Tsiropoulos (2016), we showed that emissions from production and import of biomass from regions outside the Netherlands were approximately 4 Mt CO_2, which are 4–9% compared to the emission reduction achieved across scenarios from the baseline of this study. These emissions do not affect the main conclusions drawn in this study, as they do not affect domestic emissions and distance to target. However, they are relevant when emissions at a larger geographical scope are assessed. Finally, non-CO_2 emissions (e.g. methane, nitrous oxides), from agriculture, industry and waste processing, while not related to the energy system that is assessed in the present study, accounted for about 16% (or 30 $MtCO_{2eq.}$) of national total GHG emissions in 2014 (Table S9). It is estimated that over time it is technically feasible and cost-effective for these sectors to achieve approximately 50% emission reduction (Roelofsen et al., 2016).

While the above are important to consider, this study shows results that are well-aligned with other efforts.

A study that assessed lowest-cost complementarity of integrating fossil-based capacity with predetermined RES diffusion to achieve low-carbon power systems illustrates the significance of wind turbines and CCS to achieve emission reduction (Brouwer et al., 2016). While there are key differences in scope (geographical, temporal) and modelling techniques between the present study and Brouwer et al. (2016), they both show that lowest system costs are achieved with a mix of RES and CCS in the power sector. Similar outcomes are supported by van den Broek et al. (2011) in scenarios which take ambitious climate policies into account. A key difference between the outcomes of these studies compared to the results presented here is the deployment of CCS in gas-fired instead of coal-fired plants. An explanation to this can be the recent instalment of coal-based capacity in the Netherlands, which remains operational until 2030. An additional explanation could be that in the present study, more sectors are included in the energy system. As van Vliet et al. (2011) showed, when accounting for the transport sector in the energy system, the role of BECCS in biomass-based FT-fuel production is prominent. This finding, as confirmed by the present study, is also relevant when emerging bioeconomy sectors (i.e. biochemicals and RJF) are included in the energy system. Therefore, the significance of BECCS is demonstrated not only as a longer-term emission mitigation option, which many studies support (Fischedick et al., 2011; Fuss et al., 2014), but also in earlier in the time horizon, provided that the technology can be commercialized within the assumed timeframe.

Regarding biochemicals, to our knowledge there are limited studies that provide future estimates at a systems level, as for example Daioglou et al. (2014). According to their study, in the long term biomass has the potential to supply up to 40% of total demand for nonenergy in 2100 (or about 1 Gt yr^{-1} assuming 45 GJ t^{-1} as heating value; Daioglou et al., 2014). Other studies have also performed short-term assessments of future biochemical potential (Dornburg et al., 2008; Ren

& Patel, 2009; Ren et al., 2009; nova-Institut, 2013; Saygin et al., 2013, 2014; Gerssen-Gondelach et al., 2014; Piotrowski et al., 2015) without, however, taking systems dynamics into account. European Bioplastics estimate that global production capacity of bioplastics will reach 7.85 Mt in 2019 (EuBP, 2016). Our study estimates that production output of biochemicals may reach up to 1.1 Mt in the Netherlands in 2020 depending on scenario conditions. While results of these studies cannot be directly comparable with the output of this study, they all confirm that biochemical increases over time.

Against this background, the most important observations can be summarized in the following:

The size of bioeconomy depends on developments across the supply chain and the fossil fuel price

By 2030, the contribution of biomass in renewable energy supply is higher than the approximately 50% that is anticipated according to other studies by 2020 (Rijksoverheid, 2010; Stralen et al., 2013). It ranges between 52% and 77% and corresponds to biomass consumption volumes of 183–760 PJ, depending on scenario assumptions. Biomass supply depends on intra-EU and extra-EU biomass, and based on literature, it is deemed available (Chum et al., 2011; Ganzevles, 2014; Smeets, 2014). The supply from RES observed in the decade 2020–2030 is due to technological growth and increase in the CO_2 emission tax. Other RES remain fairly constant to 2020 levels, while the bioeconomy grows. Investments across the supply chain both on the supply side, as modelled by the low-cost biomass scenario, and on the conversion side, as modelled by the high technology development scenario, lead to increased contribution of biomass in the system (Fig. 6a). Low fossil fuel prices do not lead to contraction of the RES share compared to 2020 and reduce total system costs by about 6.5 billion € yr^{-1} in 2030, however, even under high technology development no growth is observed. In the face of low fossil fuel prices, mechanisms are required to ensure bioeconomy growth such as a CO_2 tax higher than 69 € t^{-1} CO_2 by 2030.

A mixed technology portfolio is required to achieve deep emission reduction

A wide technology portfolio is required to achieve emission reduction in the mid-term, to realize long-term climate goals. The role of wind in the electricity sector, bioenergy in road transport and industrial heat, but also CCS and BECCS are significant. This finding is widely supported by literature (IPCC, 2014; Matthews et al., 2015; Winchester & Reilly, 2015). Introducing new bioeconomy sectors in the energy system, namely biochemicals and RJF does not alter it; however, the latter are supplied only under high technology development assumptions. As other RES do not increase significantly in the scenario outcomes, except when high fossil fuel prices are assumed, the post-2020 emission reduction can be attributed to biomass (20–40 Mt CO_2 or 40–60% compared to the baseline) and CCS (19–41 Mt CO_2 or 42–60% compared to the baseline). In high technology development scenarios, which among other options include BECCS, emission reduction is higher by 6–17% (7.5–20 Mt) compared to low technology development scenarios (BECCS contributes 47–83%). BECCS can have a significant role earlier in the time horizon than most studies indicate (Fischedick et al., 2011; Fuss et al., 2014), if the technology is commercialized. With demand-side improvements (e.g. on industrial and residential energy efficiency), the role of biomass heat may diminish in the longer term. This could create opportunities for other bioeconomy sectors to grow. Such an assessment requires incorporation to the model of energy efficiency measures or a longer temporal scope (e.g. 2040), which may also result in other structural changes of sectors such as electrification in transport.

Sector-specific assumptions do not compromise the potential emission reduction

A decrease in demand for chemicals in combination with other factors such as delayed retirement of steam cracking capacity and low fossil fuel prices affects the size of the biochemical sector. The latter reduce the output of biochemicals by about 70% compared to the reference, while combined with the former assumptions the reduction ranges between 85 and 99%. However, the systems' CO_2 emissions are not affected. Furthermore, dismantling all coal-based power generation capacity leads to an increase in RES (wind) and natural-gas power generation. While coal is effectively phased out entirely from the energy system of the Netherlands, the emission levels in 2030 remain the same as CCS capacity compared to reference scenarios is lower and the emission reduction is offset by wind turbines.

High technology development is a no-regrets option to achieve deep emission reduction

Post-2020, high technology development uses 313–760 PJ of biomass depending on scenario assumptions. Compared to the low technology development counterparts, it offers additional opportunities to utilize biomass in the energy system as indicated by the additional 100–270 PJ that are consumed. High technology development combined with the low-cost biomass

scenario uses approximately 100 PJ more compared to the reference. Assuming low-cost biomass does not lead to increased consumption in low technological growth scenarios. Thus, improvements early in the supply chain increase the size of the bioeconomy only under high technological growth. Furthermore, high technology development consistently leads to lower emissions and cumulative system costs than low technology development in 2030. At the same time, high technology development creates a more resilient bioeconomy even if fossil fuel prices remain low as there is continuous growth to 2030. However, this observation excludes external costs, which are required to achieve high technological growth, such as in R&D or support to 1st-of-a-kind plant. Furthermore, this comparison is sensitive to the underlying production costs of bioenergy and biochemicals as illustrated by the two technology development scenario variants. Nonetheless, to achieve deeper levels of emission reduction required to embark on low-cost trajectories that meet long-term climate targets technological development is required to reduce production costs of advanced biomass conversion technologies.

Acknowledgements

This work was carried out within the BE-Basic R&D Program, which was granted a FES subsidy from the Dutch Ministry of Economic affairs, agriculture and innovation (EL&I).

References

AEBIOM (2015) *AEBIOM Statistical Report – European Bioenergy Outlook.* European Biomass Association (AEBIOM), Brussels, Belgium.

Batidzirai B, Van Der Hilst F, Meerman H, Junginger MH, Faaij APC (2014) Optimization potential of biomass supply chains with torrefaction technology. *Biofuels, Bioproducts and Biorefining*, **8**, 253–282.

BCG (1968) *Perspectives on Experience.* Boston Consulting Group Inc (BCG), Boston, MA.

van den Broek M, Faaij A, Turkenburg W (2008) Planning for an electricity sector with carbon capture and storage. Case of the Netherlands. *International Journal of Greenhouse Gas Control*, **2**, 105–129.

van den Broek M, Veenendaal P, Koutstaal P, Turkenburg W, Faaij A (2011) Impact of international climate policies on CO2 capture and storage deployment. *Energy Policy*, **39**, 2000–2019.

Broeren MLM, Saygin D, Patel MK (2014) Forecasting Glob developments in the basic chemical industry for environmental policy analysis. *Energy Policy*, **64**, 273–287.

Brouwer AS, van den Broek M, Zappa W, Turkenburg WC, Faaij A (2016) Least-cost options for integrating intermittent renewables in low-carbon power systems. *Applied Energy*, **161**, 48–74.

CBS (2016) *Statistics Netherlands: Total Final Consumption.* Centraal Bureau voor de Statistiek (CBS), The Hague/Herleen.

Chèze B, Gastineau P, Chevallier J (2011) Forecasting world and regional aviation jet fuel demands to the mid-term (2025). *Energy Policy*, **39**, 5147–5158.

Chum H, Faaij A, Moreira J et al. (2011) Bioenergy. In: *IPCC Special Report on Renewable Energy Sources and Climate Change Mitigation* (eds Sokona Y, Seyboth K, Matschoss P, Kadner S, Zwickel T, Eickemeier P, Hansen G, Schlomer S, von Stechow C), pp. 209–332. Cambridge University Press, Cambridge, UK and New York, NY.

Daioglou V, Faaij APC, Saygin D, Patel MK, Wicke B, van Vuuren DP (2014) Energy demand and emissions of the non-energy sector. *Energy & Environmental Science*, **7**, 482.

Daioglou V, Wicke B, Faaij APC, van Vuuren DP (2015) Competing uses of biomass for energy and chemicals: implications for long-term global CO2 mitigation potential. *GCB Bioenergy*, **7**, 1321–1334.

Dammer L, Carus M, Raschka A, Scholz L (2013) *Market Developments and Opportunities for Biobased Products and Chemicals*. nova-Institut, Hürth.

Dornburg V, Hermann BG, Patel MK (2008) Scenario projections for future market potentials of biobased bulk chemicals. *Environmental Science & Technology*, **42**, 2261–2267.

EC (2003) *EU-15 Energy and Transport Outlook to 2030 (part II)*. European Commission (EC), Brussels, Belgium.

EC (2009) Directive 2009/28/EC of the European Parliament and of the Council of 23 April 2009. *Official Journal of the European Union*, **140**, 16–62.

EC (2011) *Launch of the European Advanced Biofuels Flightpath*. European Commission (EC), Brussels.

EC (2015) *Intended Nationally Determined Contribution of the EU and its Member States*, pp. 1–5. European Commission (EC), Riga, Lithuania.

ECN (2015) *MONITweb*. Energy Research Centre of the Netherlands (ECN), Petten, the Netherlands.

Elbersen B, Startisky I, Hengeveld G, Jeurissen L, Lesschen J (2015) *Outlook of spatial value chains in EU28 – Deliverable 2.3 of Biomass Policies Project*. Alterra – Part of Wageningen UR, Wageningen, the Netherlands and Laxenburg, Austria.

EuBP (2016) *Bioplastics Facts and Figures*, pp. 1–6. European Bioplastics (EUBP), Berlin, Germany.

Fischedick M, Schaeffer R, Adedoyin A et al. (2011) Mitigation potential and costs. In: *IPCC Special Report on Renewable Energy Sources and Climate Change Mitigation* (eds Sokona Y, Seyboth K, Matschoss P, Kadner S, Zwickel T, Eickemeier P, Hansen G, Schlomer S, von Stechow C), pp. 791–864. Cambridge University Press, Cambridge, UK and New York, NY.

Fuss S, Canadell JG, Peters GP et al. (2014) Betting on negative emissions. *Nature Climate Change*, **4**, 850–853.

Ganzevles J (2014) *Bijlage 1: Vraag en aanbod in Nederland in 2030*, pp. 1–17. Commissie Duurzaamheidsvraagstukken Biomassa, Nijmegen, the Netherlands.

Gerssen-Gondelach SJ, Saygin D, Wicke B, Patel MK, Faaij APC (2014) Competing uses of biomass: assessment and comparison of the performance of bio-based heat, power, fuels and materials. *Renewable and Sustainable Energy Reviews*, **40**, 964–998.

Hoefnagels R, Searcy E, Cafferty K, Cornelissen T, Junginger M, Jacobson J, Faaij A (2014a) Lignocellulosic feedstock supply systems with intermodal and overseas transportation. *Biofuels, Bioproducts and Biorefining*, **8**, 794–818.

Hoefnagels R, Resch G, Junginger M, Faaij A (2014b) International and domestic uses of solid biofuels under different renewable energy support scenarios in the European Union. *Applied Energy*, **131**, 139–157.

IPCC (2006) Energy. In: *2006 IPCC Guidelines for National Greenhouse Gas Inventories, Prepared by the National Greenhouse Gas Inventories Programme* (eds Eggleston S, Buendia L, Miwa K, Ngara T, Tanabe K), pp. 1.1–6.14. Intergovernmental Panel on Climate Change to Institute for Global Environmental Strategies, Kanagawa.

IPCC (2014) *Climate Change 2014: Synthesis Report. Contribution of Working Groups I, II and III to the Fifth Assessment Report of the Intergovernmental Panel on Climate Change* (eds Core Writing Team, Pachauri RK, Meyer LA). IPCC, Geneva, Switzerland.

Keegan D, Kretschmer B, Elbersen B, Panoutsou C (2013) Cascading use: a systematic approach to biomass beyond the energy sector. *Biofuels, Bioproducts and Biorefining*, **7**, 193–206.

Matthews R, Mortimer N, Lesschen JP et al. (2015) *Carbon Impacts of Biomass Consumed in the EU: Quantitative Assessment October 2015*. Forest Research, Farnham.

van Meijl H, Tsiropoulos I, Bartelings H et al. (2016) *Macro-economic Outlook of Sustainable Energy and Biorenewables Innovations (MEV II)*, pp. 1–168. LEI Wageningen UR (University & Research centre), Den Hague, the Netherlands.

Moss RH, Edmonds JA, Hibbard KA et al. (2010) The next generation of scenarios for climate change research and assessment. *Nature*, **463**, 747–756.

Nicola S, Andresen T (2015) Germany gives dirtiest coal plants six years for phase out. Bloomberg, Businessweek. Available at: http://www.bloomberg.com/news/articles/2015-07-02/germany-to-close-coal-plants-in-effort-to-curb-pollution (accessed 25 May 2016).

nova-Institut (2013) *Bio-based Polymers – Production Capacity will Triple from 3.5 Million Tonnes in 2011 to Nearly 12 Million Tonnes in 2020*, pp. 1–13. nova-Institut, Hürth.

OECD/IEA (2014) *World Energy Outlook 2014*. Organisation for Economic Co-operation and Development (OECD), International Energy Agency (IEA), Paris, France.

OECD/IEA (2015) *World Energy Outlook 2015*. Organisation for Economic Co-operation and Development (OECD), International Energy Agency (IEA), Paris, France.

OGJ (2012) Global ethylene capacity continues advance in 2011. *Oil & Gas Journal*, 78.

Pieters J (2016) Dutch cabinet to close two more coal plants. NLtimes.nl. Available at: http://nltimes.nl/2016/04/11/dutch-cabinet-to-close-two-more-coal-plants/ (accessed 26 May 2016).

Piotrowski S, Carus M, Essel R (2015) Global bioeconomy in the conflict between biomass supply and demand. Nova Paper #7 on Bio-based Economy, 1–13.

Ren T, Patel MK (2009) Basic petrochemicals from natural gas, coal and biomass: energy use and CO2 emissions. *Resources, Conservation and Recycling*, **53**, 513–528.

Ren T, Daniëls B, Patel MK, Blok K (2009) Petrochemicals from oil, natural gas, coal and biomass: production costs in 2030–2050. *Resources, Conservation and Recycling*, **53**, 653–663.

Rijksoverheid (2010) *National Renewable Action Plan of the Netherlands*. Government of the Netherlands, the Netherlands.

Roelofsen O, de Pee A, Speelman E (2016) *Accelerating the Energy Transition: Cost or Opportunity?*. McKinsey & Company, Amsterdam.

Rose SK, Kriegler E, Bibas R, Calvin K, Popp A, van Vuuren DP, Weyant J (2014) Bioenergy in energy transformation and climate management. *Climatic Change*, **123**, 477–493.

Saygin D, Patel MK, Tam C, Gielen DJ (2009) *Chemical and Petrochemical Sector Potential of best practice technology*. International Energy Agency (IEA), Paris, France.

Saygin D, van den Broek M, Ramírez A, Patel MK, Worrell E (2013) Modelling the future CO2 abatement potentials of energy efficiency and CCS: the case of the Dutch industry. *International Journal of Greenhouse Gas Control*, **18**, 23–37.

Saygin D, Gielen DJ, Draeck M, Worrell E, Patel MK (2014) Assessment of the technical and economic potentials of biomass use for the production of steam, chemicals and polymers. *Renewable and Sustainable Energy Reviews*, **40**, 1153–1167.

Schwartz P (1996) *The Art of the Long View: Planning for the Future in an Uncertain World*, pp. 1–24. Random House Digital Inc, New York, NY.

SER (2013) *Energy Agreement for Sustainable Growth – Summary*, pp. 6–9. Sociaal Economische Raad (SER), The Hague.

Smeets E (2014) *Bijlage 2: Beschikbaarheid van biomassa voor export naar Nederland*, pp. 1–20. Commissie Duurzaamheidsvraagstukken, The Hague.

Sterl S, Höhne N, Kuramochi T (2016) *What Does the Paris Agreement Mean for Climate Resilience and Adaptation?*. New Climate Institute, Cologne, Berlin, Germany.

Stralen JNP, Uslu A, Longa FD (2013) The role of biomass in heat, electricity, and transport markets in the EU27 under different scenarios. *Biofuels, Bioproducts & Biorefining*, **7**, 147–163.

Tsiropoulos I (ed.) (2016) Emerging bioeconomy sectors in energy systems modelling – Integrated systems analysis of electricity, heat, road transport, aviation and chemicals: a case study for the Netherlands. In: *Emerging Bioeconomy – Assessing the Implications of Advanced Bioenergy and Biochemicals with Bottom-Up and Top-Down Modelling Approaches*, pp. 89–150. Utrecht University, Utrecht.

UNFCCC (2015) *Adoption of the Paris Agreement*, 21st Conference of Parties, United Nations Framework Convention on Climate Change (UNFCCC), Paris.

Uslu A, Faaij APC, Bergman PCA (2008) Pre-treatment technologies, and their effect on international bioenergy supply chain logistics. Techno-economic evaluation of torrefaction, fast pyrolysis and pelletisation. *Energy*, **33**, 1206–1223.

van Vliet O, van den Broek M, Turkenburg W, Faaij A (2011) Combining hybrid cars and synthetic fuels with electricity generation and carbon capture and storage. *Energy Policy*, **39**, 248–268.

Vuuren D, Boot P, Ros J, Hof A, den Elzen M (2016) *Wat betekent het Parijsakkoord voor het Nederlandse langtermijn-klimaatbeleid?*. Planbureau voor de Leefomgeving (PBL), den Haag, the Netherlands.

Wicke B (2011) *Bioenergy Production on Degraded and Marginal Land: Assessing its Potentials, Economic Performance, and Environmental Impacts for Different Settings and Geographical Scales*. Utrecht University, Utrecht.

Wicke B, Verweij P, van Meijl H, van Vuuren DP, Faaij APC (2012) Indirect land use change: review of existing models and strategies for mitigation. *Biofuels*, **3**, 87–100.

Winchester N, Reilly J (2015) *The Contribution of Biomass to Emissions Mitigation under a Global Climate Policy*. MIT Joint Program on the Science and Policy of Global Change, Cambridge, MA.

de Wit M, Faaij APC (2009) European biomass resource potential and costs. *Biomass and Bioenergy*, **34**, 188–202.

de Wit M, Junginger M, Lensink S, Londo M, Faaij A (2010) Competition between biofuels: modeling technological learning and cost reductions over time. *Biomass and Bioenergy*, **34**, 203–217.

Yeo S (2015) *In-depth: UK Pledges Coal Phase Out by 2025, but Uncertainty Remains*. CarbonBrief. Available at: https://www.carbonbrief.org/in-depth-uk-pledges-coal-phase-out-by-2025-but-uncertainty-remains (accessed 25 May 2016).

10

Genetic diversity of *Miscanthus sinensis* in US naturalized populations

YONGLI ZHAO[1,†], SUMA BASAK[1,†], CHRISTINE E. FLEENER[2], MARCELINE EGNIN[1], ERIK J. SACKS[2], CHANNAPATNA S. PRAKASH[3] and GUOHAO HE[1]

[1]*College of Agriculture, Environment and Nutrition Sciences, Tuskegee University, Tuskegee, AL 36088, USA,* [2]*Department of Crop Sciences, University of Illinois, Urbana, IL 61801, USA,* [3]*College of Arts and Sciences, Tuskegee University, Tuskegee, AL 36088, USA*

Abstract

Miscanthus is increasingly gaining popularity as a bioenergy grass because of its extremely high biomass productivity. Many clones of this grass were introduced into United States over the past century from East Asia where it originated, and planted for ornamental and landscaping purposes. An understanding of the genetic diversity among these naturalized populations may help in the efficient selection of potential parents in the *Miscanthus* breeding program. Here, we report our study analyzing the genetic diversity of 228 *Miscanthus* DNA samples selected from seven sites in six states (Ohio, North Carolina, Washington D.C., Kentucky, Pennsylvania, and Virginia) across the eastern United States. Ten transferable DNA markers from other plant species were employed to amplify genomic DNA of *Miscanthus* because of the paucity of molecular markers in *Miscanthus*. There were significant genetic variations observed within and among US naturalized populations. The highest genetic diversity (0.3738) was found among the North Carolina genotypes taken from Biltmore Deer Park and Biltmore, Madison County, Cody Rd. The lowest genetic diversity (0.2776) was observed among Virginia genotypes that were diverged from those from other states, suggesting Virginia genotypes might be independently introduced into the United States from the different origin. By the cluster and structure analysis, 228 genotypes were categorized into two major groups that were further divided into six subgroups at the DNA level and the groups were generally consistent with geographic region.

Keywords: genetic marker, genetic variation, *Miscanthus*, transferability

Introduction

Miscanthus (Gramineae) is an herbaceous perennial grass, native to eastern Asia found throughout China, Korea, and Japan. Due to its high biomass yield, high ligno-cellulose content, and high photosynthesis, *Miscanthus* is among the top potential bioenergy-producing plants in Europe and North America (Somerville *et al.*, 2010; Glowacka, 2011). This 'green-energy' grass is widely cultivated as a prospective energy source for power production and liquid biofuel generation in Europe and North America. *Miscanthus* is not native to the United States and has no natural wild relatives with potential out-crossing risks. Thus, it represents an attractive candidate crop for future transgene modifications (ex. to develop clones producing *in-planta* enzymes or with targeted modification of biomass genes) because

of the reduced biosafety concern related to gene flow, with consequent low regulatory burden (Yuan *et al.*, 2008; Jakob *et al.*, 2009; Bransby *et al.*, 2010; Somerville *et al.*, 2010).

The study of genetic diversity of *Miscanthus* is not only important for the effective conservation, management, and utilization of the genetic resource, but it is also a prerequisite in breeding this grass to select desirable plants. Molecular markers are useful tools for studying genetic diversity and facilitating crop improvement programs, as well as evolutionary and conservation studies. Compared to earlier molecular marker technologies such as AFLPs, RFLPs, and RAPDs, simple sequence repeat (SSR) or microsatellite markers are more effective tools for plant genetics and genomics studies because of their abundance, hyper-variability, ease of PCR-based recognition, codominant transmission, reproducibility, transferability among species and genera, and ubiquitous distribution in the genome and high frequency of polymorphisms (Brown *et al.*, 1996; Cordeiro *et al.*, 2000). Although the advantages of SSR markers are obvious, the identification of SSRs from

[†]Contributed equally to the project.

Correspondence: Channapatna S. Prakash
e-mail: prakash@mytu.tuskegee.edu and Guohao He
e-mail: hguohao@mytu.tuskegee.edu

genomic DNA is an expensive and lengthy process, requiring library construction as well as clone sequencing (Zhao *et al.*, 2011). Because of the lack of SSR markers in *Miscanthus*, it may be prudent to explore the transferability of SSR markers from closely related species. A large number of such transferable markers have been identified from related species (Hernández *et al.*, 2001; Zhou *et al.*, 2011; Kim *et al.*, 2012; Swaminathan *et al.*, 2012; Dai *et al.*, 2013; Yu *et al.*, 2013; Tamura *et al.*, 2015), opening up access to genetic variation studies of *Miscanthus* and molecular breeding. Kim *et al.* (2012) constructed two genetic linkage maps in *Miscanthus*, in which total 261 and 303 loci were mapped in the populations of *M. sacchariflorus and M. sinensis* using sugarcane EST-SSRs, respectively. Moreover, many unigenes in sugarcane deposited in GenBank provide potential for development of gene-based markers in *Miscanthus*.

Cluster analysis and phylogenetic studies have elicited interest among scientists to study genetic identity, genetic diversity, genetic similarity and dissimilarity, and genetic distance using molecular markers on many different crops. *Miscanthus* diversity and population structure have been studied in different geographic regions, in China (Selvi *et al.*, 2003; Xu *et al.*, 2013; Zhang *et al.*, 2013), in Japan (Clark *et al.*, 2015), in Korea (Yook *et al.*, 2014), and in Asia (Clark *et al.*, 2014). *Miscanthus sinensis* has been grown throughout the United States since shortly after they were introduced from eastern Asia in the early 1870s (Anonymous, 1876; Bailey & Miller, 1901). Naturalized populations of *Miscanthus* have become established in the United States (Clark *et al.*, 2014). However, there are few studies reporting on their origin, genetic diversity, and relationships between U.S. naturalized populations.

To better understand genetic variation in naturalized populations of *Miscanthus* introduced as ornamental plants from East Asia over the past century in the United States, the transferable DNA markers could be used to analyze the diversity among *Miscanthus* genotypes based on marker amplicon data. Analysis and comparison of gene sequences would further reveal genetic variation at the DNA level in *Miscanthus* naturalized populations. Therefore, the aim of this study was to gain an insight into genetic variations of US naturalized populations of *Miscanthus* with following objectives: (i) use heterologous sugarcane DNA markers to identify genetic variation within and among naturalized populations of *Miscanthus*; (ii) conduct phylogenetic analysis of *Miscanthus* in US naturalized populations using DNA sequences.

Materials and methods

Experimental materials

Two hundred and twenty-eight seedlings of *M. sinensis* were collected from seven sites in six states (Ohio, North Carolina, Washington D.C., Kentucky, Pennsylvania, and Virginia) (Fig. 1, Table 1). *M. sinensis* is an obligate outcrossing species due to self-incompatibility; thus, each seedling is a unique genotype. Accessions collected from the same state were considered as a single naturalized population for this study. These six naturalized populations likely represent adaptation across complex topography and various climate conditions, especially cold tolerance, within the major distribution areas of

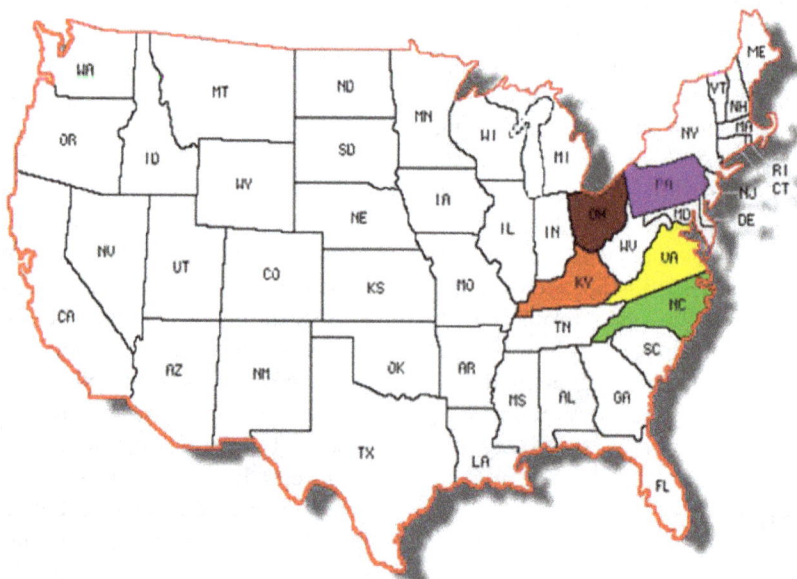

Fig. 1 Source of naturalized populations of *Miscanthus*, collected across six states of the United States.

Table 1 Descriptions of the ecological and geographic data of six natural populations of *Miscanthus sinensis* sampled

Accession ID	Location	Elevation (m)	Latitude	Longitude	Number of plants found	Number of plants sampled*	Sample size†
OH 2009-001		1094	N 39°47′	W 81°30′		19	34
NC-2010-001.5	Biltmore, Deer Park	572	N 35°33′	W 82°34′	1000	18	31
NC-2010-001	Biltmore, Madison Co, Cody Rd.	698	N 35°32′	W 82°32′	3000	64	26
DC-2010-001	Rock Creek Park	48	N 38°56′	W 77°02′	100	39	36
KY-2009-001		1293	N 37°80′	W 83°66′		20	33
PA-2010-003	Route 252	92	N 39°57′	W 75°23′	200	20	36
VA-2010-002	Prince William National Park	655	N 37°23′	W 80°21′		10	32

*Number of plants at the collection site from which seed was collected.
†Number of seedlings per population evaluated in this study.

M. sinensis (Fig. 1) in the eastern United States. The seedlings were grown in a field at the University of Illinois at Urbana Champaign.

DNA extraction and PCR amplification

About 10–15 g of fresh young leaves was collected from each seedling and freeze-dried and then ground to powder in sterile acid washed sand. Genomic DNA (gDNA) was extracted from 500 mg of powdered leaf samples using a modified protocol (Egnin *et al.*, 1998).

Seven genomic SSR markers and 29 EST-SSR markers from sugarcane and 38 gene markers from *Arabidopsis, Lolium perenne,* strawberry, and pea (Table S1) were used for PCR amplification. Polymerase chain reaction (PCR) amplification of *Miscanthus* DNA was carried out using 74 molecular markers (Table S1). The PCR conditions included an initial step of 5 min at 94 °C, followed by 35 cycles of denaturation at 94 °C for 30 s, annealing at 50–55 °C for 30 s depending on primers used, extension at 72 °C for 1 min, and final extension at 72 °C for 7 min. The total volume of each PCR reaction was 10 μL containing 25 ng template DNA, 1.0 μL of the 10× reaction buffer (MgSO$_4$ free), 1.0 μL of 25 mM Mg^{++}, 0.2 μL of 10 mM dNTP, 1.0 μL of 5 μM (0.5 μL Forward and 0.5 μL Reverse) primer, and 0.05 μL of 5 u μL^{-1} Taq polymerase. The amplified products were separated on 6.0% polyacrylamide gel using 0.5% Tris–borate–EDTA (TBE) buffer. After electrophoresis, the gel was stained by 1% ethidium bromide solution and the image of the gel was visualized under UV light. The PCR amplification was repeated one more time to ensure reproducibility.

The presence and absence of the DNA bands for each primer–genotype combination was scored as either 1 or 0. Several genetic parameters including allele frequency, genetic diversity (H_e), and Nei's genetic distance (*D*) (Nei *et al.*, 1983) were estimated using the software of POWER-MARKER 3.25 (Liu and Muse 2005). Cluster analysis was performed using the neighbor-joining method by the Molecular Evolutionary Genetics Analysis software (MEGA 6.06) (Tamura *et al.*, 2013). Unique alleles were considered as those present in one accession or one group of accessions but absent in other accessions or groups of accessions. Rare alleles were those with frequency of ≤5% in investigated materials, while those alleles with frequencies >20% were classified as frequent alleles (Zhao *et al.*, 2013).

Population structure was determined and individuals assigned to groups using the software STRUCTURE 2.3.4, Stanford University, CA (Pritchard *et al.*, 2000). The admixture model, using correlated allele frequency, was used. The program Structure was run eight times for each subpopulation value (*K*, ranging from 1 to 10) with a burn-in period of 10^4 followed by 10^5 iterations. Evanno's delta *K* (Evanno *et al.*, 2005) was chosen to determine the optimum number of subpopulations. The run with maximum likelihoods value was chosen to assign accessions with the posterior membership coefficients (Q). A graphical bar plot representing the posterior membership coefficients of each accession was then generated.

Analysis of molecular variance (AMOVA) was performed to evaluate population differentiation using GENALEX6.5 software (Peakall & Smouse, 2006, 2012). The calculation of pairwise GDs for binary data followed the method of Huff *et al.* (1993).

DNA sequence data analysis

Seventy-four DNA markers from other species were used to amplify *Miscanthus* genomic DNAs obtained from genotypes in US naturalized populations. When amplicons displayed a monomorphism among genotypes by certain marker, the amplified PCR products of such monomorphic marker were sequenced to further reveal genetic variation within and between populations. On the basis of DNA sequences of individuals, the GD among genotypes was calculated and the phylogenetic relationship was reconstructed using MEGA version 6.06 (Tamura *et al.*, 2013).

Results

Genetic diversity and population structure

Of 36 sugarcane SSR markers, 32 produced amplicons and the remaining four EST-SSR markers did not. However, only 20 from 38 gene-based markers of different species could be transferable to *Miscanthus*. Among DNA markers tested in this study, eight SSR markers and two gene markers (actin, GA30x, PF00931, PF03856, SMC226CG, SMC248CG, SMC319CG, SMC1039CG, EST-SSR29, and EST-SSR38-2) producing polymorphic bands were used for downstream analyses in this study

(Table 2). A total of 23 alleles were generated by these 10 markers, including 22 frequent alleles and one rare allele with an average of 2.3 alleles per primer pair. The rate of polymorphic alleles, Nei's genetic diversity (H_e), and polymorphism information content (PIC) were 0.7054, 0.3833, and 0.3030, respectively, among the genotypes studied (Table 3). This study presented a sizeable molecular marker dataset in a diverse panel of *M. sinensis*.

Genetic variation within and between US naturalized populations of *M. sinensis* were both significant; however, variation within populations (89%) was substantially greater than between populations (11%) in the distance-based AMOVA analysis (Table 4). The results indicated genotypes within naturalized population were genetically diverse, whereas genetic differences among populations were substantially less common.

The estimated GD indicated divergence among the populations (Table 5). The VA population was generally the most diverged from the other five populations. The largest GD (0.245) was found between VA and DC

populations (Table 5), while the smallest distance (0.035) was obtained between NC and OH, and (0.038) between KY and PA. Using the software STRUCTURE 2.3.4 (Evanno et al., 2005), we identified six groups among the US *M. sinensis* naturalized populations (Fig. 2). Genotypes within each group identified by the Structure were represented with different colors in Figure 2. To further evaluate how well the six groups identified by Structure were consistent with the six naturalized populations, cluster analysis was performed among all 228 genotypes using neighbor-joining with the software MEGA 6.06. The resulted radial tree is shown in Figure 3, where each genotype branch has the same color as that genotype had in groups in Figure 2. Although the clades in the tree were not entirely the same as the groups by the Structure analysis (Fig. 2), genotypes with the same color were typically clustered together with a few exceptions. Given the degree of admixture indicated in the Structure analysis, the groups identified were largely consistent with geographic regions, although some genotypes did not fit into major clades. Population composition of the structure groups was 88% VA for green group, 26% OH and 31% NC for blue group, 45%

Table 2 List of 10 primer pairs used in this study

Primer name	Forward primer (5′–3′)	Reverse primer (5′-3′)	Annotation
EST-SSR29	CGACTGCTGCTTCGACTACA	GACCGATCCACCGAATCTC	Pathogenesis-related protein (*Oryza sativa*)
EST-SSR38-2	GTAGTCCTGCGCGTACTTGG	TAGCAAACATGGCGTTTCTG	Type-1 pathogenesis-related protein
SMC1039CG	AGGTGAGAGTTCCTGGCTTTCCA	TGTGCTGGCAAGCCCCTACTT	
SMC226CG	GAGGCTCAGAAGCTGGCAT	ACCCTCTATTTCCGAGTTGGT	
SMC248CG	TGCGCCTAGATGTACGATATGT	TTGTGTTATCCCAACTATTATGTCA	
SMC319CG	CCTTTCATCCACAGAGGACAG	GGTTCACCGAAGCAAGAGAAC	
PF00931	GCGTCTTCATCATCTGCAAC	TAGAGAGACATGGGGTGCAT	NB-ARC domain
PF03856	TTGTTAGTTTATTGGAGGGAA	GGCACATCTCTTGCTGTC	Beta-glucosidase (SUN family)
GA3ox	CCTCACAATCATCCACCAATCC	CGCCGATGTTGATCACCAA	
Actin	AGAGATTCAGATGCCCAGAAGTCTTGTTCC	AACGATTCCTGGACCTGCCTCATCATACTC	

Table 3 Allele frequency, genetic diversity, and PIC within each naturalized *Miscanthus sinensis* population analyzed by POWER-MARKER 3.25 based on DNA marker data

Naturalized population	Number of genotypes in population	Major allele frequency	Nei's genetic diversity	Polymorphism information content (PIC)
Virginia (VA)	32	0.7911	0.2776	0.2228
Washington DC (DC)	36	0.7830	0.3005	0.2440
Kentucky (KY)	33	0.7533	0.3322	0.2660
Ohio (OH)	34	0.7174	0.3569	0.2824
Pennsylvania (PA)	36	0.7234	0.3728	0.2966
North Carolina (NC)	57	0.7058	0.3738	0.2933
All genotypes	228	0.7054	0.3833	0.3030

Table 4 Analysis of molecular variation (AMOVA) results for six naturalized U.S. *Miscanthus sinensis* populations based on the dataset of 10 DNA markers tested

Source	Degree of freedom	Sum of square	Mean of square	Est. variance component	Percentage (%)
Among pops	5	112.937	22.587	0.494	11
Within pops	222	891.744	4.017	4.017	89
Total	227	1004.681		4.511	100

Table 5 Genetic distance (*D*) between six US naturalized populations of *Miscanthus sinensis* based on the DNA marker data

Population	DC	KY	NC	OH	PA	VA
KY	0.054					
NC	0.083	0.069				
OH	0.074	0.056	0.035			
PA	0.120	0.038	0.088	0.076		
VA	0.245	0.195	0.163	0.202	0.206	
Mean distance	0.115	0.082	0.088	0.089	0.106	0.202

DC for yellow group, 62% PA for aquamarine group, 56% NC for red group, and 48% NC for dark pink group.

The blue group included 15 genotypes from OH, 18 were from NC, 11 from KY, eight from DC, five from PA, and one from VA; it included 58 genotypes and formed the biggest group. The second biggest group, yellow, contained 44 genotypes of which 20 were from DC, eight from KY, seven from OH, five from NC, and one from VA. The aquamarine group included 34 genotypes of which 21 were from PA, eight from NC, three from KY, and two from OH genotypes. The red group had 34 genotypes including 19 form NC, five from OH, four from PA, and two from KY. The green group in Figure 3 included 28 genotypes from VA, two from PA, one was from NC, and one from KY. In 'dark pink' group, there were the least number (26) of genotypes.

Most of the genotypes from VA were included in a clade, although a few genotypes were grouped in other clades, further suggesting that the VA naturalized population had the smallest genetic variation among the six populations and it was the most genetically different from the other populations.

Genetic variation in DNA sequences

Because 29 of 36 sugarcane DNA markers were monomorphic among *Miscanthus* genotypes, the PCR products of one such marker SSR38-1 (cellulose synthase gene-related marker) were sequenced to investigate the genetic variation within DNA sequences among the genotypes. Four sequences were removed due to their poor sequence quality. The genetic distance (GD) among individuals within a naturalized population and among populations was calculated using 224 sequence data by

the software MEGA 6.06. Genetic variation within a population was lowest (0.006) for VA followed by OH (0.008) (Table 6). The naturalized populations KY, PA, and DC had the same genetic variation (0.011), while NC had the greatest variation (0.012). For GD among populations, the lowest GD was between OH and VA (0.011), as well as OH and PA (0.011), whereas the greatest GD was between NC and VA (0.015) (Table 7).

Phylogenetic relationships

Genetic variation may exist in the DNA sequences even when their PCR products show monomorphism on gel electrophoresis due to the size homoplasy. Phylogenetic relationships were reconstructed using the 224 DNA sequences amplified by the marker SSR38-1 (Fig. 4). Using neighbor-joining method in MEGA 6.06 software, 224 genotypes were categorized into two major groups on the basis of genetic similarity matrix obtained using the software POWER-MARKER 3.25. One major group included four subgroups I, II, III, and IV and another one having two subgroups V and VI (Fig. 4). Genotypes from different naturalized populations were mixed within subgroups. However, genotypes from VA were clustered and concentrated in three subgroups (II, III, and IV) of one major group. Genotypes from OH were found in four subgroups (I, II, III, and VI) in both major groups.

Discussion

In this study, DNA markers from sugarcane proved the most useful tool for analyzing *Miscanthus* DNA. Ten of these markers (actin, GA30x, PF00931, PF03856, SMC226CG, SMC248CG, SMC319CG, SMC1039CG, EST-SSR29, and EST-SSR38-2) generated clear and unambiguous amplicons showing the polymorphism at each locus in *Miscanthus*. Thus, these polymorphic markers can be used in *Miscanthus* genetic and genomic studies, such as genetic linkage mapping and genetic diversity studies.

M. sinensis is an important ornamental crop in the United States and Europe and it is also one of the parental species of *M. × giganteus*, the most economically important *Miscanthus* species for bioenergy (Linde-Laursen, 1993; Lafferty & Lelley, 1994). *M. sinensis* can be used not only as a genetic resource for development of new hybrids or improvement of fertile lines, but also as an alternative biofuel crop besides *M.* × *giganteus* based on a series of studies on agronomy, productivity, and utilization (Christian & Haase, 2001; Jørgensen & Muhs, 2001; Lewandowski *et al.*, 2003; Clifton-Brown *et al.*, 2008). Therefore, it is important to understand the genetic diversity of US naturalized *M. sinensis* populations and their potential value for breeding improved *Miscanthus* cultivars that would benefit the ornamental horticulture and bioenergy industries. In the present study, we found high diversity observed among these genotypes, indicating the potential of its use in the genetic improvement of *Miscanthus*. Our study shows that the genetic diversity within each U.S. naturalized population ranged from 0.28 to 0.37 (Table 3), which was similar to the genetic diversity ranging from 0.25 to 0.32 within different provinces that was observed for

some Chinese *M. sinensis* populations although different markers were used (Zhao *et al.*, 2013). Clark *et al.* (2014, 2015) found that US naturalized *M. sinensis* were derived from ornamental cultivars that originated in portions of southern Japan. The present results suggest that although the US naturalized populations are the product of a genetic bottleneck, considerable genetic diversity remains and thus could be useful resource for enabling selection of parents in breeding to develop hybrids with high heterosis.

In this study, 228 *M. sinensis* genotypes from six naturalized populations in the United States could be divided into six groups (Fig. 2) using two statistical methods, STRUCTURE and MEGA 6.06 employing the genotyping dataset. Although differentiation among the groups was incomplete, as indicated by admixture estimates and group geographic composition, the groups were largely

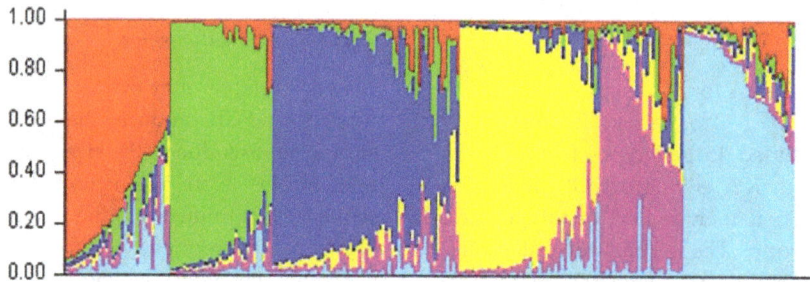

Fig. 2 Molecular classification by Structure showed six groups of US naturalized *Miscanthus sinensis*.

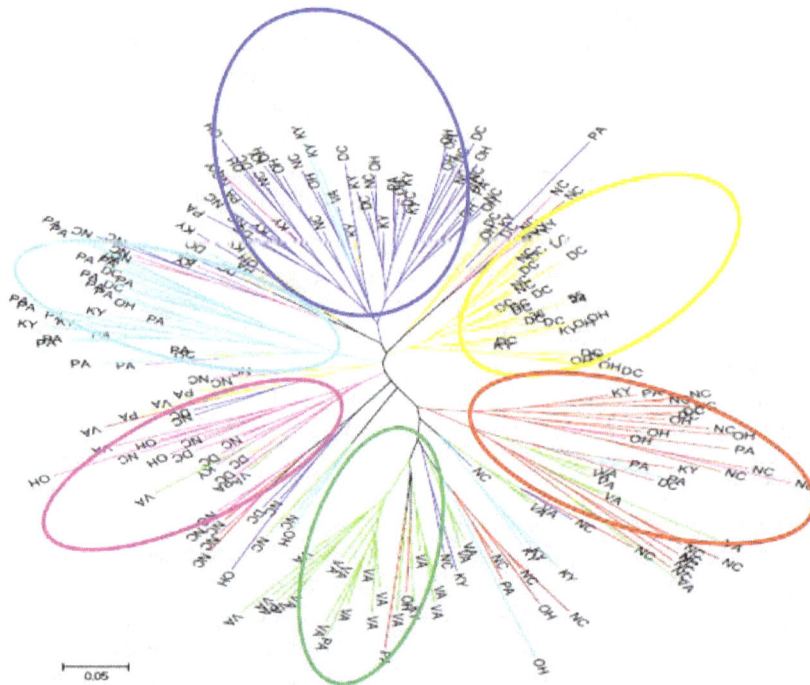

Fig. 3 Classification of US naturalized *Miscanthus sinensis* based on neighbor-joining tree clustering analysis using genotyping data by 10 polymorphic DNA markers. (Notes: the colors of the genotype branch are consistent with the colors of genotype groups in Fig. 2).

Table 6 Genetic variation in six naturalized U.S. *Miscanthus sinensis* populations based on DNA sequences

Population	Sample size	Genetic variation
VA	32	0.006
OH	34	0.008
KY	31	0.011
NC	57	0.012
PA	34	0.011
DC	36	0.011

Table 7 Genetic distance between *Miscanthus* naturalized populations by DNA sequences

Population	DC	KY	NC	OH	PA	VA
DC						
KY	0.012					
NC	0.013	0.013				
OH	0.012	0.011	0.012			
PA	0.013	0.012	0.013	0.011		
VA	0.014	0.012	0.015	0.011	0.013	
Mean distance	0.013	0.012	0.013	0.011	0.012	0.013

Fig. 4 Phylogenetic tree of US naturalized *Miscanthus sinensis* by analysis of sequences of SSR38-1 (cellulose synthase gene-related marker) based on neighbor-joining tree clustering analysis using MEGA 6.06 software. (NC = red, VA = green, OH = blue, DC = yellow, PA = aquamarine, and KY = dark pink).

consistent with geographic region (Figs 2 and 3). The result suggests that either there has been insufficient time for greater differentiation to have occurred and/or migration has allowed for gene flow between geographically distant populations. Seed of *Miscanthus* is dispersed long distances by wind, and many naturalized populations in the United States are found along highways that would further facilitate seed dispersal. In contrast, Zhao *et al.*

(2013) observed that Chinese *M. sinensis* were grouped based on their different provinces. The NC naturalized population was the most genetically diverse and also present in many of the Structure groups (Fig. 3, Table 3), which is consistent with the historical record that the Biltmore Estate in Asheville, NC, was an early grower and distributor of *M. sinensis* in the late 1800s and early 1900s (Quinn *et al.*, 2010). The largest GD between DC and VA naturalized populations was shown by both DNA marker data and DNA sequence data. Moreover, most genotypes from the VA naturalized population could be clustered in the same structure group (Fig. 3) and had the lowest genetic diversity based on DNA marker data (0.2776 in Table 3) and DNA sequence data (0.006 in Table 6), indicating they could be independently introduced from the same place or a narrow origin region differing from the origin for genotypes in other naturalized populations.

M. sinensis has great potential to be a feedstock for producing bioenergy including bioethanol (Heaton *et al.*, 2008; Hastings *et al.*, 2009a,b), and it has long been an important ornamental grass in American gardens. The information obtained in this study on genetic structure and diversity of US naturalized populations of *M. sinensis* will be useful for breeders to improve *M. sinensis* cultivars and hybrids. Both polymorphic markers and DNA sequences indicated that large genetic variation was found in NC and PA genotypes and small variation in VA genotypes. These results suggest that the higher genetic gain would be obtained within NC and PA populations for *Miscanthus* breeding. While less genetic advance would be reached within the VA population, it could be a valuable resource for new hybrids as parental lines because they are genetically diverged from those in other naturalized populations. The genetic information on *Miscanthus* genotypes in naturalized populations in this study could not only broaden the genetic knowledge of US *M. sinensis* germplasm, but also be of benefit to future association mapping studies.

There is a concern that *M. sinensis* could escape production to become a potential invasive species because it produces viable seeds (Raghu *et al.* 2006). To effectively control escape, strategies were proposed to develop complete sterility in breeding programs by inducing triploidy and functional sterility using combination of genotype by environment interactions to minimize seed-related invasiveness in *M. sinensis* (Quinn *et al.*, 2010). In this study, we focused on genetic diversity of naturalized genotypes of *M. sinensis* rather than its invasiveness. However, one rare allele with frequency of ≤5% was observed among genotypes studied. A further study is needed to understand whether this rare allele was caused by mutation or gene flow between *Miscanthus* and related species due to its nature of open pollination.

Acknowledgements

The research project conducted at Tuskegee University was financially supported by a grant from USDA/NIFA under the 1890 Institutions Capacity Building Grants Program (Award number 2012-38821-20046) and the George Washington Carver Agricultural Experiment Station at Tuskegee University.

References

Anonymous (1876) The zebra-striped Eulalia. *The American Agriculturalist*, **3**, 460.

Bailey LH, Miller W (1901) *Cyclopedia of American Horticulture*. The MacMillan Co, New York.

Bransby ID, Allen DJ, Gutterson N, Ikonen G, Richard E, Rooney W, Van Santen E (2010) Engineering advantages, challenges and grass energy crops. *Biotechnology in Agriculture and Forestry*, **66**, 125–154.

Brown M, Hopkins MS, Mitchell SE *et al.* (1996) Multiple methods for the identification of polymorphic simple sequence repeats (SSRs) in sorghum (*Sorghum bicolor* (L.) Moench). *Theoretical and Applied Genetics*, **93**, 190–198.

Christian DG, Haase E (2001) Agronomy of *Miscanthus*. In: *Miscanthus for energy and fibre* (eds Jones M, Walsh M), pp. 21–45. James and James, London, UK.

Clark LV, Brummer JE, Głowacka K, *et al.* (2014) A footprint of past climate change on the diversity and population structure of *Miscanthus sinensis*. *Annals of Botany*, **114**, 97–107.

Clark LV, Stewart JR, Nishiwaki A, *et al.* (2015) Genetic structure of *Miscanthus sinensis* and *Miscanthus sacchariflorus* in Japan indicates a gradient of bidirectional but asymmetric introgression. *Journal of Experimental Botany*, **66**, 4213–4225.

Clifton-Brown JC, Chiang YC, Hodkinson TR (2008) *Miscanthus*: genetic resources and breeding potential to enhance bioenergy production. In: *Genetic Improvement of Bioenergy Crops* (ed. Vermerris W), pp. 273–294. Springer, New York.

Cordeiro GM, Taylor GO, Henry RJ (2000) Characterisation of microsatellite markers from sugarcane (*Saccharum* sp.), a highly polyploidy species. *Plant Science*, **155**, 161–168.

Dai L, Wang B, Zhao H, Peng J (2013) Transferability of genomic simple sequence repeat and expressed sequence tag-simple sequence repeat markers from sorghum to *Miscanthus sinensis*, a potential biomass crop. *Crop Science*, **53**, 977–986.

Egnin M, Mora A, Prakash CS (1998) Factors enhancing *Agrobacterium tumefaciens*. Mediated gene transfer in peanut (*Arachis hypogea* L.). *In vitro cellular and Developmental Biology plants*, **34**, 310–318.

Evanno G, Regnaut S, Goudet J (2005) Detecting the number of clusters of individuals using the software STRUCTURE: a simulation study. *Molecular Ecology*, **14**, 2611–2620.

Glowacka K (2011) A review of the genetic study of the energy crop *Miscanthus*. *Biomass and Bioenergy*, **35**, 2445–2454.

Hastings A, Clifton-Brown J, Wattenbach M, Mitchell CP, Smith P (2009a) The development of MISCANFOR, a new *Miscanthus* crop growth model: towards more robust yield predictions under different climatic and soil conditions. *GCB Bioenergy*, **1**, 154–170.

Hastings A, Clifton-Brown J, Wattenbach M, Mitchell CP, Stampfl P, Smith P (2009b) Future energy potential of *Miscanthus* in Europe. *GCB Bioenergy*, **1**, 180–196.

Heaton EA, Dohleman FG, Long SP (2008) Meeting US biofuel goals with less land: the potential of *Miscanthus*. *Global Change Biology*, **14**, 2000–2014.

Hernández P, Dorado G, Laurie DA, Martin A, Snape JW (2001) Microsatellites and RFLP probes from maize are efficient sources of molecular markers for *Miscanthus*. *Theoretical and Applied Genetics*, **102**, 616–622.

Huff DR *et al.* (1993) RAPD variation within and among natural populations of outcrossing buffalograss *Buchloe dactyloides* (Nutt) Engelm. *Theoretical and Applied Genetics*, **86**, 927–934.

Jakob K, Zhou F, Paterson A (2009) Genetic improvement of C4 grasses as cellulosic biofuel feedstocks. *In Vitro cellular and Developmental Biology plants*, **45**, 291–305.

Jørgensen U, Muhs HJ (2001) *Miscanthus* breeding and improvement. In: *Miscanthus for Energy and Fibre* (eds Jones MB, Walsh M), pp. 68–85. James & James, London.

Kim C, Zhang D, Auckland SA *et al.* (2012) SSR based genetic maps of *Miscanthus sinensis* and *M. saccahriflorus*, and their comparison to sorghum. *Theoretical and Applied Genetics*, **124**, 1325–1338.

Lafferty J, Lelley T (1994) Cytogenetic studies of different *Miscanthus* species with potential for agricultural use. *Plant Breed*, **113**, 246–249.

Lewandowski I, Scurlockb JMO, Lindvall E *et al.* (2003) The development and current status of perennial rhizomatous grasses as energy crops in the US and Europe. *Biomass Bioenerg*, **25**, 335–361.

Liu K, Muse SV (2005) PowerMarker: an integrated analysis environment for genetic marker analysis. *Bioinformatics*, **21**, 2128–2129.

Linde-Laursen IB (1993) Cytogenetic analysis of *Miscanthus* 'Giganteus', an inter-specific hybrid. *Hereditas*, **119**, 297–300.

Nei M, Tajima FA, Tateno Y (1983) Accuracy of estimated phylogenetic trees from molecular data. *Journal of Molecular Evolology*, **19**, 153–170.

Peakall R, Smouse PE (2006) GENELEX 6: genetic analysis in Excel. Population genetic software for teaching and research. *Molecular Ecology Notes*, **6**, 288–295.

Peakall R, Smouse PE (2012) GenAlEx 6.5: genetic analysis in Excel. Population genetic software for teaching and research-an update. *Bioinformatics*, **28**, 2537–2539.

Pritchard JK, Stephens M, Donnelly P (2000) Inference of population structure using multilocus genotype data. *Genetics*, **155**, 945–959.

Quinn LD, Allen DJ, Stewart R (2010) Invasiveness potential of *Miscanthus sinensis*: implications for bioenergy production in the United States. *GCB Bioenergy*, **2**, 310–320.

Raghu S, Anderson RC, Daehler CC, Davis AS, Wiedenmann RN, Simberloff D, Mack RN (2006) Adding biofuels to the invasive species fire? *Science*, **313**, 1742–1742.

Selvi A, Balasundaram NN, Mohapatra T (2003) Evaluation of maize microsatellite markers for genetic diversity analysis and fingerprinting in sugarcane. *Genome*, **46**, 394–403.

Somerville C, Youngs H, Taylor C, Davis SC, Long SP (2010) Feedstocks for lignocellulosic biofuels. *Science*, **329**, 790–792.

Swaminathan K, Chae WB, Mitros T, *et al.* (2012) A framework genetic map for *Miscanthus sinensis* from RNAseq-based markers shows recent tetraploidy. *BMC Genomics*, **13**, 142.

Tamura K, Stecher G, Peterson D, Filipski A, Kumar S (2013) MEGA6: molecular evolutionary genetics analysis version 6.0. *Molecular Biology and Evolution*, **30**, 2725–2729.

Tamura K, Sanada Y, Shoji A *et al.* (2015) DNA markers for identifying interspecific hybrids between *Miscanthus sacchariflorus* and *Miscanthus sinensis*. *Grassland Science*, **61**, 160–166.

Xu WZ, Zhang XQ, Hung LK, Nie G, Wang JP (2013) Higher genetic diversity and gene flow in wild populations of *Miscanthus sinensis* in southwest China. *Biochemical Systematics and Ecology*, **48**, 174–181.

Yook MJ, Lim SH, Song JS, *et al.* (2014) Assessment of genetic diversity of Korean *Miscanthus* using morphological traits and SSR markers. *Biomass and Bioenergy*, **66**, 81–92.

Yu J, Zhao H, Zhu T, Chen L, Peng J (2013) Transferability of rice SSR markers to *Miscanthus sinensis*, a potential biofuel crop. *Euphytica*, **191**, 455–468.

Yuan JS, Tiller KH, Al-Ahmad H, Stewart NR, Stewart CN Jr (2008) Plants to power: bioenergy to fuel the future. *Trends Plant Science*, **13**, 421–429.

Zhang QX, Shen YK, Shao RX, *et al.* (2013) Genetic diversity of natural *Miscanthus sinensis* populations in china revealed by ISSR markers. *Biochemical Systematics and Ecology*, **48**, 248–256.

Zhao H, Yu J, You FM, Luo M, Peng J (2011) Transferability of microsatellite markers from *Brachypodium distachyon* to *Miscanthus sinensis*, a potential biomass crop. *Journal of Integrative Plant Biol.*, **53**, 232–245.

Zhao H, Wang B, He J, Yang J, Pan L, Sun D, Peng J (2013) Genetic diversity and population structure of *Miscanthus sinensis* germplasm in China. *PLoS One*, **8**, e75672.

Zhou HF, Li SS, Ge S (2011) Development of microsatellite markers for *Miscanthus sinensis* (Poaceae) and cross-amplification in other related species. *American Journal of Botany*, **98**, e195–e197.

11

The expansion of short rotation forestry: characterization of determinants with an agent-based land use model

JULE SCHULZE[1,2], ERIK GAWEL[3,4], HENNING NOLZEN[1,2], HANNA WEISE[5] and KARIN FRANK[1,2,6]

[1]Department Ecological Modelling, UFZ – Helmholtz Centre for Environmental Research, Permoserstr. 15, 04318 Leipzig, Germany, [2]Institute for Environmental System Research, Osnabrück University, Barbarastr. 12, 49076 Osnabrück, Germany, [3]Department of Economics, UFZ – Helmholtz Centre for Environmental Research, Permoserstr. 15, 04318 Leipzig, Germany, [4]Institute for Infrastructure and Resources Management, Leipzig University, Grimmaische Str. 12, 04109 Leipzig, Germany, [5]Institute of Biology, Biodiversity and Ecological Modelling, Freie Universität Berlin, Altensteinstr. 6, 14195 Berlin, Germany, [6]iDiv – German Centre for Biodiversity Research Halle-Jena-Leipzig, Deutscher Platz 5a, Leipzig, Germany

Abstract

Wood is a limited resource which is exposed to a continuously growing global demand not least because of a politically fostered bioenergy use. One approach to master the challenge to sustainably meet this increasing wood demand is short rotation forestry (SRF). However, SRF is only gradually evolving and it is not fully understood which determinants hamper its expansion. This study provides theoretical insights into economic and environmental determinants of an SRF expansion and their interplay. This assessment requires the incorporation of farmers' decision-making based on an explicit investment appraisal. Therefore, we use an agent-based model to depict the decision-making of profit-maximizing farmers facing the choice between SRF, the cultivation of conventional annual agricultural crops and abstaining from cultivation (fallow land). The land use decisions are influenced by general economic determinants, such as market prices for wood and annual crops, and by site-dependent determinants, such as the environmental site quality. We found that the willingness to pay for SRF-based products and for annual crops most strongly influences the coverage of SRF in the landscape. SRF will in most cases be established on sites with low productivity. However, a decrease in the willingness to pay for annual crops will lead to a reallocation of SRF plantations to sites with higher productivity. Furthermore, our model results indicate that the impact of the distance to processing plants on farmers' decisions strongly depends on general economic determinants and the given spatial structure of the underlying natural landscape. Analysing the relative importance of different determinants of an SRF expansion, this study gives insights into the approach of using SRF to sustainably meet the growing wood demand. Moreover, these insights are taken as a starting point for the design of effective government interventions to promote SRF.

Keywords: agent-based model, bioeconomy, energy crops, farmer, human decision-making, landscape generators, woody biomass

Introduction

Wood is a limited bio-based resource that serves as a source for material, power and heat. The global wood demand is increasing due to economic growth and demographic change (FAO, 2014). Lamers *et al.* (2012) depicted a more than tenfold increase in EU demand for wood pellets and an exponential increase in global trade of wood pellets from 0.5 to 6.6 Mt between 2000 and 2010. This increase is expected to be further pushed by the growing relevance of the bioeconomy, that is the enclosure of all economic sectors that develop, produce or use bio-based renewable resources. The European

Commission, for instance, has presented a bioeconomy strategy in 2012 that aims at a low-carbon and resource-efficient economy (European Commission, 2012). The Netherlands, Denmark, Sweden, Finland, Germany, Canada and the United States have presented national bioeconomy strategies, and other countries are expected to follow (BMEL, 2014). The associated stronger role of bio-based resources including innovative wood uses may even further increase the wood demand in the future.

As a consequence, the challenge is to meet the increasing wood demand without negative environmental effects. Woodland and natural forests provide multiple regulating ecosystem services such as carbon storage or purification of water and air. Furthermore, forests are a habitat for about 80% of world's terrestrial

Correspondence: Jule Schulze
e-mail: jule.schulze@ufz.de

biodiversity (IUCN, 2012). They are cleared at the rapid rate of about 13 million hectares per year leading to severe negative environmental impacts. Therefore, a variety of policy instruments aiming at protecting forests and avoiding such negative impacts are implemented worldwide (e.g. German Federal Forest Act, Código florestal in Brazil or REDD+). These policies set limits to the amount of wood that can be sourced from forests.

An alternative approach to meet the increasing wood demand is short rotation forestry (SRF). SRF plantations consist of fast-growing trees, whose common species include poplar and willow, which are grown as perennial energy crops on agricultural land (Faasch & Patenaude, 2012). SRF plantations can either be managed as stem plantations with rotation cycles of 10–15 years or as coppice systems using stump sprouting with rotation cycles of approximately 4 years. After several of these rotations, the land is re-cultivated. While the first group is used for fibre production, the latter practice is referred to as short rotation coppices (SRCs) and is often used for energy purposes (Mantau et al., 2010). Therewith, SRF plantations may fulfil multiple bioeconomic purposes. At the same time, several environmental advantages over conventional agriculture are being discussed (for overviews of environmental impacts of SRF see BfN 2012, Thrän et al. 2011 or Weih & Dimitriou 2012). For example, SRF is expected to have a positive effect on biodiversity (Sage et al., 2006; Rowe et al., 2011; Holland et al., 2015) as well as on soil and water quality (Makeschin, 1994; Schmidt-Walter & Lamersdorf, 2012). Nonetheless, environmental benefits of SRF are strongly dependent on site- and plantation-specific characteristics (e.g. tree species, cultivation design). Negative impacts, for example on the water balance, can also occur (e.g. Dauber et al. 2010, Thrän et al. 2011 or Strohm et al. 2012). Still, positive impacts predominate and SRF expansion is seen as promising approach to sustainably meet the growing wood demand.

However, the expansion of SRF is proceeding slowly. For example, for Germany and the year 2013, Drossart & Mühlenhoff (2013) reported an area of approximately 6500 ha SRF which only represents 0.03% of the total agricultural land (FAOSTAT, 2015). For Sweden and the year 2011, Dimitriou et al. (2011) reported an area of 14 000 ha willow SRC cultivations or 0.5% of total agricultural land. Past studies have predicted strong increases in SRF for several European countries. For example, in the 1990s, stakeholders predicted that the SRC area in Sweden would increase to several hundreds of thousands of hectares (Helby et al., 2004). Almost two decades later, in 2006, the European Environment Agency still stated that SRF would substantially increase from 2010 onwards (EEA, 2006). Given the above stated statistics on current cultivation areas, it

becomes evident that these predictions have failed so far. At the same time, EU wood pellet demand increased by 43.5% from 2008 to 2010 (Cocchi et al., 2011).

Various reasons for the slow uptake of SRF in Europe are discussed in the literature. Main barriers include high initial investment costs combined with uncertain returns on investment. The high uncertainty is caused by price volatility (Finger, 2016) as well as by uncertain yields and production costs (Strohm et al., 2012). In such a situation, it is a good strategy to postpone investment in order to wait for the occurrence of learning curve effects (Musshoff, 2012; de Wit et al., 2013). In addition, capital (especially land) is bound for a long time, leading to inflexibility to react to changing market developments (Strohm et al., 2012; Schweier & Becker, 2013). Still, the relative importance of different determinants that hamper SRF expansion in the EU is not fully understood.

Empirical analyses of spatial distributions of SRF are one approach to identify such determinants. For example, Mola-Yudego & Gonzalez-Olabarria (2010) use a geostatistical method to depict determinants of SRC establishment in Sweden. However, low SRF establishment leads to low data availability on commercial plantations, and therefore, only a few studies exist, which focus on specific regions. We believe that this issue can be tackled by considering SRF expansion as a result of land use decisions and by analysing the decision-making and its implication for the regional land use pattern within a modelling framework. Agricultural decisions as on the adoption of SRF are mostly driven by expected profits, that is expected revenues and costs. These can depend on both site conditions (e.g. soil quality or precipitation) and factors that are not site-specific (e.g. market conditions). For our analysis, we will refer to them as site-dependent determinants and general economic determinants. Mean annual temperature and precipitation, soil quality and transportation costs to the next woody biomass processing plant are important site-dependent determinants for the economic feasibility of SRF (cf. Dunnett et al. 2008 and Faasch & Patenaude 2012, respectively). Demands or prices for agricultural products are important general economic determinants. The interplay of general economic and site-dependent determinants and its effect on individual land use decisions have not been systematically analysed so far. This may be owed to the complexity of the underlying decision mechanisms which evolves from the need to compare crops with harvest cycles of different lengths.

This study investigates how the above-mentioned economic and environmental determinants affect SRF expansion in terms of the increase in land cover and spatial distribution of plantations. We focus on the

European context and analyse the relative importance of site-dependent determinants and general economic determinants. More specifically, we investigate the two site-dependent determinants 'environmental site quality' and 'distance to woody biomass processing plants' as well as seven general economic determinants such as 'willingness to pay for agricultural products' or 'investment expenditures'. In addition, we test the transferability of model results between regions by analysing to what extent these findings depend on the spatial structure of the underlying natural landscape. In particular, we assess the relevance of the explicit spatial configuration and the predictive power of aggregated spatial characteristics of the underlying landscape. For this purpose, we develop a spatially explicit agent-based model (ABM) to depict the decision-making of profit-maximizing farmers in a stylized landscape indirectly interacting via a market mechanism. This approach enables us to simulate and analyse land use decisions under different economic framework conditions and in differently structured stylized landscapes. Instead of providing quantitative predictions for a specific case study, we aim to derive a comprehensive general mechanistic understanding on the SRF expansion. We take these insights as a starting point to discuss the design of effective government interventions to promote SRF. Finally, we conclude by reflecting on the potential of the applied modelling approach.

Material and methods

In the following, we present the model INCLUDE (INdividual Cultivators' Land Use DEcisions). It is based on an ABM developed by Weise (2014): a stylized model of rational land use decisions that comprises markets and policy instruments to assess land use effects of promoting bioenergy. We expand this model to enable the incorporation of spatial heterogeneity and of an explicit investment appraisal to include crops with harvest cycles of different lengths.

General conception

The model INCLUDE is a simple ABM based on a stylized landscape. These types of models are considered particularly valuable for the purpose of system understanding, hypothesis testing and communication (Schlüter et al., 2013). In this sense, the model purpose of this study is not to provide quantitative predictions for specific case studies but to derive a comprehensive general mechanistic understanding on the SRF expansion.

The model INCLUDE considers regional land use change as result of individual land use decisions. The landscape is described as regular grid of 50 × 50 cells of approximately 45 ha each (based on Fischer et al. 2011). In each cell, there is one agent (i.e. farmer) who decides on the crop to be cultivated in the next time step. The agents are assumed to be rational profit maximizers with full knowledge over revenue and costs

of all possible land use options. We believe that profit maximization is an appropriate assumption for decisions in the European industrial agricultural sector.

In the model, agricultural markets are assumed to be endogenous and to mediate interactions among agents. Therefore, equilibrium market prices for both SRF-based products and products based on annual crops are described in the model by the ratio of exogenously given demands and the endogenously resulting supply that is determined by the agents' cultivation decisions. This price formation is in line with standard economic theory (e.g. equilibrium concept; cf. Mankiw & Taylor 2006 or Engelkamp & Sell 2007) and incorporates the critical market feedback of supply decisions that result in prices which influence again supply decisions (as also used by Lawler et al. 2014). In the result of the individual decisions of all agents and the interactions mediated by the market mechanism, land use patterns emerge and evolve over time.

We assume that the agents' land use decisions are influenced by general economic (i.e. same for all cells) and site-dependent determinants (i.e. different between cells). All determinants investigated in this study are shown in Table 1. The site quality of a cell subsumes environmental site characteristics such as mean annual precipitation and soil quality and therefore influences agricultural productivity. In the model, the determinant 'harvest costs' represents the costs for harvesting SRF plantations and no other production factors. Harvest costs of annual crops are included in the production costs of annuals which further include seed and crop protection of these crops. Therefore, and due to the extent of the landscape stated above, harvest and production costs are seen as general determinant and transport costs are the only site-dependent costs.

To address the site-dependent determinants, we need to incorporate spatial heterogeneity. Moreover, as we aim to gain general mechanistic understanding of SRF expansion, rather than exploring a specific region, we decided to investigate stylized landscapes. The underlying landscape is generated using a randomization algorithm which allows

Table 1 Determinants of land use decisions in the model

Determinants		General	Site-dependent
Economic	Aggregated willingness to pay for SRF products	x	
	Aggregated willingness to pay for annual crops	x	
	Investment expenditures	x	
	Discount rate	x	
	Recovery costs	x	
	Harvest costs	x	
	Transport price	x	
Environmental	Site quality value in cell		x
	Distance to processing plant		x

generating a variety of landscapes that coincide in certain aggregated spatial characteristics but differ in their explicit spatial configuration. This enables to test the transferability of results between landscape types. Each generated landscape consists of a grid of cells with both specific site qualities and locations of woody biomass processing plants (Fig. 1). These site-dependent determinants together with the general economic determinants influence the agents' land use decisions and hence the emerging land use pattern (Fig. 1). The approach of combining the ABM and a landscape generator enables us to systematically investigate the relative importance of the general economic and site-dependent determinants for the SRF cultivation decisions.

In addition to the spatial heterogeneity, the perennial character of SRF requires the incorporation of an explicit investment appraisal. INCLUDE runs on an annual temporal scale as annual crops are also included. To enable the comparison between land use options with different lengths of harvest cycles, the equivalent annual annuity approach from investment theory is chosen (e.g. Brigham & Houston (2006)). This approach calculates a constant annuity from an uneven cash flow for several periods. In a first step, the net present value for the investment is calculated by discounting the annual profits. In a second step, this net present value is multiplied by the annuity factor to receive a constant value per year, the equivalent annual annuity. Discount rates are seen as subjective discount rates which can vary depending on personal risk aversion (Barberis & Thaler, 2003). The equivalent annual annuity approach is appropriate as it is often recommended to farmers interested in SRF practice (for example Schweinle & Franke 2010) and has been used in several studies on the financial analysis of SRF (Kasmioui & Ceulemans, 2012).

Initialization of landscape

At the beginning of each simulation, the underlying landscape is randomly generated: (i) environmental site qualities are assigned to cells, and (ii) woody biomass processing plants are spatially allocated within the landscape.

1. The distribution of site quality for the ABM was generated using a randomization algorithm that returns uniformly distributed, spatially correlated numbers with a fixed arithmetic mean and a certain spatial correlation. For this purpose, the method of Cholesky decomposition, which considers the covariances among all cells, was used (see appendix A in Thober *et al.* 2014 for details). This enables the generation of ensembles of landscapes with varying explicit configuration but the same aggregated spatial characteristics, that is mean and spatial correlation, of the site quality distribution (Fig. 2).

2. A fixed number of woody biomass processing plants are randomly placed within the landscape. At this, the number of processing plants can be adapted to represent regions with different areal densities (see Table S1 for standard parameter values).

Model processes

At the beginning of each decision step, the current market prices $p_j(t)$ in year t for the different products j, that is annual crops (ANN) and SRF crops, are calculated based on the regional supplies $H_j(t)$ and the following pricing rule:

$$p_j(t) = \frac{D_j}{H_j(t)} \text{ with } H_j(t) = \sum_{i=1}^{n} h_j^i(t), \quad (1)$$

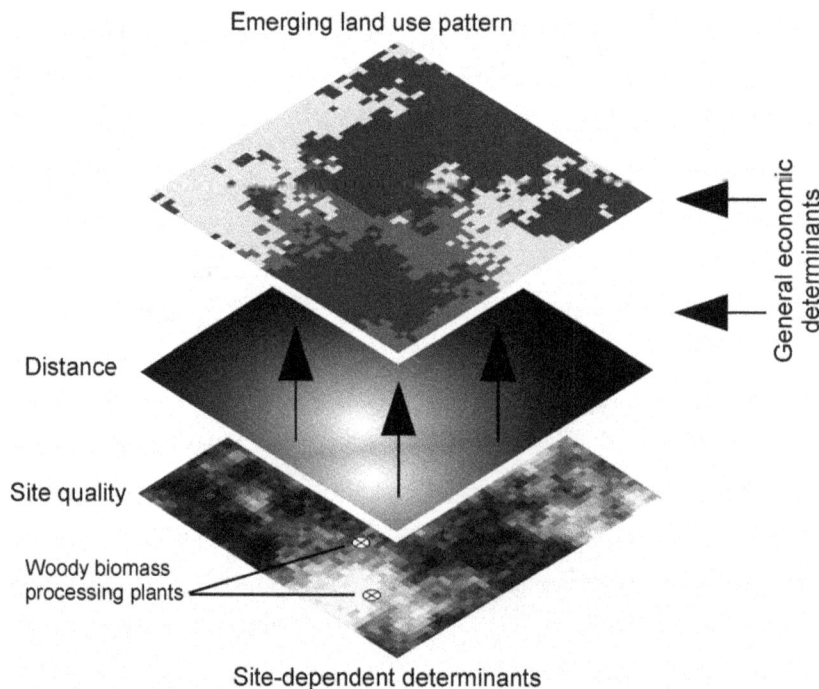

Fig. 1 Interplay of general economic and site-dependent determinants in the course of the short rotation forestry expansion.

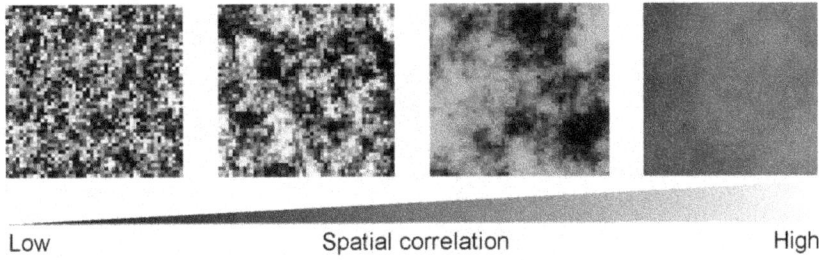

Low   Spatial correlation   High

Fig. 2 Examples of generated landscapes with increasing spatial correlation from left to right.

where D_j is the aggregated willingness to pay for product $j \in \{\text{ANN}, \text{SRF}\}$, n the number of agents and $h_j^i(t)$ the harvest amount of product j in cell i given by:

$$h_{\text{ANN}}^i(t) = \begin{cases} q^i, \text{if land use is ANN} \\ 0, \text{if land use is not ANN} \end{cases} \tag{2}$$

$$h_{\text{SRF}}^i(t) = \begin{cases} q^i \cdot 0.2 + h_{\min}, \text{if land use is SRF} \\ 0, \text{if land use is not SRF} \end{cases}, \tag{3}$$

where q^i is the site quality of the cell of agent i. Site quality subsumes factors such as mean annual precipitation and soil quality known to strongly impact the agricultural output and. Therefore, we assume that the yield of the annual crops and the site quality are linearly correlated with both factors being normalized between 0 and 1. The yield of SRF plantations is also assumed to decrease on poor sites. At this, the dependence on the site quality is less pronounced than for annual crop production because SRF is more resistant against poor site conditions than annual crops.

The land use in the cells is determined by the agents' decisions based on profit calculation. This calculation differs between the three land use options: no cultivation (NoC), ANN and SRF. If agent i abstains from cultivation, neither costs nor revenue arise and the related profit Π for agent i is therefore:

$$\Pi_{\text{NoC}}^i = 0. \tag{4}$$

For annual agricultural crop production, the following profit function applies:

$$\Pi_{\text{ANN}}^i(t) = p_{\text{ANN}}(t) \cdot h_{\text{ANN}}^i - c_{\text{ANN}}, \tag{5}$$

where $p_{\text{ANN}}(t)$ is the current market price (calculated by the pricing rule shown in Eqn 1), h_{ANN}^i the harvest of annual crops in the cell of agent i and c_{ANN} the production costs of annuals. For the profit calculation of the SRF option, the profit of agent i in year t, $\Pi_{\text{SRF}}^i(t)$, over the whole lifetime T of the SRF is calculated by Eqn (6). This stream of profits will be the basis for calculating the equivalent annual annuity (Eqn 11). In the first year, only costs accrue, followed by both profit and costs accruing after each rotation cycle:

$$\Pi_{\text{SRF}}^i(t) = \begin{cases} -c_{\text{SRF}}^i(t), \text{if } t = 0 \\ p_{\text{SRF}}(t) \cdot h_{\text{SRF}}^i - c_{\text{SRF}}(t), \text{if } t \bmod a = 0, \\ 0, \text{else} \end{cases} \tag{6}$$

where $p_{\text{SRF}}(t)$ is the current market price in year t for SRF products produced in one rotation cycle on optimal site conditions calculated by the pricing rule shown in Eqn (1), h_{SRF}^i the harvest of SRF in the cell of agent i, $c_{\text{SRF}}^i(t)$ are all incurring costs

in year t calculated by Eqn (7) and a is the number of years after which harvest takes place, that is the rotation cycle (therefore $t \bmod a = 0$ indicates the end of a rotation cycle).

Finally, all occurring costs are calculated by Eqn (7). As perennial crops are associated with higher risks than annual crops (e.g. damages from drought or pests), farmers require a compensation for accepting the higher risk (Sherrington & Moran, 2010; Rosenquist et al., 2013). To reflect this, we include yearly risk costs in the decision model as have been empirically quantified by Rosenquist et al. (2013). These risk costs are assumed to decrease with the increase in SRF coverage in the landscape due to learning effects.

In the first year, only investment expenditures v accrue. At the end of each rotation cycle (i.e. $t \bmod a = 0$), harvest costs h, transport costs to the processing plant Γ^i and risk costs k occur. Finally, at the end of the lifetime T, in addition to harvest, transport and risk costs, recovery costs of the land r have to be paid. In all other years, no treatments are needed, and therefore, only risk costs occur:

$$c_{\text{SRF}}^i(t) = \begin{cases} v, \text{if } t = 0 \\ h + \Gamma^i \cdot h_{\text{SRF}}^i + k(\Phi_{\text{SRF}}), \text{if } t \bmod a = 0 \text{ and } t < T \\ h + \Gamma^i \cdot h_{\text{SRF}}^i + r + k(\Phi_{\text{SRF}}), \text{if } t = T \\ k(\Phi_{\text{SRF}}), \text{ else} \end{cases}, \tag{7}$$

where t is the current year, v are the investment expenditures, k are the risk costs calculated by Eqn (8), Φ_{SRF} is the SRF coverage, h the harvest costs, Γ^i the transportation costs of wood produced under optimal site quality conditions calculated by Eqn (9), h_{SRF}^i the actual harvest of SRF in the cell of agent i, a the rotation cycle and r the recovery costs. The risk costs k are assumed to be linearly dependent on the current SRF coverage Φ_{SRF} (given in percentage of the whole landscape). The function has been parameterized following results of Rosenquist et al. (2013):

$$k(\Phi_{\text{SRF}}) = \max(0, k_{\max} - k_{\text{slope}} \cdot \Phi_{\text{SRF}}). \tag{8}$$

The transportation costs are calculated based on Bauen et al. (2010), including a tortuosity factor of 1.6 to model the road network. The transportation costs are assumed to be linearly dependent on the distance to woody biomass processing plants:

$$\Gamma^i = \tau + \gamma \cdot d^i, \tag{9}$$

where d^i is the distance of agent i to the processing plant, τ are fixed costs for transportation and γ the transport price per distance. We assume a homogeneous cell size f to calculate the distance d using Euclidean distance (Deza & Deza, 2013).

From the sequence of profits $\Pi^i_{\text{SRF}}(t)$, the net present value is calculated as the sum of the discounted profits:

$$N^i = \sum_{t=0}^{T} (1+s)^{-t} \cdot \Pi^i_{\text{SRF}}(t), \qquad (10)$$

where T is the lifetime of the plantation, s the discount rate and $\Pi^i_{\text{SRF}}(t)$ the profit in year t calculated by Eqn (6). Subsequently, the equivalent annual annuity E is calculated from the net present value N to enable the comparison of land use options with unequal lifespans:

$$E^i = \frac{1}{1 - (1+s)^{-T}} \cdot N^i, \qquad (11)$$

where s is the discount rate, T the lifetime of a SRF plantation and N^i the net present value calculated by Eqn (10).

Finally, the agent compares the equivalent annual value E^i with the possible profit from annual agricultural crop production $\Pi^i_{\text{ANN}}(t)$ and chooses the option with the higher profit. If both, the equivalent annual value E^i of SRF and the profit of annual agricultural crop production $\Pi^i_{\text{ANN}}(t)$, would yield negative profits, the agent decides to abstain from cultivation.

All model parameters, their values and the references for parameterization can be found in the Table S1.

Evaluation criteria and simulation experiments

In this study, we investigate how different determinants affect a possible SRF expansion after entering the market in terms of the increase in SRF coverage and their spatial distribution across the stylized landscape. We assess the relative importance of different general economic and site-dependent determinants in differently structured stylized landscapes.

For this purpose, we apply an ensemble approach and perform a spatial sensitivity analysis as follows. All landscapes belonging to the same ensemble coincide in the aggregated spatial characteristics but differ in their explicit spatial configuration. Accordingly, the variance in the outcomes for all landscapes of the ensemble indicates the sensitivity of the evaluation criteria to changes in the explicit spatial configuration. Additionally, the randomization algorithm enables us to generate ensembles with different aggregated spatial characteristics. In this study, we compare two scenarios with ensembles of different spatial correlations of site quality (Fig. 2). Therefore, we vary the spatial correlation and hold the mean site quality constant. As a consequence, the frequency of site qualities also changes with the spatial correlation because of the changing spatial variability. A low spatial correlation leads to a uniform frequency distribution because site qualities of all levels are occurring. A high spatial correlation implies a clustering of site qualities around their mean while extreme values are not occurring.

Based on this ensemble approach, we perform a systematic model analysis in two steps, which are summarized in Table 2.

In the first step, we analyse the impact of general economic determinants (see Tables 1 and 2 for the specific determinants and the respective model parameters) on the land use pattern in general and the SRF coverage in particular. At this, we vary each general economic determinant individually, while all other parameters are kept constant. To quantify how sensitive the SRF expansion reacts to these determinants, we use the sensitivity index SI (see for example Bauer & Hamby 1991) which is given by the percentage difference in model output when varying one parameter over its entire range:

$$\text{SI} = \frac{O_{\max} - O_{\min}}{O_{\max}}, \qquad (12)$$

where O represents the model output. As we are interested in SRF expansion, we chose the SRF coverage Φ_{SRF} in year 50, that is the number of cells with SRF divided by total number of

Table 2 Overview of analysis steps: evaluation measures and model parameters investigated under different scenarios for which the single analysis steps are repeated

Subject of analysis	Evaluation measure	Investigated model parameters	Scenarios for transferability test	Section
Step 1: General economic determinants	Sensitivity index of short rotation forestry (SRF) coverage in landscape	Aggregated willingness to pay for SRF-based products D_{SRF}; Aggregated willingness to pay for annual crops D_{ANN}; Investment expenditures v; Recovery costs r; Harvest costs h; Transport price γ	a) Standard b) High discount rate c) High spatial correlation of site quality	Influence of general economic determinants
Step 2: Site-dependent determinants and interplay with general economic determinants	Probability of SRF occurrence	Aggregated willingness to pay for SRF products D_{SRF}; Aggregated willingness to pay for annual crops D_{ANN}	a) Standard b) High spatial correlation of site quality	Influence of site-dependent determinants

cells in the landscape, as investigated model output. As a result, a ranking of the relative importance of general economic determinants can be derived. As stated above, the standard deviation of the SRF coverage Φ_{SRF} and of the sensitivity index over the ensemble gives insights into the importance of the explicit spatial configuration. In addition, we test the transferability of the sensitivity results between landscapes with different aggregated spatial characteristics (high and low spatial correlation of site quality) and between landscapes populated by farmers with different risk attitudes. Therefore, we repeat the gradual variation in general economic determinants for two more scenarios: a high discount rate of the agents and a high spatial correlation of site quality.

In a second step, we analyse the impact of the two site-dependent determinants 'site quality' and 'distance to processing plant'. Therefore, we determine the probability that an agent in year 50 cultivates SRFs given a certain site quality and distance to processing plant. The probability calculation is based on the ensemble of underlying landscapes. In addition, we analyse the interplay of the site-dependent determinants with general economic determinants by repeating the analysis for an increasing aggregated willingness to pay for the two agricultural products. Finally, we again test the transferability of this interplay between landscapes with different aggregated spatial characteristics of the underlying natural landscape (high and low spatial correlation of site quality).

Results

When SRF enters the market

For a better understanding of model dynamics, we first show land use patterns that emerge under the standard parameter set (see Table S1). Here, we compare the case

with and without SRF as land use option available (Fig. 3a, b respectively).

Without SRF (Fig. 3a) as agricultural option, annual crops represent the dominant land use option with coverage of approximately 94% in this example and occupation of cells with high environmental site quality. The remaining 6%, characterized by low site quality, are covered with fallow land. The parameterization of this baseline scenario was chosen based on the situation in European countries where on average, 6% of agricultural land is fallow land (Allen *et al.*, 2014). In the model, the fallow sites are not chosen for agricultural production because here the yield of annual crops is low and agricultural practice hence not profitable, given the assumed willingness to pay for annual agricultural crops D_{ANN}. With SRF as land use option available (Fig. 3b), 17% of the landscape is covered by SRF plantations, largely at the expense of fallow land. The sites where SRF is cultivated are characterized by inferior sites. The reason for SRF cultivation on inferior sites is the low profit that annual crop cultivation yields on these sites. In the following section, we will investigate how different general economic determinants affect the expansion of SRF.

Influence of general economic determinants

To investigate the relative role of different general economic determinants, we analyse their impact on the mean SRF coverage Φ_{SRF} over the ensemble of landscapes with low spatial correlation of site qualities.

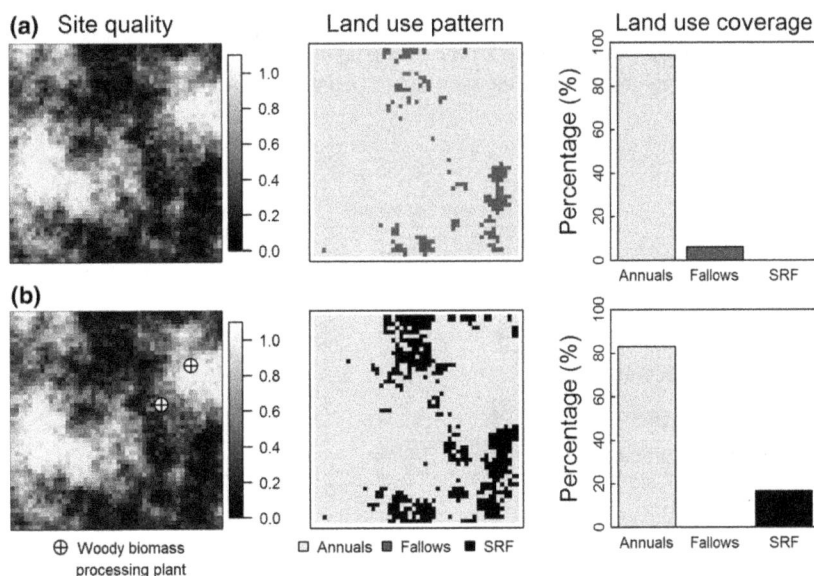

Fig. 3 Underlying landscape of site qualities and processing plants, resulting land use patterns and coverage of land use options after 50 years (a) without and (b) with short rotation forestry available as option.

Increasing aggregated willingness to pay for SRF products D_{SRF} as well as decreasing investment expenditures v positively affect the mean coverage of SRF plantations (Fig. 4a, b respectively). Triggered by a higher willingness to pay, the market price increases and positively influences the profit (see Eqns 1 and 6). The other way around, high investment expenditures represent a hurdle, which hinders the SRF cultivation decision. Note the very low standard deviation (indicated by the grey shading in Fig. 4) for the entire regarded parameter range. The landscapes within the ensemble only differ in their explicit spatial configuration. Therefore, the low standard deviation indicates that the explicit spatial configuration is not important for SRF coverage (possible reasons will be discussed in section General economic determinants). Instead, the general economic determinants strongly affect the coverage of SRF plantations in the landscape and dominate the importance of the explicit spatial configuration.

In a second step, we quantified the impact of various general economic determinants on the SRF coverage by performing a local sensitivity analysis and calculating sensitivity indices (see Eqn 12). To test the relative importance of these general economic determinants and the aggregated spatial characteristics of the underlying landscape, we performed the analysis for (i) the standard scenario, (ii) a higher discount rate and (iii) higher spatial correlation of site qualities. Therefore, we derive an indication whether general economic determinants would equally affect SRF expansion in different scenarios.

High-sensitivity indices indicate a high impact of the corresponding determinant. Under the standard scenario, the main drivers of the SRF expansion are the aggregated willingness to pay for SRF products and annual crops, the investment expenditures and the harvest costs (see Fig. 5a). The relative importance of these major determinants is influenced by the spatial correlation of site quality (Fig. 5c) and the higher discount rate (Fig. 5b). In the scenario with a higher discount rate, the impact of investment expenditures strongly increases (see Fig. 5b). With a higher discount rate, agents value profit accruing at the end of each rotation cycle less, and therefore, the initial hurdle of investment expenditures more strongly influences the SRF cultivation decision. Regarding landscapes with a different spatial structure, namely a higher spatial correlation of site qualities, the relative importance of the different economic variables is also changing. For instance, the impact of the aggregated willingness to pay for annual

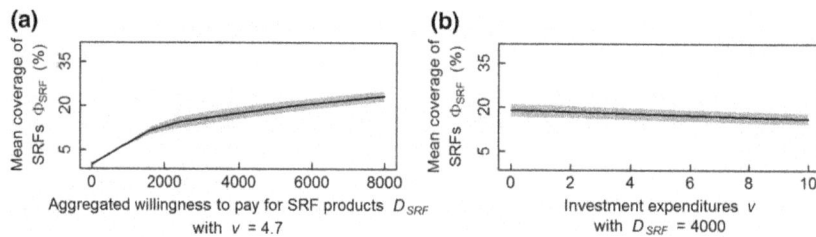

Fig. 4 Mean short rotation forestry (SRF) coverage Φ_{SRF} for increasing (a) aggregated willingness to pay for SRF products D_{SRF} and (b) investment expenditures vs. Grey shading indicates the standard deviation over the ensemble of the low spatial correlation of site qualities.

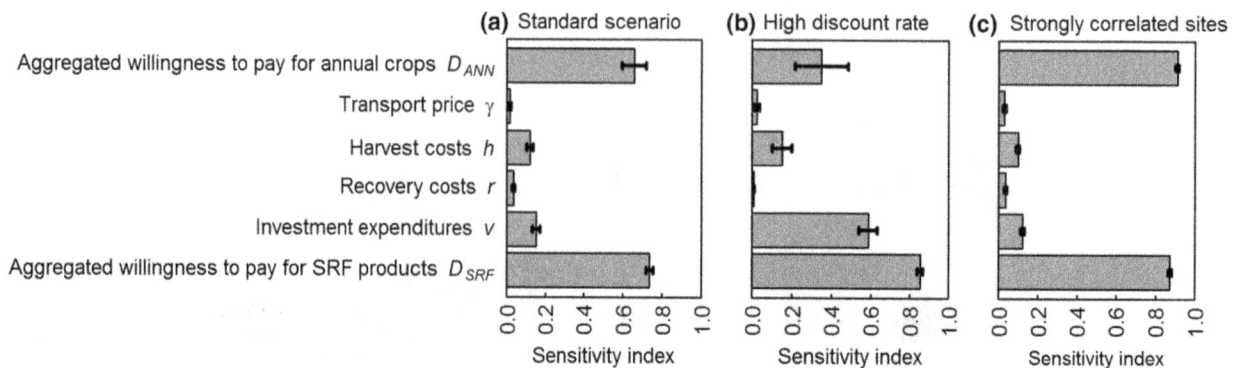

Fig. 5 Sensitivity indices of short rotation forestry coverage to general economic determinants in the three scenarios: (a) standard, (b) higher discount rate and (c) higher spatial correlation of site qualities. Error bars indicate the standard deviation over the respective ensemble.

crops increases (see Fig. 5c). The reason for this lies in the distribution of site qualities in the underlying landscape. While the spatial correlation of the distribution of site quality is higher than that under the standard scenario, the mean site quality is kept constant. As a consequence, the range of available site qualities for the landscape with high spatial correlation of site quality is narrower. The landscape contains fewer sites with low site qualities. We assume that the productivity of annual crops is more affected by low site quality than that of SRF. Therefore, fewer sites of low site quality also imply fewer sites on which the yield of annual crops is very low and the cultivation of SRF is therefore competitive. Therefore, the coverage of SRF is more strongly dependent on the economic situation of the competitive land use option. Again, the explicit spatial configuration is not influential as standard deviations are low for all parameters and scenarios. Hence, the results are transferable to regions with the same aggregated spatial characteristics but different explicit spatial configuration.

Influence of site-dependent determinants

In the second step of the analysis (cf. Table 2), the attention is shifted to the spatial pattern of SRF occurrence, its determinants and the explanatory power of certain site-dependent determinants. The focus is on the relative importance of environmental site quality and the distance to woody biomass processing plants for SRF allocation, that is two attributes which are both site-dependent, heterogeneously distributed and known to influence yield and/or costs of the various options of crop cultivation under consideration. Additionally, we

investigate the extent to which general economic determinants influence this relationship.

In all cases with the standard value for the willingness to pay for annual crops D_{ANN}, SRF occurrence is restricted to sites with low environmental site qualities (Fig. 6). On sites with high site qualities, the cultivation of SRF is economically not competitive with the high yields of annual crops.

In the scenario with standard spatial correlation of site qualities (Fig. 6a) and for a low to medium willingness to pay for SRF crops, the probability of SRF occurrence is positively correlated to the site quality, however, only up to a certain threshold of site quality above which the probability decreases abruptly. Higher site qualities increase the yield of SRF and therefore the probability of cultivating SRF. Here, higher distances to the processing plants d and therewith higher transport costs lead to a decreasing probability of SRF occurrence. Additionally, higher site qualities compensate for higher distances and, vice versa, lower distances for lower site qualities (indicated by the triangle shape of high probabilities in Fig. 6a). Yield of SRF, and therewith revenue, is higher on good sites. This compensates for higher transport costs of longer distances. Contrary, lower transport costs compensate for the lower revenue of SRF on sites with lower site quality.

The distance of the chosen SRF sites to their next processing plants d varies with the aggregated willingness to pay for SRF products D_{SRF}. For an increasing D_{SRF} (left to right column in Fig. 6a), sites with higher distances d become economically attractive and are therefore chosen for SRF cultivation. The higher willingness to pay leads to higher revenues from SRF which compensates for higher costs of longer distances.

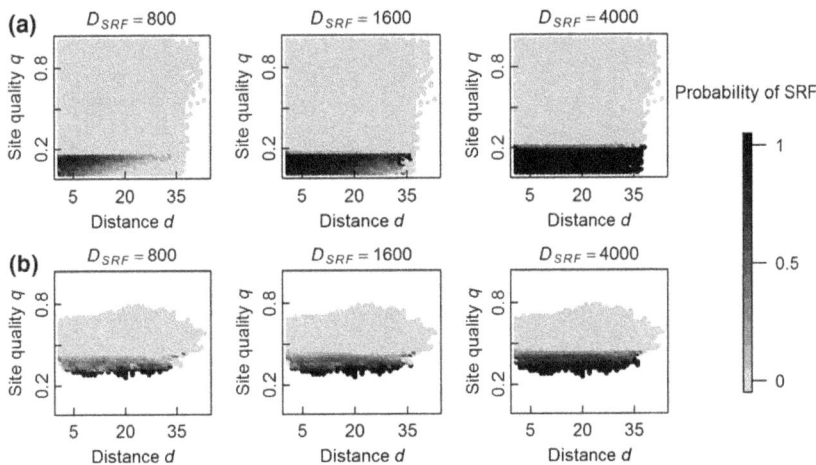

Fig. 6 Probability of short rotation forestry (SRF) occurrence for combinations of site quality q and distance d present in the underlying landscapes for an increasing aggregated willingness to pay for SRF-based products D_{SRF} and scenarios (a) standard and (b) high spatial correlations of site qualities. The willingness to pay for annual crops is set to the standard value of $D_{ANN} = 20500$.

In contrast, the importance of site quality as site-dependent determinant for SRF cultivation decision does not change with D_{SRF}. SRF plantations are cultivated on lower quality sites, independent of D_{SRF}. This is not a gradual interrelation. Instead, a threshold of site quality can be identified above which the cultivation of SRF is economically not competitive anymore. .

Finally, we investigate how the aggregated spatial characteristic of the underlying landscape affects the results (i.e. the spatial correlation of site quality; compare Fig. 6a, b). Recalling, a higher spatial correlation leads to a narrower range of available site qualities in the landscape; that is, site quality varies closely around the mean. While the site quality and SRF occurrence probability are positively correlated up to certain threshold for the standard scenario (Fig. 6a), they are negatively correlated up to a certain threshold for the high spatial correlation of site qualities (Fig. 6b). In the latter case, very low-quality sites are not available and farmers need to evade to higher site qualities to cultivate SRF. Here, the competition with annuals increases with the increase in site quality, resulting in a decrease in SRF probability. The importance of distance also changes between the two scenarios. Under the scenario with highly correlated site qualities, distance is not relevant under all of the investigated D_{SRF} (Fig. 6b). As described above, for higher correlated site qualities, fewer sites with low site qualities are available. This reduces the number of potential sites where SRF cultivation is competitive with annual crops. Therefore, farmers accept longer distances to processing plants. In other words, the comparison of the two scenarios indicates that the general economic determinant D_{SRF} alters the

importance of the site-dependent determinant 'distance' for the SRF decision. While the distance is still influential for a low and medium aggregated willingness to pay for SRF-based products D_{SRF} in the standard scenario, it is not in the landscape with high spatial correlation of site qualities. Hence, the results are not fully transferable between landscapes with different aggregated spatial characteristics.

In addition to the impact of the aggregated willingness to pay for SRF-based products D_{SRF}, we assessed the influence of the aggregated willingness to pay for annual crops D_{ANN} (Fig. 7). Again, higher distances to the processing plants d negatively influence the SRF occurrence probability. Moreover, site qualities and distances can compensate for each other (see explanation of Fig. 6). Here, these relationships are even more sensitive to the willingness to pay for annual crops D_{ANN} than they were to D_{SRF}.

For a lowered willingness to pay for annual crops D_{ANN}, sites with high site qualities are more likely to be chosen for SRF cultivation, independent of the spatial correlation of site qualities (left column of Fig. 7). Here, no competition with annuals takes place and SRF plantations are most profitable on good sites due to higher yields. As demand for annuals D_{ANN} increases, sites with low to medium site qualities are chosen for SRF cultivation.

A high willingness to pay D_{ANN} also leads to an increase in the realized distance of the chosen SCF sites to the processing plants. Due to the advantageous situation of the competitive annual crops, only sites with lower site qualities are chosen for SRF cultivation where yield of annual crops is low. These sites, however, can also be located far away from processing plants, and

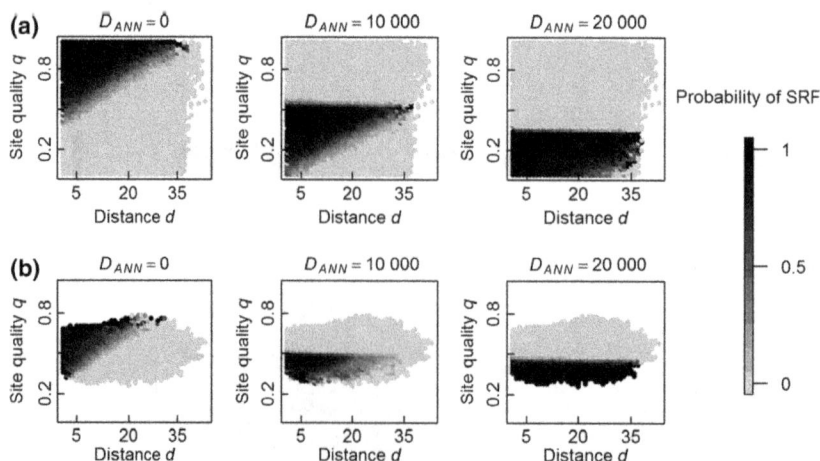

Fig. 7 Probability of short rotation forestry (SRF) occurrence for combinations of site quality q and distance d present in the underlying landscapes for an increasing aggregated willingness to pay for annual crops D_{ANN} and scenarios (a) standard and (b) high spatial correlations of site qualities. The willingness to pay for SRF-based products is set to the standard value of $D_{SRF} = 4000$.

therefore, also these sites with long distances to processing plants are chosen for SRF cultivation.

The spatial structure of the underlying landscapes again influences the impact of distance: while distance is still slightly influential for a high willingness to pay D_{ANN} in the standard scenario, it is not in the landscape with high spatial correlation of site qualities. The impact of site quality is again stable across the different spatial structures.

Discussion

In this work, we assessed the relative importance of different economic and environmental determinants for agricultural crop cultivation choice and showed how these influencing factors might affect a possible SRF expansion in terms of the SRF coverage and their spatial distribution. In the following paragraphs, we will draw conclusions from our model results, discuss advantages of the applied method and finish with an outlook on future research.

Determinants of SRF expansion

General economic determinants. Our model results indicate that general economic determinants have a strong impact on the uptake of SRF practice. This effect is relatively stable across the investigated scenarios with differently structured landscapes and different risk attitudes of farmers:

1. Independent of the investigated scenarios (i.e. spatial correlation of site quality and discount rate of farmers), the willingness to pay for SRF products showed to be one influential economic determinant of SRF expansion in the model. The reason is that the willingness to pay strongly affects the revenue of SRF.
2. Furthermore, given our model assumptions, the willingness to pay for the competitive land use option 'annual crops' and the investment expenditures represent strong determinants of SRF expansion. Therefore, the strength of their impact depends on the investigated scenario.
3. Transport price, harvest costs and recovery costs have a relatively low impact under all investigated scenarios.

These results are in accordance with empirical and model-based studies which showed the importance of electricity prices (analogue to the importance of the willingness to pay for SRF-based products in our model), establishment grants and demand for the spread of SRF cultivation (Mola-Yudego & Gonzalez-Olabarria, 2010; Alexander *et al.*, 2014; Mola-Yudego *et al.*, 2014). The low impact of the transport price is contrary to previous studies (e.g. Dunnett *et al.* 2008) and might increase when investigating a larger landscape than the one in this study.

In addition, we assessed to what extent these findings depend on the spatial structure of the underlying natural landscape. Therefore, we assessed the relevance of (i) explicit spatial configurations and (ii) aggregated spatial characteristics (i.e. the spatial correlation influencing the range of environmental site qualities present):

1. We showed that while general economic determinants have a strong impact on the SRF coverage, the importance of the explicit spatial configuration as we depicted it in the underlying landscape is negligible.
2. In contrast, the range of site qualities present in the landscape influenced the impact of the general economic determinants more strongly.

The results are therefore fully transferable between regions with different explicit spatial configurations but are not between regions with different aggregated spatial characteristics. However, further model experiments showed that with a substantial increase in transport price, the variation over the ensemble increases. This indicates that the transport price governs the relevance of the explicit spatial configuration of site quality distribution. Furthermore, in this study, we did not model the spatial allocation of the processing plants in dependence on the current feedstock supply. Modelling the two-way interaction between the establishment of processing plants and feedstock suppliers (as done by Alexander *et al.* 2013) may increase the importance of spatial configuration in our model results. In this study, the focus was on the supply side because the allocation of processing plants may be influenced by external factors such as political incentives or the proximity to consumer centres (esp. when the wood from SRF is used for heat supply).

Relevance of site-dependent determinants. Another focus of our analysis was on the impact of site-dependent determinants of SRF cultivation decisions. SRF plantations in the model will be located on sites with low productivity in most cases as annual crops are economically more competitive on sites with higher environmental site quality. This is confirmed by a survey among SRC operators in Bavaria in which SRC sites show below-average land rents (Hauk *et al.*, 2014). Skevas *et al.* (2015) showed a reduced difference in revenue between corn and bioenergy perennials on poor soils. Similarly, Helby *et al.* (2004) revealed a slight economic disadvantage for SRCs over food production on good soils. However, we showed that an intense decrease in the willingness to pay for annual crops will lead to a reallocation of SRF plantations in the model to sites with higher site quality.

In our model, sites chosen for SRF cultivation are characterized mostly by low environmental site qualities. Therefore, direct conflicts with food production are negligible because yields of annual crops would be low on these sites. This is in line with Aust et al. (2014): the authors argued that SRC on marginal agricultural land will only slightly affect food and feed production due to low yields on these sites. Similarly, various studies promote the use of marginal land as option to reduce competition with food production (Fitzherbert et al., 2008; van Dam et al., 2009; Hartman et al., 2011). On the other hand, areas with low site quality may possess high ecological value (e.g. in the case of grasslands (cf. BfN 2012). We do not model the ecological value of sites, but land which has been left fallow before the SRF expansion might have potentially built up ecological value.

The influence of the site-dependent determinant 'distance to the processing plants' was found to be more sensitive to general economic determinants such as the aggregated willingness to pay for SRF products and for annual crops, respectively.

Policy implications for promoting SRF. In this section, we take the model results as a starting point to discuss the design of effective government interventions to promote SRF. Therefore, we go beyond the model results to position them within the real-world context and focus on the political situation in Germany. Derived insights may also propose ways for other European countries, in particular as the model is not specific to the German case.

Currently, two main policy instruments to promote SRF expansion are applied in Germany. First, investment subsidies exist in some federal states and differ with respect to design (Strohm et al., 2012; Peschel & Weitz, 2013) They are important to overcome the barrier of high initial investment costs and to reduce the risk of investment (e.g. Faasch & Patenaude 2012, Strohm et al. 2012 or Wolbert-Haverkamp & Musshoff 2014). This is also supported by one of our model results: the high impact of investment expenditures. Therefore, it would be valuable to improve the subsidy design and provide coordination and harmonization of investment subsidies: requirements regarding minimal investment amount and minimal number of trees should be adjusted to allow for participation of small plantations and lower participation barriers (Strohm et al., 2012). Secondly, as of late, SRC can be accounted for as an ecological focus area under the greening component of the European Common Agricultural Policy (CAP) (Finger, 2016).

Further proposed instruments include the support of networks between SRF suppliers and demand side

actors, support for research and development and information instruments (Strohm et al., 2012). Additionally, in some studies, setting minimum wood chip prices through supply contracts is named as a measure to reduce investment uncertainty (Ridier et al., 2012; Wolbert-Haverkamp & Musshoff, 2014). This is also supported by our model results: the high impact of the willingness to pay for SRF products which substantially influences wood chip prices.

However, guaranteeing minimum wood chip prices or wood-specific quotas by public support instruments might cause market actors to choose cheapest wood or biomass resources available, not necessarily SRF. Therefore, a very technology- and feedstock-specific design of support instruments would be required to incentivize SRF (e.g. a higher substrate tariff class for SRF as implemented in the German Renewable Energies Act (EEG) 2012). However, attempting to incentivize SRF specifically through demand-sided, sectoral deployment support has high risks for steering errors. Large-scale SRF plantations may be incentivized if demand resulting from policy instruments is high enough, but it may end up not to be a competitive feedstock compared to other biomass resources nor a competitive climate change mitigation option. This would result in high public costs of errors as it was for example seen for the 'NaWaRo bonus' (renewable raw material bonus) in earlier versions of the EEG (cf. Britz & Delzeit 2013). In addition, decisions about the sectoral use of SRF wood would be distorted in favour of energetic applications as long as comprehensive bioeconomy policies are absent.

When assessing the appropriateness of policy instruments, it is important to consider that environmental benefits of SRF strongly depend on site- and plantation-specific characteristics (e.g. tree species, cultivation design) and that negative impacts are also possible (e.g. Dauber et al. 2010, Thrän et al. 2011 or Strohm et al. 2012). If SRF were supported through a demand-sided deployment support instrument, this would need to be complemented by specific spatial explicit environmental requirements or SRF-specific sustainability certification standards. This would ensure a positive environmental balance, but also increase complexity and transaction costs of demand-sided interventions.

From our model results and the discussion of current policy options, we conclude that investment subsidies in combination with information, networking, and research and development support seem to be the most promising approach to reduce barriers posed by high initial investment requirements, but should be combined with environmental minimum requirements (cf., Thrän et al. 2011 or Strohm et al. 2012). These subsidies would be only viable for the market entry phase to generate learning effects and should be phased out eventually.

Income stream risks would already be reduced by providing consistent and reliable political framework conditions, which increase planning security about future demand for woody biomass. Reliable framework conditions encompass general reliability of signals from sectoral bioenergy policies (e.g. in Germany the EEG in the electricity sector or the Renewable Heat Act (EEWärmeG) and the Market Incentive Programme in the heating sector), but also from biofuel policies (for innovative applications, e.g. wood gasification) and bioeconomy policies.

In general, the effectiveness increases with increasing specificity of intervention (ranging from instruments directed at renewable energy in general over wood in general to SRF-specific instruments), but so does the risk of inefficiency and market distortions. Whether SRF emerges as a competitive resource option should therefore be left to market actors, to reduce distortions of land, energy and material biomass markets.

Advantages of the applied methodology

The cultivation of perennial energy crops, such as SRF, resembles a long-term investments decision (Skevas et al., 2015). Modelling SRF cultivation decisions therefore requires incorporating different timescales and risk attitudes. We use approaches from investment theory which allow the comparison between land use options with different lengths of harvest cycles. Furthermore, perennial crops are associated with higher risks than annual crops (e.g. damages from drought or pests). Therefore, farmers require a compensation for accepting the higher risk (Sherrington & Moran, 2010; Rosenquist et al., 2013). To reflect this, we included risk costs in the decision model as have been empirically quantified by Rosenquist et al. (2013).

While financial barriers showed to be the most influential determinant of SRF cultivation decisions (e.g. Aylott & McDermott 2012), behavioural and nonfinancial determinants of SRF expansion were also identified significant by modelling (e.g. Sherrington & Moran 2010) as well as by empirical studies (e.g. Sherrington et al. 2008). In this context, diffusion processes driven by farmers' imitation or communication are of particular importance (Mola-Yudego & Gonzalez-Olabarria, 2010; Alexander et al., 2013). ABMs represent a strong tool to model diffusion of innovation processes compared to aggregated approaches (such as Bass' differential equation model) because they enable to depict heterogeneous agents and their interaction (Kiesling et al., 2012). Multiple application examples exist (Kiesling et al., 2012) which differ in the way decision rules

are modelled (e.g. simple rules such as threshold behaviour or utilitarian approaches) and the depiction of social networks (e.g. full networks, random networks). In the INCLUDE model, we follow a simple diffusion model of risk costs which decrease with the increase in SRF coverage due to learning effects (Rosenquist et al., 2013). In general, INCLUDE provides a reference model that could be enhanced in future research by including also noneconomic influence factors of decisions.

The chosen method of using stylized landscapes enables us to derive a general understanding beyond a specific region. Furthermore, the use of a landscape generator for the underlying landscape enables us to test the transferability of results between landscapes. We generate an ensemble of initial landscapes with fixed aggregated statistical characteristics (termed geostatistical model by Jager et al. 2005). Model evaluation was then performed using statistics over the entire ensemble. Besides statistically significant results (Dibble, 2006), this also enables the investigation into the relevance of explicit spatial configuration by quantifying the variation in model predictions due to variation in spatial structure (as proposed as spatial uncertainty analysis by Jager et al. 2005). Furthermore, the approach enables to test the transferability of results between landscapes with different aggregated spatial characteristics.

To conclude, by assessing different general economic and site-dependent determinants of SRF cultivation decisions, this study gave insights into barriers of a possible SRF expansion. The identification of determinants with strong impacts, such as investment expenditures or the willingness to pay for SRF products, can be taken as starting point for the future design of effective government interventions to promote SRF. This might contribute to sustainably meet an increasing demand for wood, especially in the context of a worldwide politically fostered bioeconomy. The analysis suggests that investment subsidies might be a promising approach to promote SRF, but should be combined with environmental minimum requirements.

Acknowledgements

This research was funded by the Research Programme 'Terrestrial Environmental Programme' (Integrated Project 'Land use aspects of transforming the energy system') of the Helmholtz Association. JS additionally acknowledges support from the graduate school 'Helmholtz Interdisciplinary Gradate School for Environmental Research (HIGRADE)' of the Helmholtz Centre for Environmental Research – UFZ. HW acknowledges funding through the German Research Foundation DFG project TI 824/2-1 Ecosystem resilience towards climate change – the role of interacting buffer mechanisms in Mediterranean-type ecosystems. The authors would like to thank Alexandra Purkus and two anonymous reviewers for valuable comments.

References

Alexander P, Moran D, Rounsevell MDA, Smith P (2013) Modelling the perennial energy crop market: the role of spatial diffusion. *Journal of the Royal Society Interface*, **10**, doi: 10.1098/rsif.2013.0656

Alexander P, Moran D, Rounsevell MDA, Hillier J, Smith P (2014) Cost and potential of carbon abatement from the UK perennial energy crop market. *GCB Bioenergy*, **6**, 156–168.

Allen B, Kretschmer B, Baldock D, Menadue H, Nanni S, Tucker G (2014) *Space for energy crops – assessing the potential contribution to Europe's energy future*. Report produced for BirdLife Europe, European Environmental Bureau and Transport & Environment. IEEP, London.

Aust C, Schweier J, Brodbeck F, Sauter UH, Becker G, Schnitzler J-P (2014) Land availability and potential biomass production with poplar and willow short rotation coppices in Germany. *GCB Bioenergy*, **6**, 521–533.

Aylott M, McDermott F (2012) *Domestic Energy Crops; Potential and Constraints Review*. NNFCC – The Bioeconomy Consultants, York, UK.

Barberis N, Thaler R (2003) A survey of behaviora finance. In: *Handbook of the Economics of Finance* (eds Constantinides GM, Harris M, Stulz R). Elsevier Science, Amsterdam, the Netherlands.

Bauen AW, Dunnett AJ, Richter GM, Dailey AG, Aylott M, Casella E, Taylor G (2010) Modelling supply and demand of bioenergy from short rotation coppice and Miscanthus in the UK. *Bioresource Technology*, **101**, 8132–8143.

Bauer LR, Hamby DM (1991) Relative sensitivities of existing and novel model parameters in atmospheric tritium dose estimates. *Radiation Protection Dosimetry*, **37**, 253–260.

BfN (2012) *Energieholzanbau auf landwirtschaftlichen Flächen – Auswirkungen von Kurzumtriebsplantagen auf Naturhaushalt, Landschaftsbild und biologische Vielfalt*. Bundesamt für Naturschutz, Leipzig.

BMEL (2014) *National Policy Strategy on Bioeconomy – Renewable Resources and Biotechnological Processes as a Basis for Food, Industry and Energy*. BMEL, Berlin, Germany.

Brigham E, Houston J (2006) *Fundamentals of Financial Management*. Cengage Learning, Boston, MA, USA.

Britz W, Delzeit R (2013) The impact of German biogas production on European and global agricultural markets, land use and the environment. *Energy Policy*, **62**, 1268–1275.

Cocchi M, Nikolaisen L, Junginger M et al. (2011) *Global Wood Pellet Industry Market and Trade Study*. IEA Bioenergy Task 40. Available at: http://www.bioenergytrade.de/downloads/t40-global-wood-pellet-market-study_final_R.pdf (accessed 11 September 2016).

van Dam J, Faaij APC, Hilbert J, Petruzzi H, Turkenburg WC (2009) Large-scale bioenergy production from soybeans and switchgrass in Argentina: part B. Environmental and socio-economic impacts on a regional level. *Renewable and Sustainable Energy Reviews*, **13**, 1679–1709.

Dauber J, Jones MB, Stout JC (2010) The impact of biomass crop cultivation on temperate biodiversity. *GCB Bioenergy*, **2**, 289–309.

Deza MM, Deza E (2013) *Encyclopedia of Distances*. Springer-Verlag, Berlin Heidelberg.

Dibble C (2006) *Handbook of Computational Economics* (eds Tesfatsion L, Judd KL). North-Holland/Elsevier, Amsterdam.

Dimitriou I, Baum C, Baum S et al. (2011) *Quantifying Environmental Effects of Short Rotation Coppice (SRC) on Biodiversity, Soil and Water*. IEA BIOENERGY Task 43. Available at: http://ieabioenergytask43.org/wp-content/uploads/2013/09/IEA_-Bioenergy_Task43_TR2011-01.pdf (accessed 11 September 2016).

Drossart I, Mühlenhoff J (2013) *Holzenergie Bedeutung, Potenziale, Herausforderungen*. Agentur für Erneuerbare Energien e. V, Berlin, Germany.

Dunnett AJ, Adjiman CS, Shah N (2008) A spatially explicit whole-system model of the lignocellulosic bioethanol supply chain: an assessment of decentralised processing potential. *Biotechnology for Biofuels*, **1**, 1–17.

EEA (2006) *How much bioenergy can Europe produce without harming the environment?* EEA Report. European Environment Agency, Copenhagen, Denmark.

Engelkamp P, Sell F (2007) *Einführung in die Volkswirtschaftslehre*. Springer, Heidelberg.

European Commission (2012) *Innovating for Sustainable Growth: A Bioeconomy for Europe*. European Commission, Brussels.

Faasch RJ, Patenaude G (2012) The economics of short rotation coppice in Germany. *Biomass and Bioenergy*, **45**, 27–40.

FAO (2014) *State of the World's Forests – Enhancing the Socioeconomic Benefits from Forests*. FAO, Rome.

FAOSTAT (2015) Food and Agriculture Organization of the United Nations – Statistics Division. Available at: http://faostat3.fao.org/browse/E/EL/E (accessed 11 September 2016).

Finger R (2016) Assessment of uncertain returns from investment in short rotation coppice using risk adjusted discount rates. *Biomass and Bioenergy*, **85**, 320–326.

Fischer C, Flohre A, Clement LW, Batáry P, Weisser WW, Tscharntke T, Thies C (2011) Mixed effects of landscape structure and farming practice on bird diversity. *Agriculture, Ecosystems & Environment*, **141**, 119–125.

Fitzherbert EB, Struebig MJ, Morel A, Danielsen F, Brühl CA, Donald PF, Phalan B (2008) How will oil palm expansion affect biodiversity? *Trends in Ecology & Evolution*, **23**, 538–545.

Hartman JC, Nippert JB, Orozco RA, Springer CJ (2011) Potential ecological impacts of switchgrass (*Panicum virgatum* L.) biofuel cultivation in the Central Great Plains, USA. *Biomass and Bioenergy*, **35**, 3415–3421.

Hauk S, Wittkopf S, Knoke T (2014) Analysis of commercial short rotation coppices in Bavaria, southern Germany. *Biomass and Bioenergy*, **67**, 401–412.

Helby P, Börjesson P, Hansen A, Roos A, Rosenqvist H, Takeuchi L (2004) *Market development problems for sustainable bio-energy systems in Sweden – the BIOMARK project*. IMES/EESS Report 38, Energy Environmental System Studies, Lund, Sweden.

Holland RA, Eigenbrod F, Muggeridge A, Brown G, Clarke D, Taylor G (2015) A synthesis of the ecosystem services impact of second generation bioenergy crop production. *Renewable and Sustainable Energy Reviews*, **46**, 30–40.

IUCN (2012) Facts and Figures on Forests. Available at: http://www.iucn.org/about/union/secretariat/offices/oceania/oceania_resources_and_publications/?9712/Facts-and-figures-on-Forests (accessed 11 September 2016).

Jager HI, King AW, Schumaker NH, Ashwood TL, Jackson BL (2005) Spatial uncertainty analysis of population models. *Ecological Modelling*, **185**, 13–27.

Kasmioui OE, Ceulemans R (2012) Financial analysis of the cultivation of poplar and willow for bioenergy. *Biomass and Bioenergy*, **43**, 52–64.

Kiesling E, Gunther M, Stummer C, Wakolbinger LM (2012) Agent-based simulation of innovation diffusion: a review. *Central European Journal of Operations Research*, **20**, 183–230.

Lamers P, Junginger M, Hamelinck C, Faaij A (2012) Developments in international solid biofuel trade—an analysis of volumes, policies, and market factors. *Renewable and Sustainable Energy Reviews*, **16**, 3176–3199.

Lawler JJ, Lewis DJ, Nelson E et al. (2014) Projected land-use change impacts on ecosystem services in the United States. *Proceedings of the National Academy of Sciences of the United States of America*, **111**, 7492–7497.

Makeschin F (1994) Effects of energy forestry on soils. *Biomass and Bioenergy*, **6**, 63–79.

Mankiw NG, Taylor MP (2006) *Economics*. Thomson Learning Services, Toronto, ON.

Mantau U, Saal U, Prins K et al. (2010) *EUwood – Real Potential for Changes in Growth and Use of EU Forests*. Final report. University of Hamburg, Centre of Wood Science, Hamburg.

Mola-Yudego B, Gonzalez-Olabarria JR (2010) Mapping the expansion and distribution of willow plantations for bioenergy in Sweden: lessons to be learned about the spread of energy crops. *Biomass and Bioenergy*, **34**, 442–448.

Mola-Yudego B, Dimitriou I, Gonzalez-Garcia S, Gritten D, Aronsson PA (2014) A conceptual framework for the introduction of energy crops. *Renewable Energy*, **72**, 29–38.

Musshoff O (2012) Growing short rotation coppice on agricultural land in Germany: a real options approach. *Biomass and Bioenergy*, **41**, 73–85.

Peschel T, Weitz M (2013) *Short Rotation Coppice Plantations – Concepts for Establishment and Operation Methods for Short Rotation Coppice (SRC) Projects for EU Bioenergy Plants. OPTFUEL – Optimized Fuels for Sustainable Transport*. Lignovis GmbH, Hamburg, Germany.

Ridier A, Chaib K, Roussy C (2012) The adoption of innovative cropping systems under price and production risks: a dynamic model of crop rotation choice. In: *123rd EAAE Seminar – Price Volatility and Farm Income Stabilisation – Modelling Outcomes and Assessing Market and Policy Based Responses*. Dublin.

Rosenqvist H, Berndes G, Borjesson P (2013) The prospects of cost reductions in willow production in Sweden. *Biomass and Bioenergy*, **48**, 139–147.

Rowe RL, Hanley ME, Goulson D, Clarke DJ, Doncaster CP, Taylor G (2011) Potential benefits of commercial willow Short Rotation Coppice (SRC) for farm-scale plant and invertebrate communities in the agri-environment. *Biomass and Bioenergy*, **35**, 325–336.

Sage R, Cunningham M, Boatman N (2006) Birds in willow short-rotation coppice compared to other arable crops in central England and a review of bird census data from energy crops in the UK. *Ibis*, **148**, 184–197.

Schlüter M, Müller B, Frank F (2013) How to Use Models to Improve Analysis and Governance of Social-Ecological Systems – The Reference Frame MORE. SSRN. Available at: http://ssrn.com/abstract=2037723 (accessed 11 September 2016).

Schmidt-Walter P, Lamersdorf NP (2012) Biomass production with willow and poplar short rotation coppices on sensitive areas-the impact on nitrate leaching and groundwater recharge in a drinking water catchment near Hanover, Germany. *Bioenergy Research*, **5**, 546–562.

Schweier J, Becker G (2013) Economics of poplar short rotation coppice plantations on marginal land in Germany. *Biomass and Bioenergy*, **59**, 494–502.

Schweinle J, Franke E (2010) *Beratunsghandbuch zu Kurzumtriebsplantagen* (ed. Skodawessely PB). Eigenverlag der TU Dresden, Dresden, Germany.

Sherrington C, Moran D (2010) Modelling farmer uptake of perennial energy crops in the UK. *Energy Policy*, **38**, 3567–3578.

Sherrington C, Bartley J, Moran D (2008) Farm-level constraints on the domestic supply of perennial energy crops in the UK. *Energy Policy*, **36**, 2504–2512.

Skevas T, Swinton SM, Tanner S, Sanford G, Thelen KD (2015) Investment risk in bioenergy crops. *GCB Bioenergy*, **8**, 1162–1177.

Strohm K, Schweinle J, Liesebach M *et al.* (2012) Kurzumtriebsplantagen aus ökologischer und ökonomischer Sicht. In: *Arbeitsberichte aus der vTI-Agrarökonomie*, pp. 1–55. Institut für Betriebswirtschaft Johann Heinrich von Thünen-Institut (vTI), Bundesforschungsinstitut für Ländliche Räume, Wald und Fischerei, Braunschweig, Germany.Available at: http://literatur.thuenen.de/digbib_extern/bitv/dn050857.pdf (accessed 11 September 2016)

Thober S, Mai J, Zink M, Samaniego L (2014) Stochastic temporal disaggregation of monthly precipitation for regional gridded data sets. *Water Resources Research*, **50**, 8714–8735.

Thrän D, Edel M, Pfeifer J, Ponitka J, Rode M, Knispel S (2011) *Identifizierung strategischer Hemmnisse und Entwicklung von Lösungsansätzen zur Reduzierung der Nutzungskonkurrenzen beim weiteren Ausbau der Biomassenutzung*. Deutsches Biomasseforschungszentrum, Leipzig, Germany.

Weih M, Dimitriou I (2012) Environmental impacts of short rotation coppice (SRC) grown for biomass on agricultural land. *BioEnergy Research*, **5**, 535–536.

Weise H (2014) Land use change in the context of bioenergy production: impact assessment using agent-based modelling. PhD thesis. University of Osnabrück, Osnabrück, Germany.

de Wit M, Junginger M, Faaij A (2013) Learning in dedicated wood production systems: past trends, future outlook and implications for bioenergy. *Renewable & Sustainable Energy Reviews*, **19**, 417–432.

Wolbert-Haverkamp M, Musshoff O (2014) Are short rotation coppices an economically interesting form of land use? A real options analysis. *Land Use Policy*, **38**, 163–174.

12

Microbial communities and diazotrophic activity differ in the root-zone of Alamo and Dacotah switchgrass feedstocks

RICHARD R. RODRIGUES[1], JINYOUNG MOON[2], BINGYU ZHAO[2] and MARK A. WILLIAMS[1,2]

[1]*Interdisciplinary Ph.D. Program in Genetics, Bioinformatics, and Computational Biology, 1015 Life Science Circle, Virginia Tech, Blacksburg, VA 24061, USA,* [2]*Department of Horticulture, 220 Ag Quad Lane, Virginia Tech, Blacksburg, VA 24061, USA*

Abstract

Nitrogen (N) bioavailability is a primary limiting nutrient for crop and feedstock productivity. Associative nitrogen fixation (ANF) by diazotrophic bacteria in root-zone soil microbial communities have been shown to provide significant amounts of N to some tropical grasses, but this potential in switchgrass, a warm-season, temperate, US native, perennial tallgrass has not been widely studied. 'Alamo' and 'Dacotah' are cultivars of switchgrass, adapted to the southern and northern regions of the United States, respectively, and offer an opportunity to better describe this plant–bacterial association. The nitrogenase enzyme activity, microbial communities, and amino acid profiles in the root-zones of the two ecotypes were studied at three different plant growth stages. Differences in the nitrogenase enzyme activity and free soluble amino acid profiles indicated the potential for greater nitrogen fixation in the high productivity Alamo compared with the lower productivity Dacotah. Changes in the amino acid profiles and microbial community structure (rRNA genes) of the root-zone suggest different plant–bacterial interactions can help to explain differences in nitrogenase activity. PICRUSt analysis revealed functional differences, especially nitrogen metabolism, that supported ecotype differences in root-zone nitrogenase enzyme activity. It is thought that the greater productivity of Alamo increased the belowground flow of carbon into roots and root-zone habitats, which in turn support the high energy demands needed to support nitrogen fixation. Further research is thus needed to understand plant ecotype and cultivar trait differences that can be used to breed or genetically modify crop plants to support root-zone associations with diazotrophs.

Keywords: 16s rRNA, Alamo and Dacotah, diazotrophs, function, interactions, ITS, microbial communities, root-zone, structure, switchgrass

Introduction

Associative nitrogen fixation (ANF) by diazotrophic bacteria provides an alternative to chemical fertilizers in support of crop yields. ANF is thought to be a loose form of *quid pro quo*, whereby energy and carbon fixed by the plant are exchanged for nitrogen fixed by bacterial diazotrophs. ANF has generally been considered a relatively small player in the annual nitrogen economy of temperate grasses; however, in some cases, it may provide up to 35% of nitrogen to agriculturally important nonlegumes and forage grasses, such as Miscanthus, energy cane, and switchgrass (Weier, 1980; Chalk, 1991; Wewalwela, 2014). ANF offers a natural solution to an immediate need for plant-available nitrogen while also reducing the footprint (Smil, 1999a,b, 2002; Galloway *et al.*, 2003; Oenema *et al.*, 2009; Sutton *et al.*, 2011a,b) of land applied synthetic chemical fertilizers. Some gaps in knowledge, however, need to be filled before using ANF for sustainable production. The natural variation in ANF that occurs across plant cultivars and plant growth stages will help to identify plant–microbial traits and interactions that underlie potentially high rates of ANF, and can be used to eventually breed and develop cultivars with greater ANF capacities. Research in this regard carried out in a model grass will likely be transferrable to other major worldwide grain crops (e.g., maize, wheat).

Switchgrass (*Panicum virgatum* L.) is a perennial, US native grass. Its biomass is used for forage for livestock and for bioenergy. Switchgrass is useful in the prevention of soil erosion and provides wildlife habitat and carbon sequestration (Ma, 1999; Skinner, 2009; Follett

Correspondence: Mark A. Williams
e-mail: markwill@vt.edu

et al., 2012). Due to the various environmental and monetary benefits of switchgrass, identifying microbes for its sustainable production has been an active area of research (Jesus *et al.*, 2010, 2016; Mao *et al.*, 2011, 2013, 2014; Chaudhary *et al.*, 2012; Hargreaves *et al.*, 2015). These studies have shown presence of bacteria that are capable of nitrogen fixation; however, studies rarely confirm the occurrence and extent of nitrogen fixation associated with switchgrass.

It is known that cultivars of plants often have differences in N requirements, abilities to support nitrogen fixation (Porter, 1966; Day *et al.*, 1975; Boddey & Dobereiner, 1988; Ledgard & Steele, 1992), unique rhizosphere microbiomes (Miller *et al.*, 1989; Li *et al.*, 2014), and thus are important sources of variation in plant–microbial traits. A recent study, for example, identified diverse sets of nitrogen-fixing bacteria associated with roots and shoots of switchgrass (Bahulikar *et al.*, 2014). Other studies focused on targeted approaches of using nitrogen-fixing bacterial endophytes to improve productivity across cultivars (Ker *et al.*, 2012; Xia *et al.*, 2013; Lowman *et al.*, 2016). Plant bacterial associations and how they differ among plant species and cultivars can help to understand the interplay of traits important for defining the drivers of a globally significant plant–bacterial function like that of ANF.

'Alamo' and 'Dacotah' are tetraploid cultivars of switchgrass, adapted to the southern and northern US ecotypes, respectively. Compared with Dacotah, Alamo shows higher biomass productivity, taller shoots, drought tolerance, and disease resistance. Nitrogen is required for the growth and maintenance of plants, so we hypothesized that a higher productivity cultivar of switchgrass, Alamo, has more nitrogen demand than Dacotah and would be associated with a different root-zone bacterial community with greater nitrogenase activity than a lower productivity cultivar, Dacotah. Here, we investigated the nitrogenase activity, amino acid composition, structure, function, and interactions of microbial communities in the root-zones of two switchgrass cultivars across multiple growth stages. The research offers novel insights into the dynamics of nitrogen-fixing communities in the root-zone of switchgrass that can help to explain the variation associated with this globally important plant–bacterial interaction.

Materials and methods

Experimental setup

Root-zone soil was collected from the Virginia Tech Agronomy Farm/Urban Horticulture Center field growing Alamo, Dacotah (~7 years), and their F1 progeny lines (~2 years) to serve as an appropriate inoculum of potential microbes associated with switchgrass. Alamo and Dacotah seeds were imbibed and allowed to germinate at 28 °C. After approximately 7 days, the germinated seeds were potted in sieved (4.75 mm) and homogenized field soil and grown in the glasshouse. The plants were kept moist, well aerated, and supplemented with light to maintain optimal growth. Sampling was performed in replicates (Table S1) at three time points from imbibition, indicative of important growth stages: stages V0 (~2.5 weeks old) and V2 (~1.5 months old) from the vegetative phase, and stage E3 (~3.5 months old) from the elongation phase (Moore *et al.*, 1991).

Nitrogenase enzyme activity

To estimate the relative rate of nitrogen fixation, nitrogenase activity was evaluated by the acetylene reduction (AR) assay, which measures the quantity of acetylene reduced to ethylene (Weaver & Danso, 1994). Briefly, the day before the nitrogenase measurement, the Alamo and Dacotah containing pots were wetted to saturation and allowed to drain gravimetrically to ~0.03 MPa. Pots were randomly sampled to obtain replicates ($n = 3$ for V0, $n = 4$ for V2, and $n = 5$ for E3) (Table S1) that received acetylene ('treatment') or remained untreated ('controls'). For the vegetative growth stages, V0 and V2, the entire pot with soil + plant was placed into the Mason jars for the AR assay. At stage E3, we gently removed the plant from the pot and placed the plant + rhizosphere soil into the Mason jar for the AR assay. The data was normalized as per the weight of the 'system' (soil + plant for V0 and V2; plant + rhizosphere soil for E3). Each replicate sample was sealed within a 500-mL to 1-L Mason glass jar. Calcium carbide (CaC_2) and water were mixed to produce acetylene gas, which was injected to create a ~10% (v/v) headspace. Mason glass jars with sample but without acetylene served as controls, to assess the natural (background) production of ethylene. At four time points (immediately and then approximately every 30 min), headspace gas was mixed using a 10-mL syringe and needle to extract gas from the container and analyzed for acetylene and ethylene concentrations using a gas chromatograph and column (GS-Carbonplot, Part # 113-3133, Agilent Technologies, Inc., Santa Clara, CA, USA, 30 m × 0.32 mm × 3 μm) in line with a flame ionization detector. The ethylene accumulation rate was used to determine nitrogenase enzyme activity.

Known ethylene standards were used to generate a linear calibration curve that was used to calculate ethylene concentrations for each sample and reported as the rate of ethylene production, $\mu g\ g_{system}^{-1}\ h^{-1}$. Following the confirmation of normality, equality of variances, and determination of outliers using Grubb's test (Grubbs, 1969) in GRAPHPAD (http://graphpad.com/quickcalcs/Grubbs1.cfm), the rates of ethylene production between Alamo and Dacotah were compared (*P*-value <0.05) using two-sample *t*-test (V0 and E3) and Mann–Whitney *U*-test (V2). Only the 24 controls (3, 4, and 5 samples each of Alamo and Dacotah from V0, V2, and E3, respectively), samples were used for the following experiments.

Soluble amino acids

Shoots were removed, and the root and the rhizosphere soil were extracted in 10 mL of 0.9% NaCl in a 50-mL falcon tube. The roots were washed to remove any attached soil and used to measure fresh and dry weight. Falcon tubes were vortexed at 200 rpm on electric orbital shaker for 20 min, allowed to sit for 5 min. Eight milliliters of solution was transferred to a new 15-mL falcon tube and centrifuged at $3500 \times g$ for 15 min. The amount of supernatant (approximately 5 mL) transferred to new smaller size falcon tubes was recorded, and 8 µL of internal standard (2.5 mm α-aminobutyric acid; AABA) was added to each sample. The volume for each sample was made to 10 mL by adding 0.9% NaCl. After shaking for 2 min, 1.5 mL of sample was transferred into a 2-mL microfuge tube through 0.22-µm polyvinylidene fluoride (PVDF; Thermo Scientific™ (Thermo Fisher Scientific, Waltham, MA, USA) Target2™ Syringe Filters, Cat# 03377155) membrane syringe filter; 500 µL of filtrate from the 2-mL tube was transferred to a new 1.5-mL microfuge tube and dried in a vacufuge (speedvac) for 2.5 h at 60 °C (function 3). The dried pellet was derivatized using the AccQ FluorTM reagent kit (fluorescent 6-aminoquinolyl-N-hydroxysuccinimidyl carbamate derivatizing reagent; Waters Corporation, Milford, MA, USA; Cat# WAT052880) following the standard protocol (Bosch et al., 2006; Hou et al., 2009). Chromatographic separation on the HPLC 1260 Infinity system (Agilent Technologies, Inc.,) was carried out on a reversed-phase column (Waters X-Terra MS C18, 3.5 µm, 2.1 m × 150 mm). The mobile phase consisted of A: an aqueous solution containing 140 mm of sodium acetate, 17 mm of triethylamine (TEA; Thermo Fisher Scientific, Cat# O4884100), and 0.1% (g L^{-1}, w/v) disodium dihydrogen ethylenediaminetetraacetate dihydrate (EDTA-2Na 2H2O; Sigma-Aldrich, St. Louis, MO, USA, CAS# 6381-92-6), pH 5.05, adjusted with phosphoric acid solution, and B: acetonitrile (ACN; HPLC grade, Thermo Fisher Scientific, Cat# A998-1): ultrapure water (60 : 40, v/v). The gradient conditions were 0–17 min 100–93% A, 17–21 min 93–90% A, 21–30 min 90–70% A, 30–35 min 70% A, 35–36 min 70–0% A, and then hold for 4 min before restoring to the initial composition at 40.5 min, with the final composition kept for 9 min. The column was thermostated at 50 °C and operated at a flow rate of 0.35 mL min^{-1}. The sample injection volume was 5 µL. The analytes detection was carried out using a fluorescence detector (λex = 250 nm and λem = 395 nm) (Bosch et al., 2006; Hou et al., 2009). Soluble amino acids in the samples were qualified and quantified by comparison with amino acid standard solutions. Each amino acid standard solution contained 19 protein amino acids including alanine (Ala), arginine (Arg), aspartic acid (Asp), asparagine (Asn), cystine (Cys–Cys; more stable form in oxidative condition than monomer cysteine), glutamic acid (Glu), glutamine (Gln), glycine (Gly), histidine (His), isoleucine (Ile), leucine (Leu), lysine (Lys), methionine (Met), phenylalanine (Phe), serine (Ser), threonine (Thr), tyrosine (Tyr), tryptophan (Trp), and valine (Val) and 1 nonprotein amino acid ornithine (Orn). Note that Asn and Ser were co-eluted and Cys–Cys and Tyr were co-eluted. A total of 20 peaks were quantified. The amino acid data was relativized per sample and used for multivariate data analyses using PC-ORD software version 6.0 (MjM Software, Gleneden Beach, OR, USA).

DNA extraction

Ten milliliters of 0.9% NaCl was added to the falcon tubes containing rhizosphere soil samples. The tubes were then turned upside down once and the soil solution was homogenized by vortexing 1 min each (at level 3). A 3 mL sample of the resuspended solution was taken immediately after vortexing in a 5-mL tube. After centrifugation of the 5-mL tubes at 10 000 g for 10 min at 20 °C, the supernatant was removed and the tubes were turned upside down on a Kim wipe (with cap open) to dry for 10 min. For each tube, the leftover soil (pellet) was mixed and 0.25 g was used for the DNA extraction using MoBio PowerSoil® (MoBio Laboratories, Carlsbad, CA, USA) DNA Isolation kit as per the manufacturer's protocol. The quality and concentration of the DNA was checked using 0.8% (w/v) agarose gel electrophoreses and NanoDrop 2000 spectrophotometry.

Real-time PCR to detect nitrogenase reductase gene copy numbers

qPCR assay was performed on an ABI 7300 system (Thermo Fisher Scientific) to quantify the abundance of nitrogenase reductase gene (nifH). The 20-µL reaction volumes contained 10 µL of 2X PowerUp™ SYBR® Green Master Mix (Thermo Fisher Scientific), 2 µL template at 2 ng µL^{-1}, 4 µL each of PolF/PolR (10 µM) primers (Poly et al., 2001), and standard conditions as per master mix's protocol. Triplicates of nuclease-free water and no-template controls were included. Technical triplicates of each biological replicate were averaged to get mean C_T value. DNA was extracted from *Sinorhizobium meliloti* 1021 (cultures provided by Dr. B. Scharf at Virginia Tech) using UltraClean® Microbial DNA Isolation Kit (Mo Bio). Dilutions of known concentrations of *Sinorhizobium meliloti* 1021 were used to generate the C_T vs. log (N_0) standard curve. The standard curve was used to calculate the qPCR efficiency ($E = 10^{(-1/slope)}$) and nifH gene copy numbers in the samples (Brankatschk et al., 2012). For each cultivar at each growth stage, biological replicates were checked for outliers using Grubb's test (Grubbs, 1969) in GRAPHPAD (http://graphpad.com/quickcalcs/Grubbs1.cfm) and boxplots were generated in BoxPlotR with Spear's criteria of whisker definition (Spitzer et al., 2014). The cultivars at each growth stage were tested for significant (P-value <0.05) differences in nifH gene copy numbers using Mann–Whitney U-test.

Sequencing and data analyses

As bacterial–fungal interactions are known to promote plant growth (Artursson et al., 2006; Gamalero et al., 2008; Frey-Klett et al., 2011), we investigated the bacterial and fungal communities to identify any possible interactions in root-zones of these switchgrass cultivars. The 16S rRNA and ITS gene amplification were performed using bacterial primer pair (S-D-Bact-0341-b-S-17 and S-D-Bact-0785-a-A-21) (Klindworth et al., 2013) and fungal primer pair (ITS1F and ITS2) (Smith & Peay, 2014), respectively. The library preparation was performed using Illumina 16S Metagenomic sequencing library preparation guide.

DNA concentration of the pool was determined by fluorometric quantification using the Qubit® 2.0 platform with Qubit dsDNA HS Assay Kit (Life Technologies, Carlsbad, CA, USA). Multiplexed 250-base pair paired-end sequencing using Illumina MiSeq was performed at the Biocomplexity Institute (formerly Virginia Bioinformatics Institute) core facility at Virginia Tech. Data are submitted under Biosamples of SAMN04917354–SAMN04917359 (16S), SAMN04917374–SAMN04917379 (ITS), BioProject PRJNA320123 at NCBI (SRA).

Barcode adapters and primers were trimmed from each read using CUTADAPT v1.8.1 (Martin, 2011) with a quality-trimming threshold of 30 and minimum read length of 100. The paired-end reads were merged based on overlapping sequences into single reads using PANDASEQ v2.8 (Masella et al., 2012). The bacterial and fungal sequencing data were analyzed using QIIME v1.8.0 (Caporaso et al., 2010) as previously described (Rodrigues et al., 2015). Briefly, reads were clustered into OTUs based on 97% sequence similarity using UCLUST (Edgar, 2010) and USEARCH61 (Edgar, 2010) for bacteria and fungi, respectively, using an open-reference OTU-picking strategy. The representative sequence of an OTU was used to assign it a taxonomy, using UCLUST against the GREENGENES reference database version 13.8 (Desantis et al., 2006; McDonald et al., 2012) for bacteria, and RDP classifier (Wang et al., 2007) against the UNITE reference database version 12.11 (Abarenkov et al., 2010) for fungi.

Postprocessing included diversity and richness analyses, identifying and summarizing the most abundant taxons, and describing the taxons that are different and indicative of cultivars at different growth stages. Briefly, the alpha diversity was calculated on the OTU table for all samples using several different indices, including PD whole tree (only for bacteria), chao1, observed species, Good's coverage, Shannon, and Simpson. After using a sequence threshold, the beta diversity for bacteria and fungi was calculated using weighted UniFrac (Lozupone et al., 2007, 2011) and Bray–Curtis (Beals, 1984) metrics, respectively, and were used for principal coordinate analysis (Gower, 2005) and visualization. The genus-level summary of communities was used for NMDS analysis using the Bray–Curtis dissimilarity in R (VEGAN). Using the collated alpha diversities from the rarefied OTU table, the PD whole tree (only for bacteria), chao1, and observed species indices were used to compare alpha diversity of groups (cultivar, stage, cultivar × stage) using a two-tail nonparametric t-test with FDR correction (q-value <0.05) using collated files. Multivariate data analysis methods of MRPP (Mielke, 1984), ADONIS (Anderson, 2001), and analysis of similarity (ANOSIM) (Clarke, 1993) were used to identify whether the cultivars and plant growth stage had an effect on the microbial communities.

Indicator species analysis (ISA) (Dufrene & Legendre, 1997) was performed in R (VEGAN) using the top 50% most abundant genera to identify taxa significantly (indicator value >50 and P-value <0.05) indicative of cultivars at the respective growth stage.

PICRUST (Langille et al., 2013) analysis using the bacterial OTU abundance was performed before and after ISA analysis to identify whether functions differed between cultivars at the different growth stages. OTUs not part of the closed reference OTU picking were filtered out, and the metagenomes were collapsed into KEGG pathways. Using Statistical Analysis of Metagenomic Profiles (STAMP) (Parks et al. 2014), principal component analysis was used to visualize differences in KEGG pathways of bacteria between the cultivars at different growth stages. A two-sided t-test was performed to check whether N-metabolism pathway was significantly different (P-value <0.05) between the cultivars.

Network analysis

The OTU table, at the class level for bacteria and order level for fungi, was summarized, relativized, and separated to create files as per samples belonging to cultivar irrespective of the plant growth stage, that is, Alamo (V2, E3) and Dacotah (V2, E3). Phylogenetic ecological networks were generated for each group (cultivar × growth stage) using MENAP (Deng et al., 2012) with an RMT threshold of 0.31 and module detection using leading eigenvector of the community matrix (Newman, 2006) and random walk methods (Pons & Latapy, 2005). A 'module' is a group of OTUs that were highly connected (positively or negatively correlated via abundance) among themselves, but had much fewer connections with OTUs outside the group. For each network, connections between OTUs between different modules were discarded to help focus on interactions among OTUs within a module. We refer these subnetworks as 'same-module networks' which show the connections between OTUs belonging to the same module. For each cultivar, the networks from leading eigenvector of the community matrix and random walk methods were compared to generate edges common between the two methods. These 'COMmon Edge between Two methods' (COMET) networks help to remove any potential biases due to the module detection algorithms and represent high confidence edges. Finally, the COMET networks were compared between cultivars.

Cultivable, free-living diazotrophs

For samples from stage V2, the NaCl solutions containing rhizosphere soil were (100x, 1000x, and 5000x) serial diluted, spread on yeast-mannitol (YM) agar (YMA) petri plates, and incubated at 28 °C for 3–5 days to selectively grow diazotrophs. Bacterial colonies were differentiated based on morphology and color. Colonies were counted and will be reported as colony forming unit per gram of root–soil. Representative colonies were sampled and isolation plated on new YMA plates to obtain pure cultures. After 3–5 days of incubation, the isolation-plated samples were grown in yeast-mannitol broth and DNA was extracted using UltraClean® Microbial DNA Isolation Kit (Mo Bio). The quality of the DNA was estimated using 0.8% (w/v) agarose gel electrophoreses and NanoDrop 2000 spectrophotometry. The 16S rRNA genes were amplified using polymerase chain reaction using primers 27F (5′-AGAGTTTGATCMTGGCTCAG-3′) and 1492R (5′-TACGGYTACCTTGTTACGACTT-3′) (Lane, 1991; Fredriksson et al., 2013) and KAPA2G Robust PCR kit (Kapa Biosystem, Wilmington, MA, USA), and cleaned using Agencourt AMPure XP PCR product cleanup kit (Beckman Coulter, Brea, CA, USA). High-quality, cleaned PCR products, as determined by 1.2% (w/v) agarose gel electrophorese and NanoDrop 2000 spectrophotometry, were sent to Biocomplexity Institute (Virginia Bioinformatics Institute) for Sanger sequencing. The 'ab1' files obtained from Sanger sequencing were converted to sequence files using 4Peaks and quality trimmed with a Phred score of 25. We used the taxonomy of the top hit from nucleotide BLAST (MEGABLAST) on the sequences to identify the bacteria. LIBSHUFF (Singleton et al., 2001; Schloss et al., 2004) analysis on the sequences and Pearson's chi-squared test on the genus-level CFU abundance were performed to identify significant differences (P-value <0.05) between the cultivable diazotrophic communities associated with Alamo and Dacotah.

Results

Alamo showed higher rates of biological nitrogen fixation (BNF)

Quality control analysis using the Grubb's test (Grubbs, 1969) in GRAPHPAD identified an outlier data point of rate of ethylene production in Dacotah at stage V2 and hence was omitted from further statistics. The rate of ethylene production, which indicates the nitrogenase enzyme activity or BNF rates, was highest at stage V2 than those at stages V0 and E3 (Fig. 1). At stages V2 and E3, Alamo had significantly greater nitrogenase activity than Dacotah (Mann–Whitney U-test P-value = 0.017 at stage V2 and t-test with equal variance P-value = 0.030 at stage E3). Both Alamo and Dacotah showed (>22 µg ethylene per g system per h) reduction in BNF rates at E3 as compared to those at V2. We quantified the nifH gene in the rhizosphere samples of switchgrass cultivars (Fig. S1) using Sinorhizobium meliloti 1021 as the standard (C_T = 22.064–4.0825 log(N_0); R^2 = 0.985; E_{std} = 1.758), however, found no significant differences between cultivars.

The cultivars showed differences in the free soluble amino acid profiles

Plant growth stages had a significant effect on the relative abundance of free soluble amino acids in the root-zone of the cultivars (adonis P-value = 0.001). Hence, further analysis was performed for each growth stage comparing the soluble amino acid profiles of Alamo and Dacotah. Adonis indicated that at stages V0 (P-value = 0.8) and E3 (P-value = 0.108), there were no differences in the relative content of root-zone soluble amino acids of cultivars. However, at stage V2 (Fig. S2), the relative abundance of soluble amino acids in the root-zones between Alamo and Dacotah was significantly different (adonis P-value = 0.04). The Mann–Whitney U-test showed that the cultivars had significantly (P-value <0.05) different levels of the following amino acids in their root-zones (Fig. 2): aspartic acid (Asp), glutamic acid (Glu), glutamine (Gln), glycine (Gly), and serine + asparagine (Ser + Asn) at V2; and arginine (Arg), histidine (His), and ornithine (Orn) at E3.

Sequencing data

Removal of barcodes, adapters, and primers using cutadapt and stitching with PANDAseq gave ~ 1.9 million and ~ 1.7 million high-quality 16S rRNA (bacterial) and ITS (fungal) gene sequence reads, respectively, for the 24 samples. The bacterial and fungal sequences, respectively, had average lengths of 333.2 and 254.3 bases with standard deviations of 57.7 and 48.7 bases. The raw data has been submitted to NCBI Sequence Read Archive (BioProject PRJNA320123) according to MIMS standard.

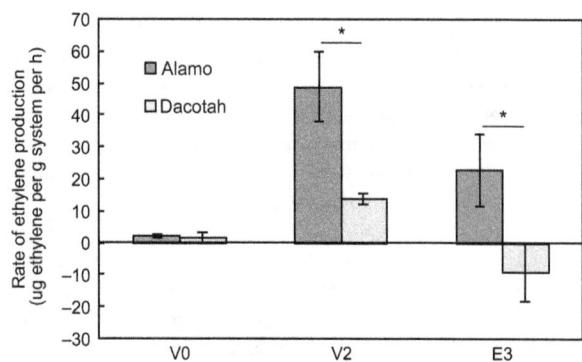

Fig. 1 Rate of ethylene production (µg ethylene per g system h^{-1}) in Alamo and Dacotah at three plant growth stages. The asterisk (*) indicates significant differences (P-value <0.05) between the ethylene production rates of Alamo and Dacotah at the particular growth stage.

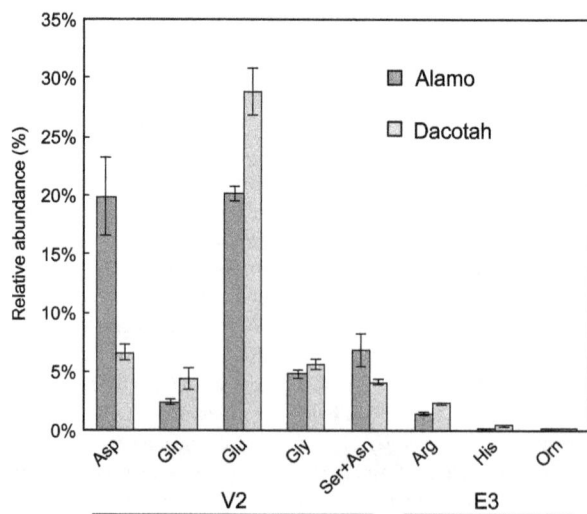

Fig. 2 Relative abundance of amino acids that were significantly different (P-value <0.05) between Alamo and Dacotah at stages V2 and E3.

The bacterial dataset had total of 72 371 distinct OTUs with a total of 1 735 730 sequences (counts) assigned to these OTUs. The fungal dataset had total of 35 986 distinct OTUs with a total of 1 604 221 sequences (counts) assigned to these OTUs. The average Good's coverage of all samples was 92.5% and 97.0% for bacteria and fungi, respectively (Table S2). Table S2 contains the relevant statistics of the counts per sample and other alpha diversity metrics.

Alpha diversity

Using samples from all three growth stages and a sequence threshold (of 21 500 for bacteria and 16 000 for fungi), we compared the alpha diversity of groups (plant growth stage, cultivar, plant growth stage × cultivar) with a two-tail nonparametric *t*-test with FDR correction using the collated alpha diversity file. The bacterial alpha diversity at stage E3 was greater than that at stage V2 (PD whole tree *q*-value = 0.003, chao1 *q*-value = 0.003). Furthermore, results from the interaction of plant growth stage and cultivar showed that the bacterial alpha diversity in Dacotah at stage E3 was greater than that at stage V2 (observed species *q*-value = 0.045, chao1 *q*-value = 0.045).

The patterns of fungal alpha diversity resembled that of bacteria. The fungal alpha diversity at stage V2 was lesser than those at stages V0 (observed species *q*-value = 0.0135) and E3 (chao1 *q*-value = 0.003, observed species *q*-value = 0.003). Also, results from the interaction of plant growth stage and cultivar showed that the fungal alpha diversity in Dacotah at stage E3 was greater than that at stage V2 (chao1 *q*-value = 0.03, observed species *q*-value = 0. 045). Certain patterns, however, were only present in the results from the interaction of plant growth stage and cultivar in the fungal data. The alpha diversity in Alamo at stage E3 was greater than those at stage V2 (chao1 *q*-value = 0.045, observed species *q*-value = 0.0225) and stage V0 (chao1 *q*-value = 0.02).

Root-zone microbial communities structure (beta diversity) different between stage and cultivars

Multivariate data analyses in QIIME using MRPP, ADONIS, and ANOSIM on the weighted UniFrac distance between samples, using all OTUs, detected significant differences (*P*-value <0.05) in the bacterial communities as per stage and interaction of cultivar and stage (cultivar × stage), but not as per cultivar. These results were consistent when the analysis was performed on samples from (i) all three growth stages (V0, V2, E3), (ii) vegetative growth stages (V0 and V2), and (iii) vegetative (V2) and elongation (E3) stages (Table S3a).

PCoA (Fig. S3a) and NMDS (Fig. S3b) analyses of bacterial communities using weighted UniFrac and Bray–Curtis distances, respectively, showed that samples from stages V2 and E3 clustered as per cultivars, however, samples from V0 did not. Hence, following analyses to compare cultivars were only performed at stages V2 and E3. *Proteobacteria*, *Actinobacteria*, and *Acidobacteria* were the most abundant phyla across samples (Fig. 3). Kruskal–Wallis test with Benjamini–Hochberg correction was performed on samples from stages V2 and E3 to identify significantly different (*q*-value <0.05) phyla. For each growth stage, the significant phyla, except 'Unassigned', were tested for differences (*P*-value <0.05) in relative abundance between cultivars using a bootstrap Mann–Whitney *U*-test in QIIME. For easier illustration, the 'Unassigned' phylum and phyla with total <1% abundance in samples across all growth stages were grouped as 'Other' taxa and phyla showing differential relative abundance between cultivars from the above test are shown by stars (Fig. 3). Stage V2 had more phyla with differential relative abundance between cultivars. *Proteobacteria* was significantly different only at stage V2; however, it consistently showed higher relative abundance in Alamo across all plant growth stages.

Similar to the bacterial data, multivariate data analyses in QIIME using MRPP, ADONIS, and ANOSIM on the Bray–Curtis distance between samples, using all OTUs, detected significant differences (*P*-value <0.05) in the fungal communities as per stage and interaction of cultivar and stage (cultivar × stage), but not as per cultivar (Table S3b). PCoA (Fig. S4) analyses of fungal communities using Bray–Curtis distances showed that samples from stage V2 clustered as per cultivars; however, samples from V0 and E3 did not. Sequences derived from the phylum *Ascomycota* were the most abundant across all growth stages (Fig. S5).

Root-zone bacterial communities showed functional differences between cultivars

Indicator species analysis (ISA) was performed for the bacterial and fungal data to identify statistically significant indicator taxa (Table S4). Functional profiles were predicted for bacteria using PICRUSt for the cultivars at the V2 and E3 stages before and after ISA. The predicted KEGG pathways in the metagenomes were different between cultivars at the corresponding growth stages for both before (Fig. S6) and after (Fig. S7) ISA. Before ISA, the KEGG pathway 'N-metabolism' showed significant differential abundance between cultivars at stage V2 (*P*-value = 0.039) but not at stage E3 (Fig. 4). However, after ISA, the pathway was observed to have significant differential abundance between cultivars at

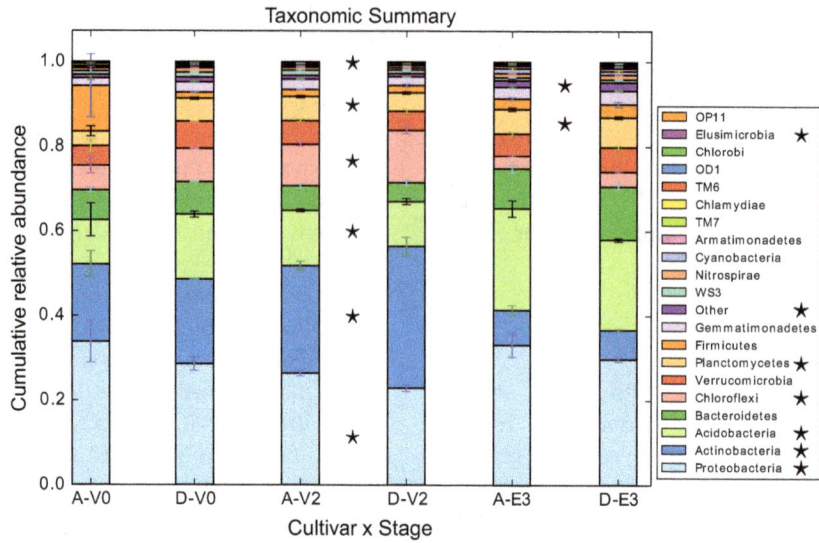

Fig. 3 Phylum-level taxonomic summaries of bacterial communities in cultivars at the different growth stages. 'Unassigned' and less abundant taxa were grouped in 'Other'. Taxa in bars and legend from bottom to top are sorted as per decreasing abundance across all samples. Stars show phyla that are significantly different (q-value <0.05) between cultivars at a specific growth stage. The significance was only calculated for stages V2 and E3.

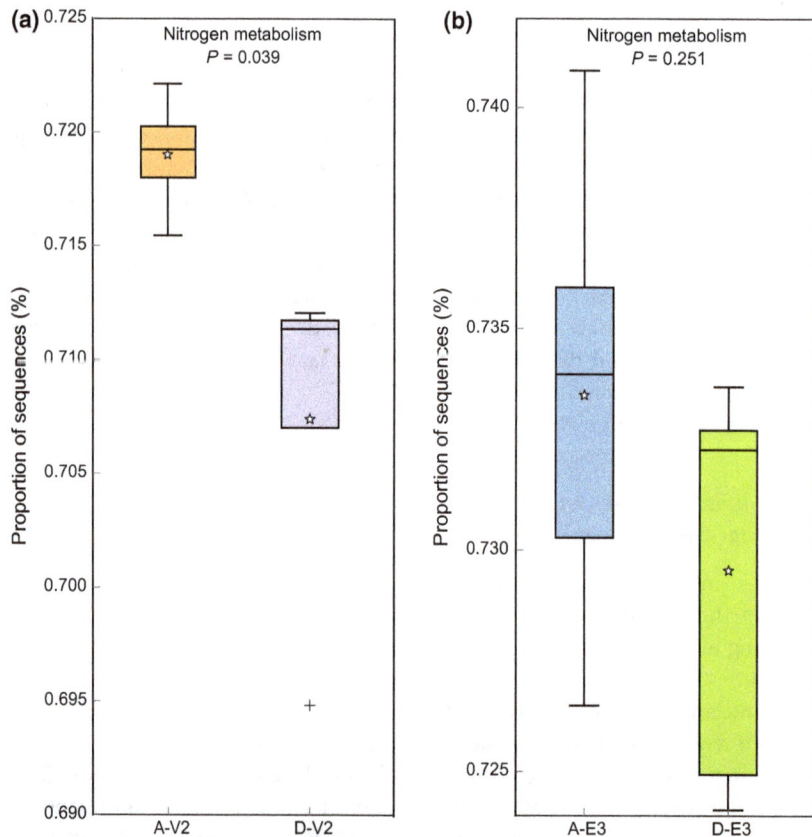

Fig. 4 Pre-ISA abundance of the 'N-metabolism' pathway in the cultivars at stages (a) V2 and (b) E3. Two-sided t-test (P-value <0.05) was used to estimate differential abundance between cultivars.

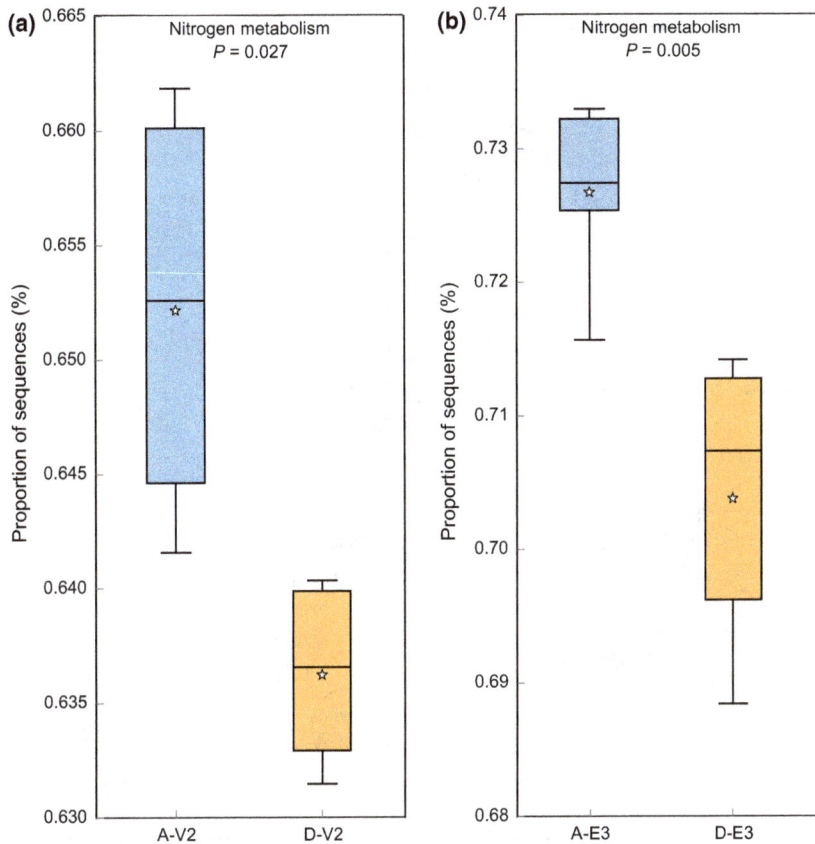

Fig. 5 Post-ISA abundance of the 'N-metabolism' pathway in the cultivars at stages (a) V2 and (b) E3. Two-sided *t*-test (*P*-value <0.05) was used to estimate differential abundance between cultivars.

stages V2 (p-value = 0.027) and E3 (*P*-value = 0.005) (Fig. 5). These indicate that the N-metabolism potential of the cultivar-specific root-zone-associated bacteria is different between Alamo and Dacotah.

Cultivars differed in the root-zone bacterial network correlation analysis

For the bacterial data, V2 and E3 samples showed differences as per cultivar and hence were used to generate bacterial COMET networks. In the bacterial COMET networks (Fig. 6), the number of classes (nodes) in Alamo and Dacotah was equal, whereas the total number of bacterial interactions in Alamo (72) and Dacotah (129) was different, with 24 interactions present in both networks (colored in black). Alamo had a higher number of modules (6) and percentage of positively correlated edges (56%) compared with the 3 and 46%, respectively, in Dacotah. For the fungal data, only samples from the V2 stage showed differences as per cultivar. As MENAP software needs at least eight samples per group, fungal COMET networks could not be generated.

Cultivable diazotrophs

After DNA extraction, 16S rRNA amplification, and Sanger sequencing of representative colonies cultivated on YM media, we obtained 45 and 42 high-quality reads for root-zone bacteria from Alamo and Dacotah, respectively. At stage V2, as per Libshuff analysis (Dacotah–Alamo P-value = 0.0054) and Pearson's chi-squared test (*P*-value <2.2e^{-16}) (Fig. S8), we observed significant differences in the cultivable nitrogen-fixing bacteria (diazotrophs) between Alamo and Dacotah.

Discussion

The results of this study support the working *quid pro quo* model of plant–microbial interactions and the hypothesis that a higher productivity cultivar of switchgrass would be associated with a different root-zone bacterial community with greater nitrogenase activity than a lower productivity cultivar. The potential for greater carbon flow in exchange for bacterial fixed nitrogen provides a consistent framework to describe plant–bacterial interactions associated with switchgrass.

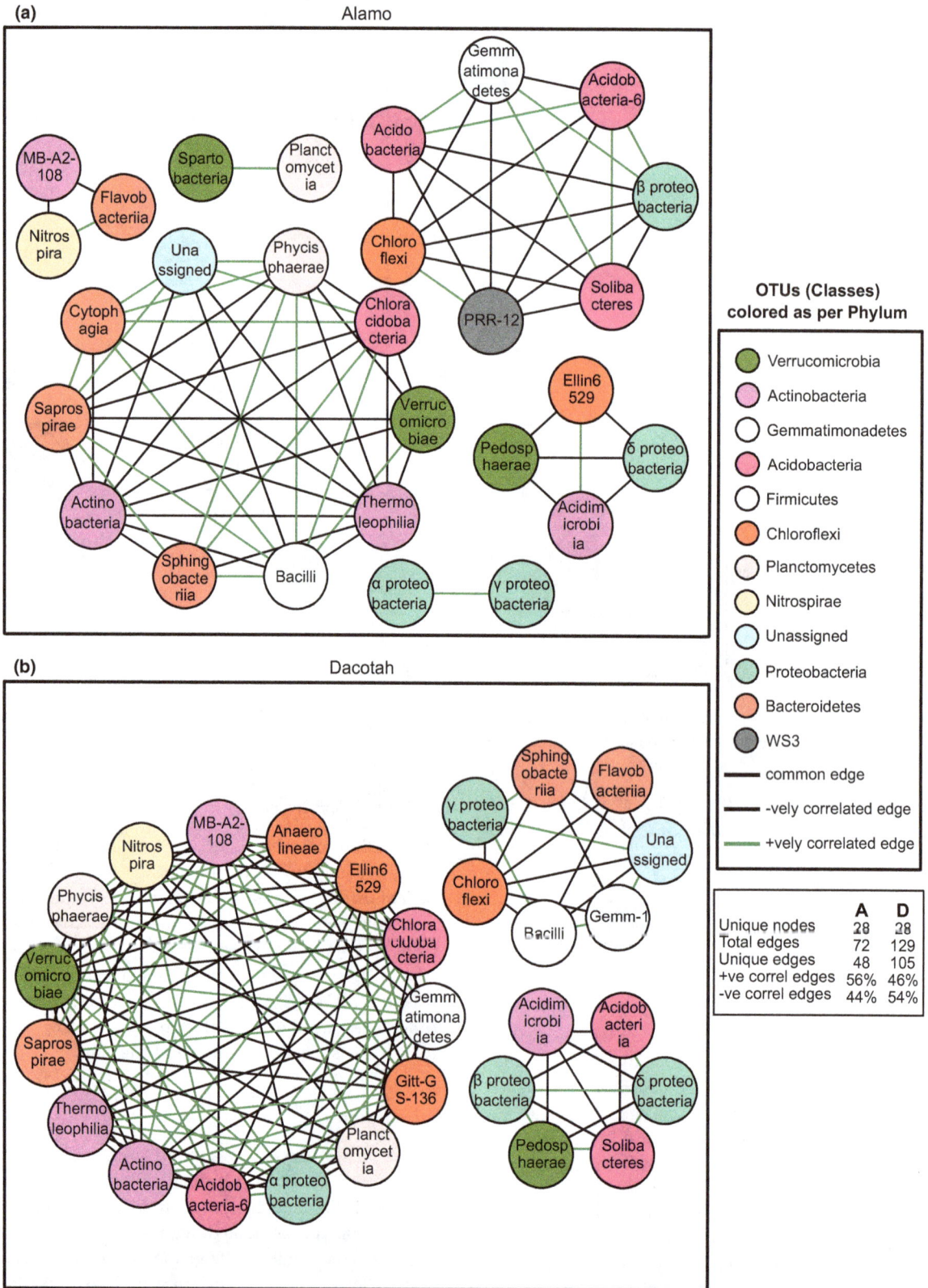

Fig. 6 The COMET networks of bacterial sequence abundance correlations from root-zones of (a) Alamo and (b) Dacotah. The nodes represent classes and they are colored as per the phyla. The red and green colored edges represent negative and positive interactions, respectively. A black colored edge represents that the interaction is present in both the Alamo and Dacotah networks.

Differences observed in the composition of amino acids, structure, cultivable diazotrophs, correlation networks, and the general functional capacity (e.g., nitrogen metabolism) of the bacterial communities associated with the two cultivars further support the above hypothesis. The results reinforce the idea that plant–diazotroph interactions and feedbacks (communications) are important descriptors of the greater nitrogen-fixing potential of the high productivity Alamo, compared with the Dacotah ecotype. Root-zone diazotroph communities, therefore, may help to meet the nitrogen demands of feedstocks and other grasses to support greater plant biomass production. The experimental outcomes support the idea that molecular breeding of switchgrass can be used to manipulate plant–bacterial interaction.

Plant growth-related N demand as a driver of switchgrass–diazotrophic interaction

Biological nitrogen fixation rates changed with growth stage and may reflect the temporal nitrogen demands of cultivars (Dong et al., 2001; Scagel et al., 2007). The increased BNF rates during the vegetative stages might be to meet the N demands of increasing biomass. On the other hand, the reduction in BNF rates during the elongation stage might be due to the nearing of maturation and the reduction in plant nitrogen demand, especially with its ability to translocate nitrogen to its roots for conservation during winter and reuse during regrowth (Monti, 2012). While it cannot be unequivocally concluded that the higher activities of nitrogen-fixing bacteria are a result of Alamo having greater productivity than Dacotah, the results are consistent with this idea and track expected growth stage changes in plant nitrogen demand and carbon flow to below ground roots (Qian et al., 1997; Voisin et al., 2003).

Plant–microbial amino acids in the root-zone

Plant root and microbial exudates, especially amino acids, are used as signals of communication between plants and the root-zone microbiome (Morgan et al., 2005; Badri et al., 2009). The amino acids glycine, glutamate, alanine, aspartate, glutamine, and asparagine were found to be the dominant soluble amino acids in relatively mature switchgrass root-zone soil habitats (Figs 2 and S2); however, the contributions of these amino acids were altered with plant growth stage. The flux rates of these same molecules as exudates from plant roots have previously been shown to be relatively high (Lesuffleur et al., 2007; Carvalhais et al., 2011, 2013) but can vary due to fertilization (e.g., N, Fe) and plant maturation. The relative concentration of glycine, in particular, was much higher in the root-

zone of mature (E3) relative to young switchgrass plants (V0 and V2). There may thus be important changes in amino acid profiles that arise during plant maturation. The observations of amino acids in the root-zone of switchgrass represent a more complex habitat than studies that focus on axenic, microbial-free root exudation. It is thus difficult to come to firm conclusions regarding the biological role that amino acid dynamics might play in the root-zone. The variation across cultivar and growth stage, nevertheless, is consistent with the idea that root-zone amino acids reflect the contributions of two different but interactive plant–microbial systems.

Asparagine and glutamine tend to be relatively low in the rhizosphere, but can vary between <1 and ~6% of the amino acid pool (Paynel et al., 2001; Lesuffleur et al., 2007). Intracellular concentrations within the roots of numerous plants (Paynel et al., 2001; Lodwig et al., 2003; Lesuffleur et al., 2007) of glutamine and asparagine tend to be relatively high, especially within phloem and xylem. Because of the importance of nitrogen-rich asparagine and the plant N shuttle glutamine, there is likely to be conservation and influx of these amino acids into organism biomass rather than efflux, particularly under N-limiting growth conditions. Hence, the relatively low concentrations of these two amino acids in the root-zone of switchgrass, relative to their constituent acids, aspartic and glutamic acids, may be the result of N-limiting growth conditions.

An amino acid feedback model has been proposed from the extensive study of legume (family Fabaceae) nodules (Dong et al., 2001). However, the application of this model to plant roots and associative diazotrophic bacteria in the rhizosphere remains to be determined. The rhizosphere (associative BNF system) is a different habitat than that of a nodule (symbiotic BNF system); however, application of this model provides a useful starting point for understanding interactions and exchanges between plants and bacteria in the root-zone. Glutamic and aspartic acids and the associated bases of glutamine and asparagine appear to provide a feedback signal that helps to control nitrogen fixation in nodules of legumes (Lodwig et al., 2003). These amino acids may thus be an important 'amino-stat' that helps to regulate the exchange of plant–bacterial carbon and nitrogen.

We used this nitrogen quid pro quo model to describe the dominant and dynamic amino acid species in the root-zone of the switchgrass ecotypes. We observed that aspartic acid was ~threefold greater in Alamo than Dacotah, and conversely, glutamic acid was ~1.4-fold greater in Dacotah than in Alamo. These differences could reflect different regulatory interactions between the plant and bacteria, perhaps related to differences in plant N demand and the availability of reduced nitrogen from diazotrophs (Lodwig *et al.*, 2003). Assuming that the different amino acid concentrations reflect differences in the flows of these molecules between plant and bacterial biomass, the high levels of aspartic acid in the root-zone of Alamo could be interpreted as the result of higher rates of nitrogen fixation by diazotrophs. The greater flow of this bacterial-supplied N transporter could help meet the higher plant nitrogen demand of Alamo (Lodwig *et al.*, 2003). In contrast, the relatively greater glutamic acid concentrations in the root-zone of Dacotah may reflect a buildup of this amino acid as a consequence of relatively low rates of nitrogen fixation. The results are also concordant with the observation that glutamic acid is secreted at lower levels by nitrogen-fixing bacteria under diazotrophic compared with adiazotrophic conditions (Gonzalez-Lopez *et al.*, 1995). Reductions in glutamine and asparagine, and elevated aspartate-to-asparagine ratios, on the other hand, have been shown to be a response to low N availability that is translated as deficiency or stress within the plant (Amarante & Sodek, 2006) and thus help to describe plant N demand (Pate *et al.*, 1980, 1981, 1984; Atkins *et al.*, 1983; Lea *et al.*, 2007). Although we did not measure concentrations of amino acids within plant tissues, differences in these ratios in the root-zone may help to provide information explaining the greater nitrogen fixation potential associated with Alamo relative to Dacotah.

Changes in the structure, functions, and interactions of rhizosphere microbiome

The broad differences in cultivable (Fig. S8 and Libshuff results) and noncultivable bacterial (Figs 3, 6 and S3) and fungal (Figs S4 and S5) communities (Table S3) in the root-zone of Alamo and Dacotah are consistent with the above described functional indicators in nitrogen fixation potential (Fig. 1), root-zone exudates (e.g., amino acids, Fig. 2), KEGG analysis (Figs 4, 5, S6, and S7), and plant–microbial interactions (Bürgmann *et al.*, 2005; Mao *et al.*, 2014). Different plant cultivars and growth stages have previously been shown to impact microbial community structure in plant root-zones (Berg & Smalla, 2009; Mao *et al.*, 2011; Chaudhary *et al.*, 2012; Chaparro *et al.*, 2014). The research reported in this manuscript build upon these differences to show specific changes in microbial communities, potentially mediated through plant–microbial signals (amino acids), and concomitant functional changes in the root-zone.

Bacterial nitrogen fixation potential tracked changes in the structure and metagenomic functional potential of the root-zone microbial community. The increased N-metabolism potential in Alamo compared with Dacotah may reflect the overall differences in the nitrogen status of the ecotypes. The phylogenetic dominance of *Proteobacteria*, *Actinobacteria*, *Acidobacteria*, and *Firmicutes* observed in our results from culture-dependent and culture-independent methods across samples (Fig. 3) has been observed previously in roots of switchgrass (Jesus *et al.*, 2010, 2016; Mao *et al.*, 2011; Plecha *et al.*, 2013; Bahulikar *et al.*, 2014). At stage V2, these phyla are among those that show differential abundances between ecotypes (Fig. 3), suggesting the potential importance of phylogenetic changes for driving functional level (Figs 4 and 5) shifts, such as that observed for the relative increase in N-metabolism potential in Alamo relative to Dacotah. The relative abundances of *Proteobacteria* and *Actinobacteria*, for example, are indicators of Alamo and Dacotah (Table S4), respectively. The observations from our study (Table S4, Figs 4 and 5) are consistent with the differential role that microbial communities and N-cycling bacteria may play in the growth of switchgrass ecotypes (Mao *et al.*, 2011, 2013).

The ecological niches of the two switchgrass cultivars also provide a framework for understanding the potential for different types of root-zone bacterial interactions in relation to the greater nitrogen fixation rates. The differences in bacterial correlations of COMET networks between cultivars point out significantly different root-zone community dynamics among the ecotypes (Fig. 6). In addition, COMET networks summarize community differences using a different analytical framework than the multivariate pattern and statistical analysis, but nevertheless complement those findings (Tables S3 and S4, Figs S3, S4, S6, and S7). The common edges between Alamo and Dacotah COMET networks can be interpreted as the core set of bacterial associations (possible interactions) that are associated with switchgrass, and with the unique edges being cultivar specific. The lower number of modules but greater proportion of positive relative to negative correlations, if interpreted as associations, in Alamo relative to Dacotah, hints at the potential of bacteria–bacteria and plant–bacteria interactions that could serve to help the plant achieve higher productivity. Although it cannot be known whether these correlations reflect different interactions in the diazotroph root-zone communities of the ecotypes, the network model is useful for identifying differences between communities and the potential positive (or negative) feedbacks between microorganisms.

Positive and negative feedbacks between plants and microbes are often driven by environmental and nutrient limitations and thus provide a working hypothesis to explain how two ecotype root-zone habitats with different energy and nutrient conditions may develop different microbial communities and plant–microbial interactions (Morgan *et al.*, 2005; Badri *et al.*, 2009; Berg, 2009). The *Proteobacteria*, for example, have diazotrophic bacterial members and could help to explain their greater relative abundance and greater nitrogen fixation potential in Alamo relative to Dacotah. Similarly, the positive association of bacilli with many other taxa within the bacterial network, and its greater abundance in the cultivable community in Alamo compared with Dacotah is consistent with a growth supportive role played by these bacteria (Lopez-Bucio *et al.*, 2007; Nautiyal *et al.*, 2013; Xia *et al.*, 2013; Gagne-Bourque *et al.*, 2015). Presumably, the surrounding taxa that benefit would in return provide resources that support the growth of the bacilli. While the diversity of potential interactions in root-zone and soil habitats is staggering, and description of them is still in scientific infancy, the networks described herein are useful for understanding, predicting, and confirming possible positive or negative interactions among organisms.

In conclusion, although there is a scientific agreement regarding the importance of microbial interactions in the root-zone, the research presented is relatively unique in that it simultaneously studies the structure, potential for interaction, and functions of microbial communities in the root-zone of switchgrass. The research highlights possible linkages in plant nitrogen cycling (KEGG, amino acids) and demand associated with changes in the structure (COMET networks, PCoA, NMDS) and diazotroph activity (nitrogenase) of root-zone microbial communities. Understanding the interplay between the diazotroph communities with associated plants offers opportunities for genetic engineering and breeding of feedstock grasses and grain crops to select varieties that can better manage microbial communities to support growth and satisfy nitrogen demands of plants.

Acknowledgements

We thank Virginia Tech's Genetics, Bioinformatics, and Computational Biology, Department of Horticulture, Multicultural Academic Opportunities, and Graduate Student Assembly for providing research, travel, and personnel funding. Research was also partially funded by a grant from USDA-NIFA (2011-03815). The authors sincerely thank Amanda Karstetter, Bronte Lantin, Brandi Edwards, Haley Feazel-Orr, Kelsey Weber, Deepak Poudel, Solmaz Eskandari, and Samantha Fenn for help with sample preparation for microbial analyses. Thanks to Dr. Li Ma for help with the soluble amino acid experiments. The authors acknowledge Virginia Polytechnic Institute and State University's Open Access Subvention Fund. Sincere thanks to Dr. Birgit Scharf for providing *S. melliloti* 1021.

References

Abarenkov K, Henrik Nilsson R, Larsson KH *et al.* (2010) The UNITE database for molecular identification of fungi-recent updates and future perspectives. *New Phytologist*, 186, 281–285.

Amarante L, Sodek L (2006) Waterlogging effect on xylem sap glutamine of nodulated soybean. *Biologia Plantarum*, 50, 405–410.

Anderson MJ (2001) A new method for non-parametric multivariate analysis of variance. *Austral Ecology*, 26, 32–46.

Artursson V, Finlay RD, Jansson JK (2006) Interactions between arbuscular mycorrhizal fungi and bacteria and their potential for stimulating plant growth. *Environmental Microbiology*, 8, 1–10.

Atkins CA, Pate JS, Peoples MB, Joy KW (1983) Amino Acid transport and metabolism in relation to the nitrogen economy of a legume leaf. *Plant Physiology*, 71, 841–848.

Badri DV, Weir TL, Van Der Lelie D, Vivanco JM (2009) Rhizosphere chemical dialogues: plant-microbe interactions. *Current Opinion in Biotechnology*, 20, 642–650.

Bahulikar RA, Torres-Jerez I, Worley E, Craven K, Udvardi MK (2014) Diversity of nitrogen-fixing bacteria associated with switchgrass in the native tallgrass prairie of northern Oklahoma. *Applied and Environment Microbiology*, 80, 5636–5643.

Beals EW (1984) Bray-curtis ordination: an effective strategy for analysis of multivariate ecological data. *Advances in Ecological Research*, 14, 1–55.

Berg G (2009) Plant-microbe interactions promoting plant growth and health: perspectives for controlled use of microorganisms in agriculture. *Applied Microbiology and Biotechnology*, 84, 11–18.

Berg G, Smalla K (2009) Plant species and soil type cooperatively shape the structure and function of microbial communities in the rhizosphere. *FEMS Microbiology Ecology*, 68, 1–13.

Boddey RM, Dobereiner J (1988) Nitrogen fixation associated with grasses and cereals: recent results and perspectives for future research. *Plant and Soil*, 108, 53–65.

Bosch L, Alegria A, Farre R (2006) Application of the 6-aminoquinolyl-N-hydroxysccinimidyl carbamate (AQC) reagent to the RP-HPLC determination of amino acids in infant foods. *Journal of Chromatography. B, Analytical Technologies in the Biomedical and Life Sciences*, 831, 176–183.

Brankatschk R, Bodenhausen N, Zeyer J, Burgmann H (2012) Simple absolute quantification method correcting for quantitative PCR efficiency variations for microbial community samples. *Applied and Environment Microbiology*, 78, 4481–4489.

Bürgmann H, Meier S, Bunge M, Widmer F, Zeyer J (2005) Effects of model root exudates on structure and activity of a soil diazotroph community. *Environmental Microbiology*, 7, 1711–1724.

Caporaso G, Kuczynski J, Stombaugh J *et al.* (2010) QIIME allows analysis of high-throughput community sequencing data. *Nature Methods*, 7, 335–336.

Carvalhais LC, Dennis PG, Fedoseyenko D, Hajirezaei M-R, Borriss R, Von Wirén N (2011) Root exudation of sugars, amino acids, and organic acids by maize as affected by nitrogen, phosphorus, potassium, and iron deficiency. *Journal of Plant Nutrition and Soil Science*, 174, 3–11.

Carvalhais LC, Dennis PG, Fan B *et al.* (2013) Linking plant nutritional status to plant-microbe interactions. *PLoS ONE*, 8, e68555.

Chalk PM (1991) The contribution of associative and symbiotic nitrogen fixation to the nitrogen nutrition of non-legumes. *Plant and Soil*, 132, 29–39.

Chaparro JM, Badri DV, Vivanco JM (2014) Rhizosphere microbiome assemblage is affected by plant development. *ISME Journal*, 8, 790–803.

Chaudhary D, Saxena J, Lorenz N, Dick L, Dick R (2012) Microbial profiles of rhizosphere and bulk soil microbial communities of biofuel crops switchgrass (*Panicum virgatum* L.) and Jatropha (*Jatropha curcas* L.). *Applied and Environmental Soil Science*, 2012, 1–6.

Clarke KR (1993) Non-parametric multivariate analyses of changes in community structure. *Australian Journal of Ecology*, 18, 117–143.

Day JM, Neves MCP, Döbereiner J (1975) Nitrogenase activity on the roots of tropical forage grasses. *Soil Biology and Biochemistry*, 7, 107–112.

Deng Y, Jiang YH, Yang Y, He Z, Luo F, Zhou J (2012) Molecular ecological network analyses. *BMC Bioinformatics*, 13, 113.

Desantis TZ, Hugenholtz P, Larsen N *et al.* (2006) Greengenes, a chimera-checked 16S rRNA gene database and workbench compatible with ARB. *Applied and Environment Microbiology*, 72, 5069–5072.

Dong S, Scagel CF, Cheng L, Fuchigami LH, Rygiewicz PT (2001) Soil temperature and plant growth stage influence nitrogen uptake and amino acid concentration of apple during early spring growth. *Tree Physiology*, 21, 541–547.

Dufrene M, Legendre P (1997) Species assemblages and indicator species: the need for a flexible asymmetrical approach. *Ecological Monographs*, 67, 345–366.

Edgar RC (2010) Search and clustering orders of magnitude faster than BLAST. *Bioinformatics*, 26, 2460–2461.

Follett RF, Vogel KP, Varvel GE, Mitchell RB, Kimble J (2012) Soil carbon sequestration by switchgrass and no-till maize grown for bioenergy. *BioEnergy Research*, 5, 866–875.

Fredriksson NJ, Hermansson M, Wilen BM (2013) The choice of PCR primers has great impact on assessments of bacterial community diversity and dynamics in a wastewater treatment plant. *PLoS ONE*, **8**, e76431.

Frey-Klett P, Burlinson P, Deveau A, Barret M, Tarkka M, Sarniguet A (2011) Bacterial-fungal interactions: hyphens between agricultural, clinical, environmental, and food microbiologists. *Microbiology and Molecular Biology Reviews*, **75**, 583–609.

Gagne-Bourque F, Mayer BF, Charron JB, Vali H, Bertrand A, Jabaji S (2015) Accelerated growth rate and increased drought stress resilience of the model grass brachypodium distachyon colonized by *Bacillus subtilis* B26. *PLoS ONE*, **10**, e0130456.

Galloway JN, Aber JD, Erisman JW, Seitzinger SP, Howarth RW, Cowling EB, Cosby BJ (2003) The nitrogen cascade. *BioScience*, **53**, 341.

Gamalero E, Berta G, Massa N, Glick BR, Lingua G (2008) Synergistic interactions between the ACC deaminase-producing bacterium Pseudomonas putida UW4 and the AM fungus Gigaspora rosea positively affect cucumber plant growth. *FEMS Microbiology Ecology*, **64**, 459–467.

Gonzalez-Lopez J, Martinez-Toledo MV, Rodelas B, Pozo C, Salmeron V (1995) Production of amino acids by free-living heterotrophic nitrogen-fixing bacteria. *Amino Acids*, **8**, 15–21.

Gower JC (2005) Principal coordinates analysis. In: *Encyclopedia of Biostatistics* (ed. Armitage P, Colton T), John Wiley and Sons Ltd, Chichester, West Sussex, UK.

Grubbs FE (1969) Procedures for detecting outlying observations in samples. *Technometrics*, **11**, 1–21.

Hargreaves SK, Williams RJ, Hofmockel KS (2015) Environmental filtering of microbial communities in agricultural soil shifts with crop growth. *PLoS ONE*, **10**, e0134345.

Hou S, He H, Zhang W, Xie H, Zhang X (2009) Determination of soil amino acids by high performance liquid chromatography-electro spray ionization-mass spectrometry derivatized with 6-aminoquinolyl-N-hydroxysuccinimidyl carbamate. *Talanta*, **80**, 440–447.

Jesus SE, Susilawati E, Smith S et al. (2010) Bacterial communities in the rhizosphere of biofuel crops grown on marginal lands as evaluated by 16S rRNA gene pyrosequences. *BioEnergy Research*, **3**, 20–27.

Jesus EDC, Liang C, Quensen JF, Susilawati E, Jackson RD, Balser TC, Tiedje JM (2016) Influence of corn, switchgrass, and prairie cropping systems on soil microbial communities in the upper Midwest of the United States. *GCB Bioenergy*, **8**, 481–494.

Ker K, Seguin P, Driscoll BT, Fyles JW, Smith DL (2012) Switchgrass establishment and seeding year production can be improved by inoculation with rhizosphere endophytes. *Biomass and Bioenergy*, **47**, 295–301.

Klindworth A, Pruesse E, Schweer T, Peplies J, Quast C, Horn M, Glockner FO (2013) Evaluation of general 16S ribosomal RNA gene PCR primers for classical and next-generation sequencing-based diversity studies. *Nucleic Acids Research*, **41**, e1.

Lane DJ (1991) 16S/23S rRNA sequencing. In: *Nucleic Acid Techniques in Bacterial Systematics* (eds Stackebrandt E, Goodfellow M), pp. 115–175. John Wiley and Sons, New York, NY.

Langille MG, Zaneveld J, Caporaso JG et al. (2013) Predictive functional profiling of microbial communities using 16S rRNA marker gene sequences. *Nature Biotechnology*, **31**, 814–821.

Lea PJ, Sodek L, Parry MAJ, Shewry PR, Halford NG (2007) Asparagine in plants. *Annals of Applied Biology*, **150**, 1–26.

Ledgard SF, Steele KW (1992) Biological nitrogen fixation in mixed legume/grass pastures. *Plant and Soil*, **141**, 137–153.

Lesuffleur F, Paynel F, Bataillé M-P, Le Deunff E, Cliquet J-B (2007) Root amino acid exudation: measurement of high efflux rates of glycine and serine from six different plant species. *Plant and Soil*, **294**, 235–246.

Li X, Rui J, Mao Y, Yannarell A, Mackie R (2014) Dynamics of the bacterial community structure in the rhizosphere of a maize cultivar. *Soil Biology and Biochemistry*, **68**, 392–401.

Lodwig EM, Hosie AH, Bourdes A et al. (2003) Amino-acid cycling drives nitrogen fixation in the legume-Rhizobium symbiosis. *Nature*, **422**, 722–726.

Lopez-Bucio J, Campos-Cuevas JC, Hernandez-Calderon E, Velasquez-Becerra C, Farias-Rodriguez R, Macias-Rodriguez LI, Valencia-Cantero E (2007) Bacillus megaterium rhizobacteria promote growth and alter root-system architecture through an auxin- and ethylene-independent signaling mechanism in *Arabidopsis thaliana*. *Molecular Plant-Microbe Interactions*, **20**, 207–217.

Lowman S, Kim-Dura S, Mei C, Nowak J (2016) Strategies for enhancement of switchgrass (*Panicum virgatum* L.) performance under limited nitrogen supply based on utilization of N-fixing bacterial endophytes. *Plant and Soil*, **405**, 47–63.

Lozupone CA, Hamady M, Kelley ST, Knight R (2007) Quantitative and qualitative beta diversity measures lead to different insights into factors that structure microbial communities. *Applied and Environment Microbiology*, **73**, 1576–1585.

Lozupone C, Lladser ME, Knights D, Stombaugh J, Knight R (2011) UniFrac: an effective distance metric for microbial community comparison. *ISME Journal*, **5**, 169–172.

Ma Z (1999) Carbon sequestration by switchgrass. Unpublished PhD Auburn University, 107 pp.

Mao Y, Yannarell A, Mackie R (2011) Changes in N-transforming archaea and bacteria in soil during the establishment of bioenergy crops. *PLoS ONE*, **6**, e24750.

Mao Y, Yannarell A, Davis S, Mackie R (2013) Impact of different bioenergy crops on N-cycling bacterial and archaeal communities in soil. *Environmental Microbiology*, **15**, 928–942.

Mao Y, Li X, Smyth E, Yannarell A, Mackie R (2014) Enrichment of specific bacterial and eukaryotic microbes in the rhizosphere of switchgrass (*Panicum virgatum* L.) through root exudates. *Environmental Microbiology Reports*, **6**, 13.

Martin M (2011) Cutadapt removes adapter sequences from high-throughput sequencing reads. *EMBnet. Journal*, **17**, 10.

Masella A, Bartram A, Truszkowski J, Brown D, Neufeld J (2012) PANDAseq: paired-end assembler for illumina sequences. *BMC Bioinformatics*, **13**, 31.

McDonald D, Price MN, Goodrich J et al. (2012) An improved Greengenes taxonomy with explicit ranks for ecological and evolutionary analyses of bacteria and archaea. *ISME Journal*, **6**, 610–618.

Mielke PW (1984) Meteorological applications of permutation techniques based on distance functions. In: *Handbook of Statistics: Nonparametric Methods* (eds Krishnaiah PR, Sen PK), pp. 813–830. North-Holland, Amsterdam.

Miller HJ, Henken G, Veen JAV (1989) Variation and composition of bacterial populations in the rhizospheres of maize, wheat, and grass cultivars. *Canadian Journal of Microbiology*, **35**, 656–660.

Monti A (2012) *Switchgrass: A Valuable Biomass Crop for Energy*. Springer-Verlag London, UK.

Moore KJ, Moser LE, Vogel KP, Waller SS, Johnson BE, Pedersen JF (1991) Describing and quantifying growth-stages of perennial forage grasses. *Agronomy Journal*, **83**, 1073–1077.

Morgan JA, Bending GD, White PJ (2005) Biological costs and benefits to plant-microbe interactions in the rhizosphere. *Journal of Experimental Botany*, **56**, 1729–1739.

Nautiyal CS, Srivastava S, Chauhan PS, Seem K, Mishra A, Sopory SK (2013) Plant growth-promoting bacteria Bacillus amyloliquefaciens NBRISN13 modulates gene expression profile of leaf and rhizosphere community in rice during salt stress. *Plant Physiology and Biochemistry*, **66**, 1–9.

Newman MEJ (2006) Finding community structure in networks using the eigenvectors of matrices. *Physical Review E*, **74**, 036104.

Oenema O, Witzke HP, Klimont Z, Lesschen JP, Velthof GL (2009) Integrated assessment of promising measures to decrease nitrogen losses from agriculture in EU-27. *Agriculture, Ecosystems & Environment*, **133**, 280–288.

Parks DH, Tyson GW, Hugenholtz P, Beiko RG (2014) STAMP: statistical analysis of taxonomic and functional profiles. *Bioinformatics*, **30**, 3123–3124.

Pate JS, Atkins CA, White ST, Rainbird RM, Woo KC (1980) Nitrogen nutrition and xylem transport of nitrogen in ureide-producing grain legumes. *Plant Physiology*, **65**, 961–965.

Pate JS, Atkins CA, Herridge DF, Layzell DB (1981) Synthesis, storage, and utilization of amino compounds in White Lupin (*Lupinus albus* L.). *Plant Physiology*, **67**, 37–42.

Pate JS, Atkins CA, Layzell DB, Shelp BJ (1984) Effects of N2 deficiency on transport and partitioning of C and N in a nodulated legume. *Plant Physiology*, **76**, 6.

Paynel F, Murray PJ, Cliquet JB (2001) Root exudates: a pathway for short-term N transfer from clover and ryegrass. *Plant and Soil*, **229**, 235–243.

Plecha S, Hall D, Tiquia-Arashiro SM (2013) Screening for novel bacteria from the bioenergy feedstock switchgrass (*Panicum virgatum* L.). *Environmental Technology*, **34**, 1895–1904.

Poly F, Monrozier LJ, Bally R (2001) Improvement in the RFLP procedure for studying the diversity of nifH genes in communities of nitrogen fixers in soil. *Research in Microbiology*, **152**, 95–103.

Pons P, Latapy M (2005) Computing communities in large networks using random walks. *Computer and Information Sciences*, **3733**, 284–293.

Porter CL (1966) An analysis of variation between upland and lowland switchgrass, *Panicum virgatum* L., in Central Oklahoma. *Ecology*, **47**, 980.

Qian JH, Doran JW, Walters DT (1997) Maize plant contributions to root zone available carbon and microbial transformations of nitrogen. *Soil Biology and Biochemistry*, **29**, 1451–1462.

Rodrigues RR, Pineda RP, Barney JN, Nilsen ET, Barrett JE, Williams MA (2015) Plant invasions associated with change in root-zone microbial community structure and diversity. *PLoS ONE*, **10**, e0141424.

Scagel CF, Bi G, Fuchigami LH, Regan RP (2007) Seasonal variation in growth, nitrogen uptake and allocation by container-grown evergreen and deciduous rhododendron cultivars. *HortScience*, **42**, 1440–1449.

Schloss PD, Larget BR, Handelsman J (2004) Integration of microbial ecology and statistics: a test to compare gene libraries. *Applied and Environment Microbiology*, **70**, 5485–5492.

Singleton DR, Furlong MA, Rathbun SL, Whitman WB (2001) Quantitative comparisons of 16S rRNA gene sequence libraries from environmental samples. *Applied and Environmental Microbiology*, **67**, 4374–4376.

Skinner H (2009) *Sequestration Potential of Switchgrass Managed for Bioenergy Production*. ARS Pasture Systems and Watershed Management Research Unit, University Park, PA.

Smil V (1999a) Detonator of the population explosion. *Nature*, **400**, 415–415.

Smil V (1999b) Nitrogen in crop production: an account of global flows. *Global Biogeochemical Cycles*, **13**, 647–662.

Smil V (2002) Nitrogen and food production: proteins for human diets. *Ambio*, **31**, 126–131.

Smith DP, Peay KG (2014) Sequence depth, not PCR replication, improves ecological inference from next generation DNA sequencing. *PLoS ONE*, **9**, e90234.

Spitzer M, Wildenhain J, Rappsilber J, Tyers M (2014) BoxPlotR: a web tool for generation of box plots. *Nature Methods*, **11**, 121–122.

Sutton MA, Howard CM, Erisman JW *et al.* (2011a) *The European Nitrogen Assessment: Sources, Effects and Policy Perspectives*. Cambridge University Press, Cambridge, UK, New York.

Sutton MA, Oenema O, Erisman JW, Leip A, Van Grinsven H, Winiwarter W (2011b) Too much of a good thing. *Nature*, **472**, 159–161.

Voisin AS, Salon C, Jeudy C, Warembourg FR (2003) Root and nodule growth in *Pisum sativum* L. in relation to photosynthesis: analysis using 13C-labelling. *Annals of Botany*, **92**, 557–563.

Wang Q, Garrity GM, Tiedje JM, Cole JR (2007) Naive Bayesian classifier for rapid assignment of rRNA sequences into the new bacterial taxonomy. *Applied and Environment Microbiology*, **73**, 5261–5267.

Weaver RW, Danso SKA (1994) Dinitrogen fixation. pp 1019–1045. In: *Methods of Soil Analysis, Part 2*, (eds Weaver RW, Angle JS, Bottomley PS). American Society of Agronomy, Madison, WI.

Weier KL (1980) Nitrogen fixation associated with grasses. *Tropical Grasslands*, **14**, 194–201.

Wewalwela J (2014) Associative nitrogen-fixing bacteria and their potential to support the growth of bioenergy grasses on marginal lands. Unpublished PhD Mississippi State University.

Xia Y, Greissworth E, Mucci C, Williams MA, De Bolt S (2013) Characterization of culturable bacterial endophytes of switchgrass (*Panicum virgatum* L.) and their capacity to influence plant growth. *GCB Bioenergy*, **5**, 674–682.

13

Landscape control of nitrous oxide emissions during the transition from conservation reserve program to perennial grasses for bioenergy

DEBASISH SAHA[1,2], BENJAMIN M. RAU[3,a], JASON P. KAYE[1], FELIPE MONTES[2], PAUL R. ADLER[3] and ARMEN R. KEMANIAN[2]

[1]*Department of Ecosystem Science and Management, The Pennsylvania State University, University Park, PA 16802, USA,* [2]*Department of Plant Science, The Pennsylvania State University, University Park, PA 16802, USA,* [3]*Pasture Systems and Watershed Management Research Unit, USDA Agricultural Research Service, University Park, PA 16802, USA*

Abstract

Future liquid fuel demand from renewable sources may, in part, be met by converting the seasonally wet portions of the landscape currently managed for soil and water conservation to perennial energy crops. However, this shift may increase nitrous oxide (N_2O) emissions, thus limiting the carbon (C) benefits of energy crops. Particularly high emissions may occur during the transition period when the soil is disturbed, plants are establishing, and nitrate and water accumulation may favor emissions. We measured N_2O emissions and associated environmental drivers during the transition of perennial grassland in a Conservation Reserve Program (CRP) to switchgrass (*Panicum virgatum* L.) and *Miscanthus x giganteus* in the bottom 3-ha of a watershed in the Ridge and Valley ecoregion of the northeastern United States. Replicated treatments of CRP (unconverted), unfertilized switchgrass (switchgrass), nitrogen (N) fertilized switchgrass (switchgrass-N), and *Miscanthus* were randomized in four blocks. Each plot was divided into shoulder, backslope, and footslope positions based on the slope and moisture gradient. Soil N_2O flux, soil moisture, and soil mineral nitrogen availability were monitored during the growing season of 2013, the year after the land conversion. Growing season N_2O flux showed a significant vegetation-by-landscape position interaction ($P < 0.009$). Switchgrass-N and *Miscanthus* treatments had 3 and 6-times higher cumulative flux respectively than the CRP in the footslope, but at other landscape positions fluxes were similar among land uses. A peak N_2O emission event, contributing 26% of the cumulative flux, occurred after a 10.8-cm of rain during early June. Prolonged subsoil saturation coinciding with high mineral N concentration fueled N_2O emission hot spots in the footslopes under energy crops. Our results suggest that mitigating N_2O emissions during the transition of CRP to energy crops would mostly require a site-specific management of the footslopes.

Keywords: conservation reserve program, energy crops, land use change, landscape position, *Miscanthus*, switchgrass

Introduction

Renewable fuels are an increasing portion of the liquid fuels portfolio of the United States and some analyses suggest that to fulfill federal mandates, the nation should produce ≈80 billion liters of ethanol from cellulosic sources by 2022 (Kim *et al.*, 2009). This will require planting up to 21 million ha with cellulosic crops (McLaughlin *et al.*, 2002) like switchgrass (*Panicum virgatum*) and *Miscanthus* (M. × *giganteus*). Land seemingly suitable for energy crops is that currently enrolled in the Conservation Reserve Program (CRP), which amounts to ≈12 million hectares of agriculturally marginal and environmentally sensitive land (Farm Service Agency, 2012).

Energy crops like switchgrass and *Miscanthus* can provide crucial ecosystem services when placed in environmentally sensitive lands across the landscape. When planted in floodplains or in lower landscape positions within watersheds, these grasses can serve as riparian buffers to reduce nutrient and sediment load to surface water and groundwater (Perez-Suarez *et al.*, 2014; Smith *et al.*, 2014). Furthermore, due to the low soil disturbance and perennial rooting systems, they can store C in the soil due to their large below-ground C allocation (Lemus & Lal, 2005; Follett *et al.*, 2012).

A critical component of the societal benefit of energy crops is that they produce fuel with a low C footprint (Tilman *et al.*, 2006; Adler *et al.*, 2007; Gelfand *et al.*, 2013). The United States Environmental Protection

[a]Present address: Savannah River Forestry Sciences Lab, USDA Forest Service, 241 Gateway Drive, Aiken, SC 29803, USA

Correspondence: Debasish Saha
e-mails: dxs1005@psu.edu and debasish992@gmail.com

Agency's revision to the National Renewable Fuel Standard requires that to qualify as renewable, advanced biofuels lifecycle greenhouse gas (GHG) emissions must be 50% of those from fossil fuel (EPA, 2010). To achieve this goal, emissions of nitrous oxide (N_2O) must be low during the feedstock production phase, because N_2O is the largest source of GHG for feedstock production (Adler *et al.*, 2012). There are arguments to posit that converting historically managed CRP lands to agricultural crops may increase N_2O emissions (Ruan & Robertson, 2013).

In the Ridge and Valley physiographic province of the Allegheny Plateau in the northeastern United States, many lands are considered marginal due to being seasonally wet. These landscapes have characteristically shallow, coarse, and rocky ridge top soils as well as fractured subsoils that drain water to the back and footslope positions (Fig. S1). The footslope soils are derived from mixed colluvial sandstone and shale, often with fragipans at shallow depth that limit drainage (Ciolkosz *et al.*, 1995). Restricted drainage in the footslope causes a shallow, temporary water table, and extended periods of soil water content above field capacity or near saturation during spring snowmelt and rainstorm events (Buda *et al.*, 2009). The steep soils with a semi-impermeable subsoil favor nitrate (NO_3^-) transport to stream water (Kleinman *et al.*, 2006; Zhu *et al.*, 2011). The simultaneous occurrence of well drained upper lands and poorly drained footslope positions may create biogeochemical hot spots for denitrification and N_2O emissions (Vilain *et al.*, 2010).

The N_2O emissions from soil predominantly originate from nitrification and denitrification (Firestone & Davidson, 1989). Soil oxygen, NO_3^-, ammonium (NH_4^+), and labile organic C concentration determine the contribution of nitrification and denitrification to the total N_2O flux (Weier *et al.*, 1993; Gillam *et al.*, 2008). We hypothesize that in seasonally wet lands of the northeastern United States, N_2O emissions could be severe during the transition from CRP to energy crops due to accelerated C and nitrogen (N) cycling of above and below-ground residues from the former CRP vegetation, soil disturbance during the land conversion (Zenone *et al.*, 2011; Nikiema *et al.*, 2012; Ruan & Robertson, 2013), and relatively low crop growth rate in the first two years of stand establishment. Furthermore, if the energy crops are fertilized, co-locating N fertilizer with decomposable organic residues along water flow paths will create conditions for increased N_2O emissions. Thus, managing N and N_2O emissions from energy crops in this region requires understanding the interactive controls of landscape, crop growth, and hydrology on soil N cycling.

In this research, we measured soil N_2O emission during land transition from CRP to energy crops *Miscanthus*

and switchgrass in a watershed in the Ridge and Valley region. The lower part of the watershed was under CRP since 1999. In 2012, we divided the landscape into replicated plots of: (i) unconverted CRP, (ii) *Miscanthus*, (iii) switchgrass, and (iv) fertilized switchgrass. The N_2O fluxes from the CRP and establishing switchgrass and *Miscanthus* were monitored during the 2013 growing season, the second year after land conversion. In June of 2013, 10.8 cm of rain from hurricane 'Andrea' caused sudden soil saturation in parts of the watershed. This event gave us an opportunity to measure the event-based response of N_2O emissions during the transition from CRP to energy crops. To the best of our knowledge, this is the first time that N_2O emissions from CRP, switchgrass, and *Miscanthus* have been simultaneously measured to assess the landscape by land management interaction during land use transition. Our research questions were: (i) what is the effect of converting CRP lands to switchgrass and *Miscanthus* on soil N_2O emissions; (ii) what is the effect of N fertilization in switchgrass on N_2O emissions; and (iii) how do landscape heterogeneity and land conversion interact to control N_2O emissions?

Materials and methods

Site description

The experimental area, hereafter called Mattern (40°42′N, 76°36′W), is located near the town of Leck Kill in east-central Pennsylvania (PA). It is part of a long-term monitoring site of the USDA-Agricultural Research Service (Sharpley *et al.*, 2008). The site has a temperate humid climate with annual mean temperature and precipitation of ≈9.2° C and ≈106 cm, respectively. Mattern is an 11-ha sub-watershed of the 726-ha watershed (WE-38) that drains to Mahantango Creek, a tributary of the Susquehanna River (Sharpley *et al.*, 2008). The site is representative of the Ridge and Valley physiographic province.

Mattern has mixed land use with 57% cropland, 30% forest, 4% pasture, 9% meadow, and <1% buildings. The upper valley lands have rotations of soybean (*Glycine max* L.), wheat (*Triticum aestivum* L.), and corn (*Zea mays* L.). Slope ranges from 1% to 20%. The elevation above sea level varies from 267 m in the valley floor to 285 m near the summit. The topsoil typically has a silt loam texture with 20–40% rock volume (Table 1) that grades to a loam and/or silty clay loam texture at depth (≈100 cm). Albrights soils are distributed along stream and valley floor and have a fragipan and argillic horizon beginning at a depth of 50–70 cm (Needelman *et al.*, 2004). These soils experience prolonged soil saturation during spring snowmelt and after rain events. In contrast, Berks soils in the shoulder and backslope positions are relatively shallow and well drained.

The plots were established in the lower portion of the watershed, in an area of approximately 3-ha that includes the watershed outlet (Fig. 1). Mattern has an ephemeral stream that is

Table 1 Soil properties in the top 20 cm soil layer at each landscape position

Landscape position	Total C g kg^{-1} soil	Total N g kg^{-1} soil	Bulk density g cm^{-3}	Rock fraction m^3 m^{-3}	Clay g kg^{-1} soil	Silt g kg^{-1} soil
Shoulder	20 ± 5	2.2 ± 0.4	1.18 ± 0.12	0.35 ± 0.05a	170 ± 30b	480 ± 28
Backslope	19 ± 5	2.1 ± 0.3	1.15 ± 0.11	0.32 ± 0.05b	190 ± 34b	480 ± 27
Footslope	19 ± 5	2.0 ± 0.4	1.21 ± 0.10	0.28 ± 0.04c	210 ± 29a	490 ± 30

Each value represents mean ± standard deviation ($n = 16$). Different letters within a column indicate a statistically significant difference among the landscape positions at $P < 0.05$.

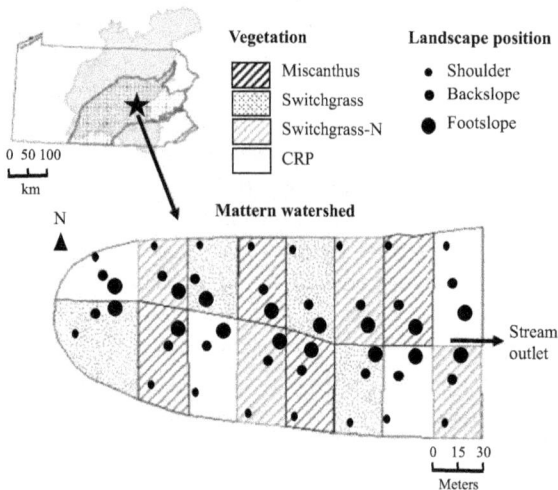

Fig. 1 Upper panel showing the location of the Mattern watershed (black star) within the Susquehanna River basin (grey shaded) and Ridge and Valley physiographic province (dotted region) of Pennsylvania. The lower panel shows treatment plots and measurement locations in shoulder, backslope, and footslope (increasing size of black circle) positions within each plot.

active most of the year except during dry summers. Both the lowland and upland were under conventionally tilled corn, soybean, wheat, and alfalfa (*Medicago* sp.) rotation with poultry manure applied at the rate of 5 Mg ha^{-1} yr^{-1}, corresponding to 205 kg N ha^{-1} yr^{-1} until 1999 (Kleinman et al., 2006). The lowland was considered marginal due to seasonal wetness and high slope that impairs annual cropping, and was brought under CRP in 1999. The CRP vegetation comprises the perennial cool-season grasses orchard grass (*Dactylis glomerata* L.), tall fescue (*Schedonorus arundinaceus* (Schreb.) Dumort.), timothy (*Phleum pratense* L.), and legumes including alfalfa (*Medicago sativa* L.) and clovers (*Trifolium* spp.).

Experimental design

The lower portion of the watershed was divided in four blocks with relatively uniform slope and aspect. Each block contained four plots measuring ≈60 × 30 m each (Fig. 1) one each of the following treatments: (i) unconverted CRP, (ii) N-fertilized switchgrass (switchgrass-N), (iii) unfertilized switchgrass (switchgrass), and (iv) unfertilized *Miscanthus*. In the 12 plots designated for energy crops, CRP vegetation was killed with

glyphosate [N-(phosphonomethyl) Glycine] in the summer of 2011, and the aboveground biomass baled and removed from the field. The glyphosate-treated area was no-till planted with winter rye (as a cover crop) in the fall of 2011, which was in turn killed with glyphosate prior to planting the energy crops in the spring of 2012. The switchgrass plots were no-till drilled in rows 20-cm apart. Rhizomes of *Miscanthus* plants were hand planted 76 cm apart in chisel-made furrows. Thus, the *Miscanthus* plots had a higher level of soil disturbance. Throughout 2012, the plots of both species were hand replanted, using plants, in areas with poor establishment, especially in wet footslope positions. By 2013, plot establishment was satisfactory except in the footslopes, especially in *Miscanthus*. In 2012, broadleaf weeds and annual grasses were controlled with herbicide; weed pressure was low in 2013. The switchgrass-N plots received 50 kg ha^{-1} of N as broadcasted urea on May 29th, 2013, a year after planting.

The upland portion of each plot borders cropland; the bottom portion merges into the ephemeral stream that drains the watershed. Based on the increasing soil wetness from top to bottom, each plot was divided into three segments: shoulder, backslope, and footslope (Fig. 1). This is customary when studying topographic effects on soil N$_2$O emissions (e.g. Pennock et al., 1992; Vilain et al., 2010). The combination of four treatments, three landscape positions, and four replications (blocks) yielded a total of 48 monitoring points.

Measurement of soil water content and air filled soil volume

We continuously monitored volumetric soil water content (θ_V, m^3 m^{-3}) with CS-616 soil moisture sensors (Campbell Scientific Inc., Logan, UT, USA). The sensors were installed at three depths (0–20, 20–40, and 40–60 cm) in each of the 48 monitoring points, for a total of 144-sensors. Each sensor was connected to one of four dataloggers through a network of buried cables.

The θ_V was used to calculate the volumetric air content in the soil layer (θ_A, m^3 m^{-3}):

$$\theta_A = \theta_S - \theta_V \qquad (1)$$

where θ_S is the total porosity of a layer after correcting for rock volume (m^3 m^{-3}).

Measurement of N$_2$O emissions

We measured N$_2$O emissions from May to September of 2013. The sampling frequency varied from weekly to biweekly, and increased after fertilization and precipitation events. The soil-atmosphere N_2O flux was measured by the static, non-steady state, vented, aluminum-foil insulated chamber method (Hutchinson & Mosier, 1981). In 2012, PVC collars of 30-cm diameter were inserted 5 cm in the soil in each of the 48 monitoring points. Each collar supports a gas-tight chamber of 10-cm height. The chamber interior was free of any vegetation.

At measurement time, chambers were placed on top of the collars and left there for 45 min. Gas samples of 20 ml were drawn from each chamber through a rubber septum connected to a manifold inside the chamber. Samples were taken 15, 30, and 45 min after chamber closure. Soil temperature near the chamber was measured during gas sampling.

The 20-ml gas samples were transferred to 12-ml pre-evacuated Labco exetainer vials (Labco Limited, Lampeter, UK). The N_2O concentration in the gas samples was measured with a Varian CP3800 (Varian, Walnut Creek, CA, USA) gas chromatograph (with Compi-Pal autosampler) equipped with a [63]Ni electron capture detector that operates at 300 °C to detect N_2O. The ideal gas law was used to calculate the μg N_2O-N and the flux was calculated from the rate of increase in N_2O concentration in the chamber headspace.

Cumulative N_2O flux (May 9th to September 13th) was calculated by interpolation and linear integration. We refer to this as the growing season flux as it roughly corresponds to the period when the grasses were green and soils were not frozen. The N_2O flux on the days in between two sampling days was estimated as:

$$F_L = F_o + \frac{(F_f - F_o) \times (d_f - d_i)}{d_f - d_o} \qquad (2)$$

where, F_L is the estimated flux of N_2O, F_o and F_f are the measured fluxes bracketing the days interpolated, d_o and d_f are the days corresponding to F_o and F_f, and d_i is the ith day in between d_o and d_f.

Soil sampling

Soil samples (0–20 cm) were taken on every other gas flux measurement day to determine NH_4^+ and NO_3^- concentration (five sampling days during the growing season). We were not able to sample soil for mineral N on the day immediately after the cessation of 'Andrea' on June 11th because the soils were flooded. Fresh soil samples were extracted with 100 ml 2M KCl and the inorganic N species were determined by colorimetric analysis (Lachat Quick Chem 8000). All values were corrected for gravimetric water content measured on a 10 g subsample oven dried at 105 °C for 48 h.

Soil cores of 50–120 cm in length, depending on the location, were collected from each sampling point in spring 2012. The cores were taken with a SIMCO Earthprobe 200 (SIMCO Drilling Equipment, Inc. Osceola, IL, USA) using plastic tube liners of 4.5 cm in diameter and a tip diameter of 3.8-cm. The soil core was used for geomorphic description, determination of bulk density, particle size distribution (USDA-pipette method; Day, 1965), total soil C and N concentration (Elemental Combustion System auto-analyzer, Costech Analytical Technologies, Inc., Valencia, CA, USA), and rock volume (water displacement method).

Statistical analysis

We used parametric and non-parametric statistical methods for data analysis. All statistical analyses were performed with the R statistical software (R Development Core Team, 2012). First, we used an analysis of variance (ANOVA) with block, vegetation type, and landscape position effects. The residuals of the daily N_2O fluxes were not always normally distributed and were log-transformed for ANOVA. The residuals of the cumulative N_2O fluxes were not normally distributed and the original fluxes were transformed using the reciprocal square root transformation based on the ladder of powers method (Tukey, 1977) to achieve normality (Shapiro-Wilk, $W = 0.96$, $P = 0.17$) and variance homogeneity. Data were back transformed for tabular presentation. Marginal means were compared using Tukey-adjusted P-values at a significance level of $P < 0.05$. The main effect and the interactions on daily and cumulative N_2O flux were assessed with the following linear model:

$$Y_{ijk} = \mu + \alpha_i + \gamma_k + (\alpha\gamma)_{ik} + \beta_j + (\alpha\beta)_{ij} + (\beta\gamma)_{jk} + e_{ijk} \qquad (3)$$

where, Y_{ijk} represents the response variable N_2O flux; α, β, and γ represent the main effect of vegetation type, landscape position, and block, respectively; e is the error term, and i, j, k denote the respective levels of the main factors.

Second, we used Random Forest method to analyze which variables best explain the observed variation in log_{10} transformed N_2O flux (Breiman, 2001). Random Forest was applied to the pooled data including only the days when soil mineral N was measured along with soil N_2O flux, θ_A, and weather variables. We used the function randomForest from the package randomForest in R (Liaw & Wiener, 2002). The control parameters for random forest were $seed = 500$ (set random number) and $ntree = 1000$ (number of trees). The control parameter $mtry$ indicates the number of variables available for splitting at each node and was calculated as square root of total number of variables (Strobl et al., 2009). The variable importance was plotted by using the function varImpPlot. The variables used for this analysis are: vegetation type, landscape position, soil NO_3^- and soil NH_4^+ concentration in the top 20 cm layer (mg N kg^{-1} dry soil), volumetric soil air content at the 0–20, 20–40, and 40–60 cm soil depth (θ_{A20}, θ_{A40}, and θ_{A60}; m^3 m^{-3}), and cumulative precipitation in the last two days (PPT_2, cm).

Finally, using the tree package in R ($seed = 500$), we constructed a conditional inference tree of N_2O emissions using observations from the five dates in which mineral N was measured. Each tree terminal node has an average N_2O flux and a number of observations (n).

Results

Weather, soil properties, and biomass yield

During the study period, the mean daily air temperature was 19.1 °C and the cumulative precipitation was 40.2 cm, with a large, 10.8 cm rain event during hurricane 'Andrea' on June 10th. The landscape positions had similar C and N distribution and bulk density in the top 20 cm soil layer ($P > 0.05$); all the treatments started with the same organic C and N in the top 20 cm of soil (Table 1). Soils had mean C and N concentration of 19 and 2.1 g kg^{-1} soil. The soil bulk density (free of rocks) varied from 1.0 to 1.5 g cm^{-3}, with a mean of 1.2 g cm^{-3}. Rocks occupy a substantial soil volume (mean 0.32 m^3 m^{-3}, range 0.19 to 0.43 m^3 m^{-3}), with a significantly higher rock fraction in the shoulder position that decreases down the slope ($P < 0.05$). The clay concentration varied from 130 to 290 g kg^{-1}, with the highest figures in the footslopes (Table 1).

The aboveground biomass yield of the energy crops harvested in fall 2013, after the second year of growth, exceeded 10 Mg ha^{-1}, and was 2–4 times higher than that of the CRP (Table 2). While the figure for the CRP reflects the harvestable biomass in fall, it does not represent total aboveground biomass produced because a large part is produced in spring, senesces in early summer and is not harvestable in fall. There was no statistically significant effect of the landscape position and N-fertilization on biomass yield.

Soil ammonium and nitrate concentration

Mean soil NH$_4^+$ concentration during the growing season was 10 ± 12 mg N kg^{-1} soil (Fig. 2), lowest in *Miscanthus* (6 ± 5 mg N kg^{-1} soil) and highest in switchgrass-N (14 ± 20 mg N kg^{-1} soil). The variation in NH$_4^+$ concentration among landscape positions was very narrow (8.8 to 10.4 mg N kg^{-1} soil). The N-fertilization in switchgrass-N increased the mean NH$_4^+$ concentration from 0.8 ± 0.5 mg N kg^{-1} soil before the fertilization to 48 ± 22 mg N kg^{-1} soil on June 7th, a week after fertilization (Fig. 2). However, this peak NH$_4^+$ concentration faded away after the hurricane. A substantial increase in NH$_4^+$ concentration was also observed in unfertilized grasses (and higher in CRP than energy crops) during summer and tended to decrease over the growing season. No plot showed N stress as vegetation growth was thick and vigorous, but the N-fertilized plots had a darker green color and the plants lodged in mid-July.

Mean growing season soil NO$_3^-$ concentrations were 19 ± 14 mg N kg^{-1} soil (Fig. 2). In contrast to NH$_4^+$, the NO$_3^-$ concentrations were highest in *Miscanthus* plots (25 ± 17 mg N kg^{-1} soil), which were twice as high as that of CRP (13 ± 13 mg N kg^{-1} soil). Again, the variation among landscape positions was narrow (footslope 17 and shoulder 20 mg N kg^{-1} soil, respectively). In the switchgrass-N plots, the N-fertilization had a minor effect on NO$_3^-$ dynamics in shoulder and backslope positions, while the footslope position showed an increase in soil NO$_3^-$ concentration (30 mg N kg^{-1} soil) during early June. This NO$_3^-$ concentration was five-times higher than that in CRP (Fig. 2). On June 7th, just before the hurricane, *Miscanthus* plots had the highest NO$_3^-$ concentration in all landscape positions (mean 43 ± 20 mg N kg^{-1} soil), while CRP plots had the lowest NO$_3^-$ concentration (8 ± 3 mg N kg^{-1} soil). These differences were not discernible after the hurricane and NO$_3^-$ concentration became comparable among treatments.

Soil air content

Soil profile air content ranged from 0 m^3 m^{-3} (soil water saturation) to 0.32 m^3 m^{-3} (Fig. 3). The greatest variation in θ_A among the treatments was in the footslope. Mean θ_A in the profile was lowest in the 20–40 cm layer (0.14 ± 0.03 m^3 m^{-3}) and highest in the top 20 cm (0.19 ± 0.04 m^3 m^{-3}). The start of the 2013, θ_A flux measurements coincided with a rain event that reduced θ_A in the soil profile. There was a steady increase in θ_A from May 9th onwards until the onset of the hurricane in June 10th.

The hurricane driven rain saturated the soil and decreased θ_A in the top soil layer of the footslope ($0.02 + 0.02$ m^3 m^{-3}), which was significantly lower than that in the backslope (0.08 ± 0.07 m^3 m^{-3}) and shoulder (0.12 ± 0.06 m^3 m^{-3}) positions on June 11th ($P < 0.05$). A prolonged period of low θ_A was noticeable in the footslope positions, especially in the subsurface

Table 2 Above-ground dry biomass production by CRP and energy crops in different landscape positions

Landscape position	CRP	Switchgrass	Switchgrass-N	*Miscanthus*	Mean
			Mg ha^{-1}		
Shoulder	3.5 ± 0.1	10.7 ± 0.4	10.5 ± 1.4	9.7 ± 1.4	8.6
Backslope	3.8 ± 0.2	11.1 ± 0.4	10.3 ± 0.6	12.5 ± 3.2	9.4
Footslope	5.2 ± 0.8	11.2 ± 0.8	11.6 ± 1.1	10.2 ± 2.1	9.6
Mean	4.2b	11.0a	10.8a	10.8a	

Each value represents mean ± standard error of mean ($n = 4$). Mean biomass yields followed by same letter within a row are not significantly different at $P < 0.05$. The above-ground biomass yields are based on post-senescence manual harvesting of 3 m × 1 m quadrats at 10 cm height from the ground surface at each monitoring point. The CRP biomass was harvested in September, while switchgrass and *Miscanthus* were harvested in early November.

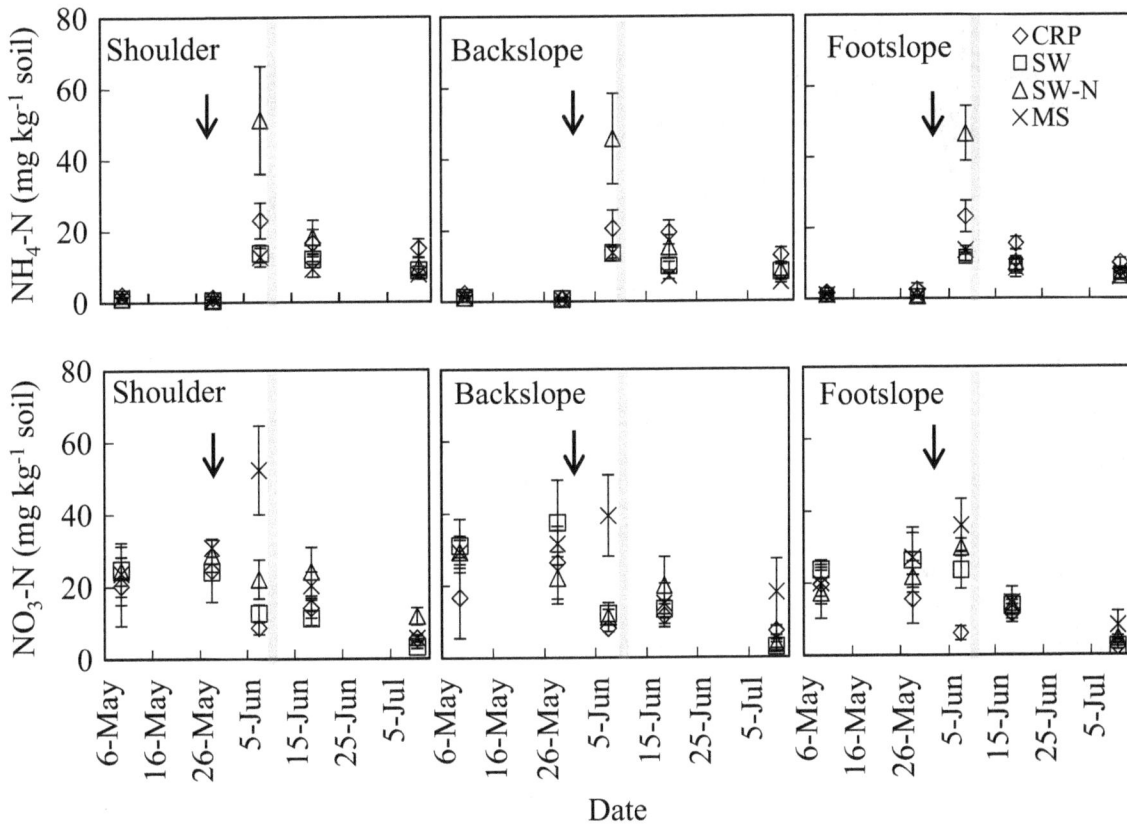

Fig. 2 Dynamics of soil NH_4-N (upper panel) and NO_3-N (lower panel) concentration in the top 20 cm soil layer under CRP and energy crops (SW, switchgrass; SW-N, fertilized switchgrass; MS, *Miscanthus*) at different landscape positions during 2013 growing season. Each point is a mean of four replicates. Error bars represent ± standard error of the mean. The solid arrow indicates the timing of N-fertilization in the switchgrass-N treatment and the vertical gray bar indicates hurricane 'Andrea'.

soil layers (Fig. 3c). Soil aeration started to increase after June 14th as drainage progressed. Subsequent rain events decreased soil aeration, but never to the extent or length of the hurricane driven storm.

N₂O emissions

The N_2O flux from soil to atmosphere widely varied among the vegetation and landscape positions, ranging from 4 to 305 g N ha^{-1} day^{-1} among vegetation types (Fig. 4). The mean flux was lowest in CRP and highest in *Miscanthus* (8 vs. 16 g N ha^{-1} day^{-1}). The footslope positions were the hot spots, with greater variability of N_2O emissions than other landscape positions, and a mean flux (18 g N ha^{-1} day^{-1}, range of 60 g N ha^{-1} day^{-1}) that was more than two-times higher than the shoulder positions (8 g N ha^{-1} day^{-1}). Footslope positions under *Miscanthus* had the highest mean growing season N_2O emission (26 g N ha^{-1} day^{-1}) followed by switchgrass-N, switchgrass, and CRP (23, 16, and 7 g N ha^{-1} day^{-1}, respectively).

During early May, emissions from the footslope positions were significantly higher than that from the shoulder positions (55 vs. 19 g N ha^{-1} day^{-1}, $P < 0.05$); emissions from backslope positions were intermediate and not significantly different from the other landscape positions (30 g N ha^{-1} day^{-1}, Fig. 4). In addition, during the same period, the average N_2O emission from energy crops of 41 g N ha^{-1} day^{-1} was 2.6-times higher than that from CRP. The N-fertilization did not have an immediate effect on N_2O flux from switchgrass-N in the shoulder and backslope positions; however, a small peak of 35 g N ha^{-1} day^{-1} occurred in the footslope position (Fig. 4).

Peak N_2O emissions were triggered by the hurricane rains on June 10th (Fig. 4). The rain saturated the soil and reduced θ_A (Fig. 3), coinciding with a period of high mineral N concentration in soil (Fig. 2). Averaged over all plots, the N_2O flux was 84 g N ha^{-1} day^{-1} on June 11th, a day after the large precipitation event. This flux was roughly four-times higher than the pre-hurricane flux of 22 g N ha^{-1} day^{-1} on June 7th. Furthermore, emissions from the footslope (156 g N ha^{-1} day^{-1}) and backslope (76 g N ha^{-1} day^{-1}) positions were significantly higher than those from the shoulder

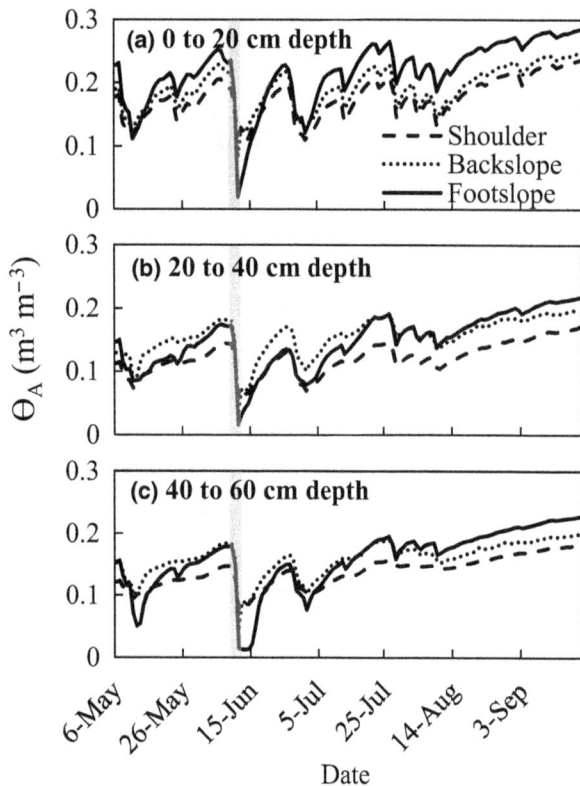

Fig. 3 Temporal and spatial variation of soil air content (θ_A) in (a) 0–20, (b) 20–40, and (c) 40–60 cm soil depths in different landscape positions during 2013 growing season N_2O flux monitoring period. Values were averaged across vegetation types (n = 16) at each depth in each landscape position. The vertical gray bar indicates hurricane 'Andrea'.

positions (20 g N ha^{-1} day^{-1}, $P < 0.05$). Immediately after the hurricane rains, the N_2O emissions from the grasses were in the order of: Miscanthus > switchgrass-N > switchgrass > CRP (130, 103, 69, and 34 g N ha^{-1} day^{-1}, respectively). A significant interaction between vegetation type-by-landscape position was only observed during this hot moment ($P = 0.03$). The emissions from lower landscape positions of Miscanthus (305 g N ha^{-1} day^{-1}) and switchgrass-N (233 g N ha^{-1} day^{-1}) was significantly higher than that in CRP (20 g N ha^{-1} day^{-1}, $P < 0.05$). The hurricane induced peak N_2O emission waned to background level after June 18th as the water drained and oxic conditions prevailed in the landscape (Fig. 3).

Growing season cumulative N_2O flux

The ANOVA of cumulative N_2O flux shows a significant vegetation-by-landscape position interaction ($P = 0.009$, Table 3). The differences in cumulative N_2O emissions between the CRP and energy crops only expressed in

the footslope positions (Table 3). While in CRP emissions were similar across landscape positions ($\approx$1 kg N ha^{-1}), these emissions were higher in the footslope for the energy crops, in particular for Miscanthus ($\approx$5.5 kg N ha^{-1}).

Analysis of conditions leading to N_2O emissions

The Random Forest analysis identified landscape position, cumulative precipitation in the two days before N_2O flux measurement (PPT$_2$), and soil NO$_3^-$ concentration as the most important factors influencing N_2O emissions (Fig. 5). It is the soil aeration (θ_A) in subsurface layers (20–40 and 40–60 cm) that seem to be related to N_2O emission, rather than the aeration in the top layer. Nonetheless, the Random Forest model explained only 28% of the variation in N_2O flux. Clearly, there are other controls of N_2O emissions that we did not measure and can include microsite properties that are difficult to characterize.

Using a regression tree, the predictor variables as identified by the Random Forest were used to classify the N_2O fluxes in groups that can be identified by specific properties. The regression tree contains seven terminal nodes. The primary node shows a split based on $\theta_{A40} = 0.03$ m^3 m^{-3}, with the highest emission when $\theta_{A40} < 0.03$ m^3 m^{-3} (Fig. 6). The observations that belong to the primary node are associated with high NO$_3$-N concentration (mean 19 mg N kg^{-1} soil) and poor aeration status ($\theta_{A40} < 0.03$ m^3 m^{-3}) in the soil profile, and are mostly found in the footslope positions. The shoulder positions never sustained $\theta_{A40} < 0.03$ m^3 m^{-3}. Accordingly, node 4, which is the node with lowest average emission, is composed only of measurements in the shoulder positions. When $\theta_{A40} > 0.03$ m^3 m^{-3}, moderately high emissions (mean 28 g N ha^{-1} day^{-1}) occur only in the footslope positions when soil NO$_3$-N concentration is >10 mg N kg^{-1} soil (node 7). Since the construction of the regression tree excluded the highest N_2O emission period after the hurricane event, the potential emissions can be much higher than the values predicted by the tree. However, the tree clearly points to the landscape position and the subsoil air content as the drivers of N_2O emissions.

Discussion

Converting CRP lands to energy crops only increased N_2O emissions significantly in the footslopes during the second year of the land use transition (Table 3). At that position, N_2O emissions from switchgrass, switchgrass-N, and Miscanthus were 2×, 3×, and 6× larger than emissions from CRP. For comparison, Gelfand et al. (2011) reported 4.5 times higher N_2O emissions from

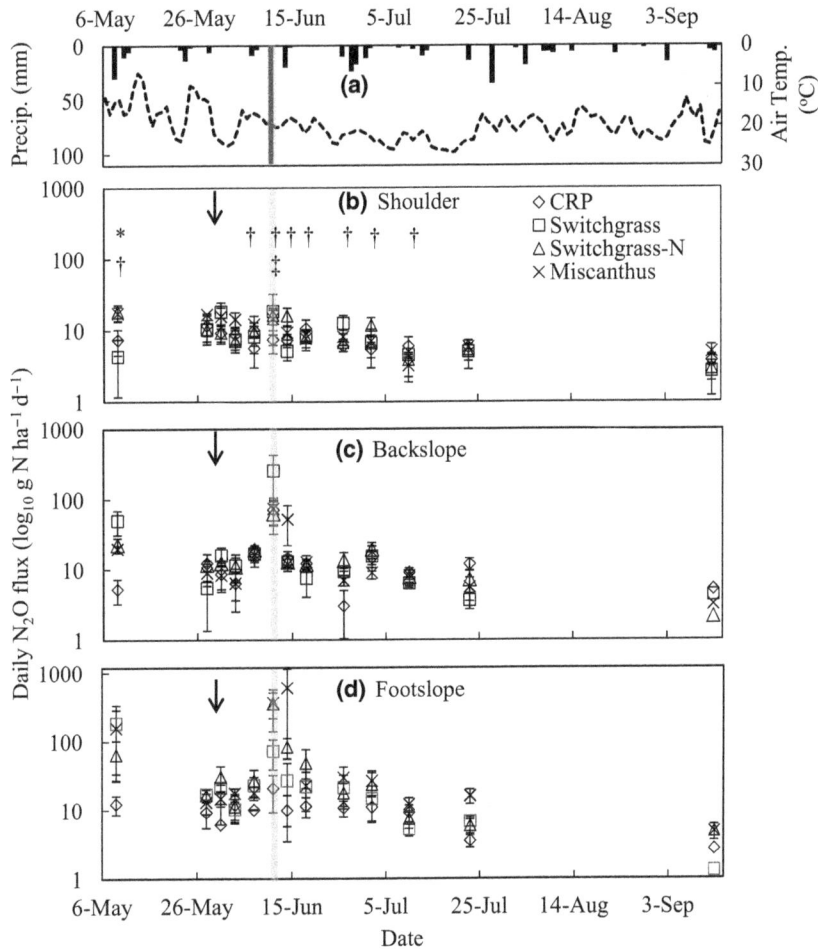

Fig. 4 Daily precipitation, air temperature (a) and $\log_{10}$ transformed daily N_2O fluxes from CRP and energy crops under (b) shoulder, (c) backslope, and (d) footslope position during 2013 growing season. Each point is a mean of four replicates. Error bars represent ± standard error of mean ($n = 4$). The solid arrow indicates the timing of N-fertilization in the switchgrass-N treatment and the vertical gray bar indicates hurricane 'Andrea'. Significant ($P < 0.05$) effects of vegetation (*), Landscape position (†), and/or their interaction (‡) on the measurement day were presented in top of panel b.

no-till soybean converted from CRP grassland. In our case, the lower landscape positions occupy at most a third of the lower part of the watershed, and therefore at least more than two thirds of the area had N_2O emissions comparable to those of CRP, hereby answering our first research question.

The N_2O emission differences among treatments built up over three distinctive periods: during early season growth prior to the hurricane, the week of the hurricane in mid-June, and the post hurricane period until the end of the growing season. These three periods have distinctive evolutions of soil mineral N and soil water in the different treatments. In the early season, the cool-season CRP vegetation starts growing before the warm season energy crops, which would reflect in comparatively more mineral N depletion in CRP. Accordingly, in early June, the soil NO_3^- concentration in CRP was lower than

that of *Miscanthus* in all landscape positions; however, the switchgrass soil NO_3^- concentration was similar to *Miscanthus* in the footslope, but similar to CRP in the shoulder and backslope (Fig. 2). Furthermore, with the obvious exception of fertilized switchgrass, CRP had slightly higher NH_4^+ than unfertilized switchgrass and *Miscanthus* at all landscape positions. Thus, the behavior of mineral N in the early season is more nuanced than expected.

Several factors may have contributed to relatively high NO_3^- accumulation under energy crops, particularly *Miscanthus*. First, in the years prior to planting the CRP in 1999, the soils received > 200 kg N ha^{-1} yr^{-1} as manure. In addition, the CRP vegetation had a visible, but not dominant, proportion of legumes. Killing the pre-existing CRP vegetation during land conversion created fresh above and belowground pools of dead grass

Table 3 Analysis of variance for significance of differences in growing season cumulative N_2O emission in different landscape positions under CRP and energy crops. The cumulative N_2O flux was reciprocal square root transformed for ANOVA, and the means were back-transformed for presentation in the table

Analysis of variance Sources of variance	df	Mean square	F value	P
Vegetation (V)	3	0.14	6.3	0.004
Block (B)	3	0.02	1.2	0.333
Landscape position (LP)	2	0.59	26.9	<0.001
V × LP	6	0.09	4.1	0.009
B × LP	6	0.02	0.9	0.514
V × B	9	0.04	2.0	0.106
Error	18	0.02		

Cumulative N_2O flux					
	CRP	Switchgrass	Switchgrass-N	Miscanthus	Mean
Landscape position			kg N ha^{-1}		
Shoulder	0.8	0.8	0.9	1.1	0.9
Backslope	1.4	1.6	1.4	1.4	1.4
Footslope	**0.9**	**2.2**	**2.9**	**5.5**	2.1
Mean	1.0	1.3	1.5	1.8	

The bold figures imply significantly higher cumulative N_2O fluxes from the energy crops than the CRP in the footslope positions only.

and legume root biomass that may have mineralized N during decomposition, adding to N mineralization from soil organic matter (Zenone *et al.*, 2011; Ruan &

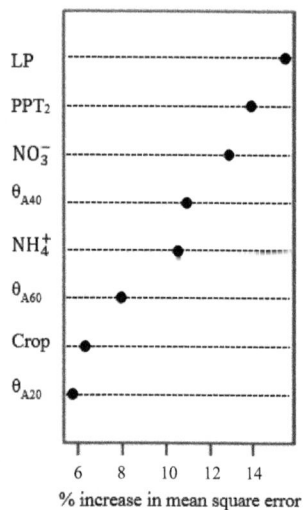

Fig. 5 Random forest based variable importance plot for N_2O flux ($\log_{10}$ transformed). The % increase in the mean square error represents the mean increase in classification error due to random permutation of the variable indicated. Higher values of percent increase in mean squared error indicate higher importance of the variable to explain N_2O emissions in the random forest model (variance explained: 28%). LP, landscape position; PPT$_2$, cumulative precipitation in preceding two days; θ_{A20}, θ_{A40}, and θ_{A60}, volumetric soil air content (m^3 m^{-3}) at 0–20, 20–40, and 40–60 cm depth, respectively.

Robertson, 2013). Killing the rye cover crop added additional decomposable residues, albeit in low quantities (<2 Mg ha^{-1}). Second, chisel plowing prior to rhizome establishment caused a greater level of soil disturbance in *Miscanthus* compared to no-till switchgrass seeding, which may have accelerated soil organic matter decomposition (Grandy & Robertson, 2006; Mazzilli *et al.*, 2015). Third, due to the different techniques used to establish the energy crops, the switchgrass stands were denser than that of *Miscanthus* early in the growing season, allowing switchgrass an earlier canopy closure and higher N uptake. This might have resulted in more N available for losses in *Miscanthus* plots. In short, killing the established CRP vegetation in C and N rich soils, can cause high concentration of soil mineral N, favoring N_2O emissions even under no-till (switchgrass); the effect was more pronounced under *Miscanthus* possibly due to additional factors such as chisel plowing and slower establishment of the wide-spaced *Miscanthus* rhizomes (Table 3). A similar case was reported by Gauder *et al.* (2012), who found higher N_2O emission from *Miscanthus* compared with shrub willow (*Salix* spp.) during early summer and attributed the difference to the partial soil cover by *Miscanthus*, in addition to high soil mineral N after fertilization. An alternative hypothesis for the lack of NO_3^- accumulation in CRP can be a limited nitrification potential due to the low disturbance, as suggested by the moderate accumulation of NH_4^+ (Fig. 2); it has been shown that the nitrification potential remarkably increased when long-term undisturbed

$\theta_{A40} < 0.03$ m³ m⁻³ — Yes (left) / No (right)

Left branch: NH_4–N < 1 mg N kg⁻¹ soil — right child: NO_3-N < 14 mg N kg⁻¹ soil

Right branch: NO_3-N < 10 mg N kg⁻¹ soil — left child: LP = Shoulder; right child: LP = Shoulder, Backslope

	Node 1	Node 2	Node 3	Node 4	Node 5	Node 6	Node 7
Average flux (g N ha⁻¹ d⁻¹)	146	22	37	15	19	21	28
Frequency	$n = 5$	$n = 6$	$n = 5$	$n = 27$	$n = 51$	$n = 101$	$n = 45$
Average node properties							
NO_3 (mg N kg⁻¹ soil)	19.5	9.2	19.6	6.3	4.7	27	24
NH_4 (mg N kg⁻¹ soil)	0.84	9.7	8.3	10.1	10.5	9.5	9.4
θ_{A20} (m³ m⁻³)	0.06	0.13	0.13	0.17	0.20	0.17	0.21
θ_{A40} (m³ m⁻³)	0.02	0.02	0.01	0.12	0.16	0.12	0.15
θ_{A60} (m³ m⁻³)	0.03	0.03	0.01	0.13	0.15	0.13	0.15
Landscape position	Backslope (1) Footslope (4)	Backslope (1) Footslope (5)	Backslope (1) Footslope (4)	Shoulder (27)	Backslope (29) Footslope (22)	Shoulder (53) Backslope (48)	Footslope (45)

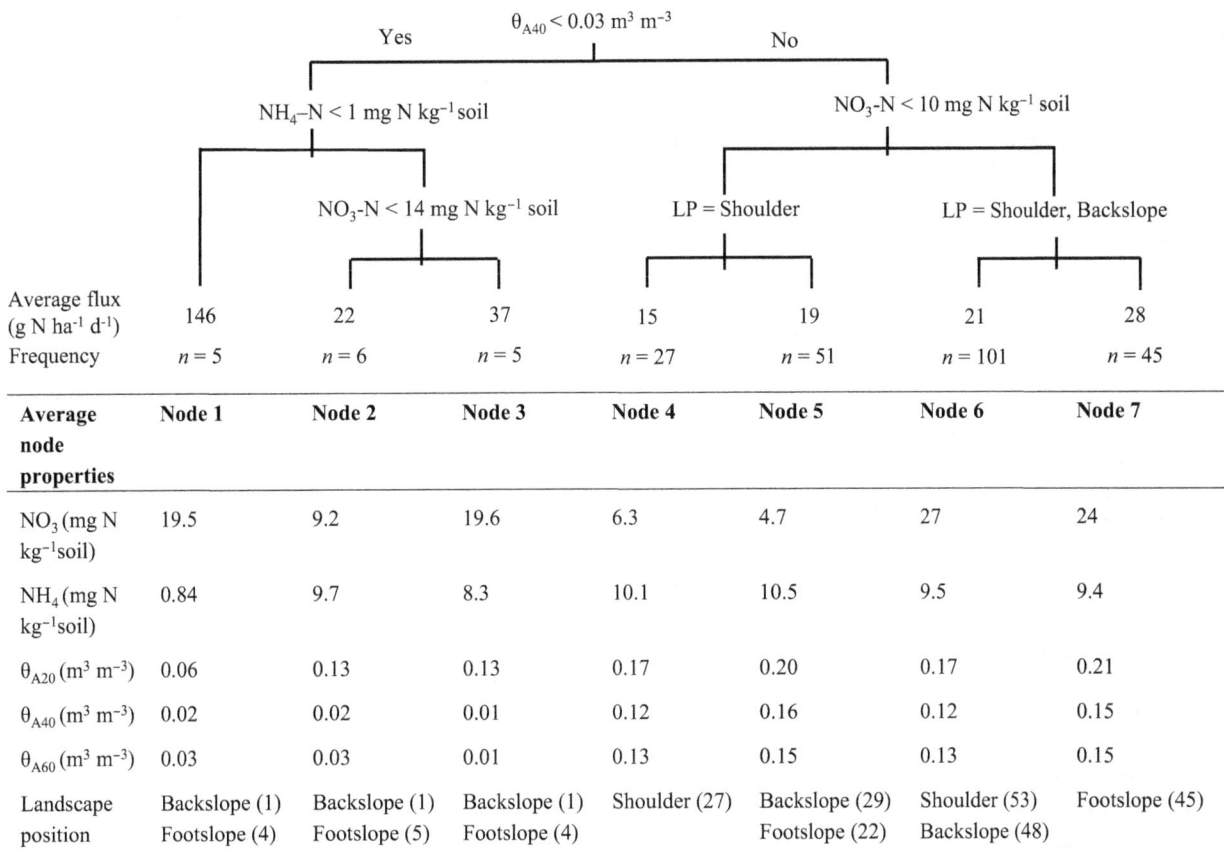

Fig. 6 Regression tree to group N₂O emissions during the transition from CRP to energy crops. Upon satisfaction of the splitting condition, the tree is routed to the left. Values in parentheses beside each landscape position (LP) show the number of observations (n) from each landscape position.

pasture land was cleared and cultivated for short-rotation poplar (*Populus* spp.) and shrub willow (Nikiema et al., 2012; Palmer et al., 2014).

According to our non-parametric statistical analysis, potentially higher N₂O emissions from these energy crops are realized only when high mineral N and soil saturation converge in the landscape, which occurs predominantly in the footslopes (Figs 5 and 6). Furthermore, the analysis identified greater influence of subsoil aeration on N₂O emissions in these landscapes, a fact that is often overlooked. At our site, limited drainage through the fragipans and argillic subsoil layers within the colluvial footslopes (Ciolkosz et al., 1995; Needelman et al., 2004) causes extended periods of water saturation, even when the top layer is relatively well drained (Fig. 3). Thus, we hypothesize that the high N₂O emission are explained by the juxtaposition of an oxic top layer above an anoxic subsurface layer that creates a hot spot for N₂O production from denitrification when NO_3^- is not limiting.

The hot moment and the hot spots for N₂O emissions were triggered by hurricane Andrea (June 10th to 13th),

when soils under energy crops had high soil NO_3^-, and the landscape was conducive to accumulation of water in the footslopes. In this time period differences among treatments built up significantly, mostly in the footslopes (Fig. 4). Given the high level of soil NO_3^- before the event, denitrification was most likely the dominant contributor of peak N₂O emissions under low soil aeration conditions following the hurricane (Weier et al., 1993; Gillam et al., 2008). This single peak emission event contributed on average 26% of the growing season cumulative N₂O flux in 2013. This is similar to the findings of Parkin & Kaspar (2006), who reported that two peak events of 29 days-long accounted for 45% of annual N₂O flux from corn in the Midwest. The remainder N₂O emissions, which are grouped mostly in the right branch of the regression tree (Fig. 6), are either from nitrification or denitrification in anoxic microsites.

The comparison of unfertilized and N-fertilized switchgrass in this period also reflects the concurrence of hot spots and hot moments. The response of N₂O flux to N fertilization was not very prominent or immediate because the soil was dry for more than a week after

N-fertilization (May 29th to June 9th) (Fig. 3), limiting N transformations (Bateman & Baggs, 2005; Davidson et al., 2008). However, a small peak from the footslopes may be attributed to favorable soil water content for nitrification. A similar lag period between fertilization and peak emission due to a dry period has been reported by Baggs et al. (2003). Addressing our second research question regarding the effect of N-fertilization on N_2O emission, we found that N_2O emissions in this N-rich soils are similar for both switchgrass-N and switchgrass treatments (Table 3); nonetheless, switchgrass-N had significantly higher N_2O emissions than CRP in the footslopes.

During and after the hurricane, N_2O emissions were more responsive in both Miscanthus and switchgrass-N in the footslopes (Fig. 4). The hurricane mobilized NO_3^- in all treatments (Fig. 2). In addition, shortly after the hurricane, N uptake and transpiration due to vigorous growth decreased the high NO_3^- and anoxia co-occurrences, which likely decreased N_2O emissions later in the growing season.

Previous studies of landscape scale N_2O emissions also identified the footslope positions as the hot spots for N_2O emissions (van Kessel et al., 1993; Castellano et al., 2010; Vilain et al., 2010). The same processes seem to operate in this watershed, with localized zones of water accumulation causing spatially variable N_2O emissions. Even after receiving a 10.8-cm of rainfall during the hurricane, fast drainage and redistribution of water from the convex shoulder positions precluded aeration limitations that favor denitrification. In fact, the critical θ_A of 0.03 m^3 m^{-3} threshold identified through the regression tree analysis is rarely, if ever, reached in the backslope and shoulder positions. However, the sources of NO_3^- for denitrification and N_2O emission in the footslope may originate from upslope, with NO_3^- transferred by interflow to the footslopes. Regardless, and answering our third research question, while footslopes in the Ridge and Valley region are at risk of high N_2O emissions, backslope and shoulder positions seem to posit a much lesser risk of emissions, even in the transitional years and in N-rich soils. Because the biomass yield did not vary significantly across the landscape, the N_2O emission per unit of biomass harvested was higher in the footslopes (0.41 ± 0.15 kg N_2O-N Mg biomass^{-1}), in particular under Miscanthus (0.86 ± 0.09 kg N_2O-N Mg biomass^{-1}).

Interannual variability of N_2O emissions as reported by many studies (Burchill et al., 2014; Oates et al., 2016), and the representativeness of the study period, limit the generalization of these results. The study period had a wet early summer (Fig. 4a) coinciding with high mineral N availability in the energy crop plots (Fig. 2), the perfect storm for a transitional system when N_2O emissions are the concern. Thus, our results may represent an upper estimate of N_2O emissions during a transition from CRP to energy crops in this landscape.

Implications for management of the transition to energy crops

Our results suggest that N_2O emissions from energy crops during the transitional period, i.e. the first and second year after killing CRP, are contingent on land conversion method, landscape properties, and management strategies. The major insight of this research is that in the Ridge and Valley region, transitioning CRP land to energy crops presents a minor risk of increasing N_2O emissions in well drained portions of the watersheds, except on the footslopes. As shown in Figs 2 and 4, once the mineral N concentrations drop due to crop uptake (and N losses), N_2O emissions are low and comparable to that of CRP.

A potential outline of N management emerges from this research. First, to minimize N_2O fluxes in N-rich environments, soil disturbance should be minimized to prevent an increase in organic matter mineralization. When switchgrass was no-till established and not fertilized, N_2O emissions were comparable to CRP. Second, Miscanthus had higher N_2O emissions during the establishment phase on footslopes when compared to switchgrass. This is likely due to a combination of tillage disturbance and slow establishment from wide-spaced plants. Higher biomass yield from Miscanthus in subsequent years (Arundale et al., 2014) may compensate for initial N_2O emissions when considering lifecycle emissions of the crop. Thus, while our data point to the transition as a hot moment, longer term studies are also needed. Third, while it is not common to apply N fertilizer to Miscanthus or switchgrass in the first year of planting due to slow growth during establishment and to prevent favoring weed growth, later fertilization requires monitoring soil mineral N in spring to avoid over fertilization. In the second growing season of our study (2013), fertilizing switchgrass did not increase biomass yield. While our original concern was the need to increase N supply through fertilization, we did not foresee that soils at Mattern, which is rocky and marginal for agriculture, would continue to mineralize substantial amounts of N upon land conversion (Fig. 2).

Fourth, the risk of soil N_2O emissions is particularly large only in footslope positions. Drainage from shoulders and backslopes is swift, which minimizes the risk of anoxic conditions. Detailed site specific delineation of seasonally saturated areas will be required to avoid management (e.g. fertilization) that could exacerbate N_2O fluxes from footslope hotspots. However, while well drained landscape positions can be fertilized with lesser risks of enhancing emissions, NO_3^- can be transported to poorly drained areas prone to higher emissions, and it is unknown how much of the N_2O emission in the footslope positions originated from the N transported from upslope areas.

In conclusion, managing the transition from CRP to energy crops maintaining a low N_2O emission per unit of biomass produced in relation to the original vegetation might be optimized by designing a transition process that minimizes the co-occurrence of high mineral N and wet soils (e.g. no-till, minimum tillage, and minimum fertilization), particularly wet subsoils. Our research suggests that a relatively small portion of the watershed is prone to significant N_2O emissions in the Ridge and Valley landscapes, and that perennial energy crops planted in well-drained portions of the landscape in current CRP land can be an important component of the portfolio of sustainable production options.

Acknowledgements

Funding for this research was provided by the USDOT Sungrant #52110-9601, the USDA AFRI grant #2012-68005-19703, and the Richard King Mellon Foundation. Mention of trade names or commercial products in this publication is solely for the purpose of providing specific information and does not imply recommendation or endorsement by the U.S. Department of Agriculture. USDA is an equal opportunity provider and employer.

References

Adler PR, Del Grosso SJ, Parton W (2007) life-cycle assessment of net greenhouse-gas flux for bioenergy cropping systems. *Ecological Applications*, **16**, 675–691.

Adler PR, Del Grosso SJ, Inman D, Jenkinson RE, Spatan S, Zhang Y (2012) Mitigation opportunities for life-cycle greenhouse gas emissions during feedstock production across heterogeneous landscapes. In: *Managing Agricultural Greenhouse Gases* (eds Liebig MA, Franzluebbers AJ, Follett RF), pp. 203–219, Coordinated agricultural research through GRACEnet to address our changing climate. Elsevier Inc., New York. DOI: 10.1016/B978-0-12-386897-8.00012-7.

Arundale RA, Dohleman FG, Heaton EA, Mcgrath JM, Voigt TB, Long SP (2014) Yields of *Miscanthus x giganteus* and *Panicum virgatum* decline with stand age in the Midwestern USA. *Global Change Biology Bioenergy*, **6**, 1–13.

Baggs EM, Stevenson M, Pihlatie M, Regar A, Cook H, Cadisch G (2003) Nitrous oxide emissions following application of residues and fertilizer under zero and conventional tillage. *Plant and Soil*, **254**, 361–370.

Bateman EJ, Baggs EM (2005) Contributions of nitrification and denitrification to N_2O emissions from soils at different water-filled pore space. *Biology and Fertility of Soils*, **41**, 379–388.

Breiman L (2001) Random forests. *Machine Learning*, **45**, 5–32.

Buda AR, Kleinman PJA, Srinivasan MS, Bryant RB, Feyereisen GW (2009) Factors influencing surface runoff generation from two agricultural hillslopes in central Pennsylvania. *Hydrological Processes*, **23**, 1295–1312.

Burchill W, Li D, Lanigan GJ, Williams M, Humphreys J (2014) Interannual variation in nitrous oxide emissions from perennial ryegrass/white clover grassland used for dairy production. *Global Change Biology*, **20**, 3137–3146.

Castellano MJ, Schmidt JP, Kaye JP, Walker C, Graham CB, Lin H, Dell CJ (2010) Hydrological and biogeochemical controls on the timing and magnitude of nitrous oxide flux across an agricultural landscape. *Global Change Biology*, **16**, 2711–2720.

Ciolkosz EJ, Waltman WJ, Thurman NC (1995) Fragipans in Pennsylvania soils. *Soil Survey Horizons*, **36**, 5–20.

Davidson EA, Nepstad DC, Ishida FY, Brando PM (2008) Effects of an experimental drought and recovery on soil emissions of carbon dioxide, methane, nitrous oxide, and nitric oxide in a moist tropical forest. *Global Change Biology*, **14**, 2582–2590.

Day PR (1965) Particle fractionation and particle-size analysis. In: *Methods of Soil Analysis*. (ed. Black CA), Part I. *Agronomy* **9**, 545–567.

EPA, 2010. Available at: http://www.epa.gov/oms/renewablefuels/420f09023.htm

Farm Service Agency (FSA) (2012), CRP Contract Summary and Statistics. Available at: http://www.fsa.usda.gov/FSA/webapp?area=home&subject=copr&topic=rns-cs (accessed December, 2012).

Firestone MK, Davidson EA (1989) Microbiological basis of NO and N_2O production and consumption in soil. In: *Exchange of Trace Gases between Terrestrial Ecosystems and the Atmosphere* (eds Andreae MO, Schimel DS), pp. 7–21. John Wiley and Sons, New York.

Follett RF, Vogel KP, Varvel GE, Mitchell RB, Kimble J (2012) Soil carbon sequestration by switchgrass and no-till maize grown for bioenergy. *Bioenergy Research*, **5**, 866–875.

Gauder M, Butterbach-Bahl K, Graeff-Honninger S, Claupein W, Wiegel R (2012) Soil-derived trace gas fluxes from different energy crops – results from a field experiment in Southwest Germany. *Global Change Biology Bioenergy*, **4**, 289–301.

Gelfand I, Zenone T, Jasrotia P, Chen J, Hamilton SK, Robertson GP (2011) Carbon debt of Conservation Reserve Program (CRP) grasslands converted to bioenergy production. *Proceedings of National Academy of Science*, **108**, 13864–13869.

Gelfand I, Sahajpal R, Zhang X, Izaurralde RC, Gross KL, Robertson GP (2013) Sustainable bioenergy production from marginal lands in the US Midwest. *Nature*, **493**, 514–517.

Gillam KM, Zebarth BJ, Burton DL (2008) Nitrous oxide emissions from denitrification and the partitioning of gaseous losses as affected by nitrate and carbon addition and soil aeration. *Canadian Journal of Soil Science*, **88**, 133–143.

Grandy AS, Robertson GP (2006) Initial cultivation of a temperate-region soil immediately accelerates aggregate turnover and CO_2 and N_2O fluxes. *Global Change Biology*, **12**, 1507–1520.

Hutchinson GL, Mosier AR (1981) Improved soil cover method for field measurement of nitrous oxide fluxes. *Soil Science Society of America Journal*, **45**, 311–316.

van Kessel C, Pennock DJ, Farrell RE (1993) Seasonal variations in denitrification and nitrous oxide evolution at the landscape scale. *Soil Science Society of America Journal*, **57**, 988–995.

Kim H, Kim S, Dale BE (2009) Biofuels, land use change, and greenhouse gas emissions: some unexplored variables. *Environmental Science and Technology*, **43**, 961–967.

Kleinman PJA, Srinivasan MS, Dell CJ, Schmidt JP, Sharpley AN, Bryant RB (2006) Role of rainfall intensity and hydrology in nutrient transport via surface runoff. *Journal of Environmental Quality*, **35**, 1248–1259.

Lemus R, Lal R (2005) Bioenergy crops and carbon sequestration. *Critical Review in Plant Science*, **24**, 1–21.

Liaw A, Wiener M (2002) Classification and regression by randomForest. *R News*, **2**, 18–22.

Mazzilli SR, Kemanian AR, Ernst OR, Jackson RB, Piñeiro G (2015) Greater humification of belowground than aboveground biomass carbon into particulate soil organic matter in no-till corn and soybean crops. *Soil Biology and Biochemistry*, **85**, 22–30.

McLaughlin SB, de la Torre Ugarte DG, Garten CT, Lynd LR, Sanderson MA, Tolbert VR, Wolf DD (2002) High-value renewable energy from prairie grasses. *Environmental Science and Technology*, **30**, 2122–2129.

Needelman BA, Gburek WJ, Peterson GW, Sharpley AN, Kleinman PJA (2004) Surface runoff along two agricultural hillslopes with contrasting soils. *Soil Science Society of America Journal*, **68**, 914–923.

Nikiema P, Rothstein DE, Miller RO (2012) Initial greenhouse gas emissions and nitrogen leaching losses associated with converting pastureland to short-rotation woody bioenergy crops in northern Michigan, USA. *Biomass and Bioenergy*, **39**, 413–426.

Oates LG, Duncan DS, Gelfand I, Millar N, Robertson GP, Jackson R (2016) Nitrous oxide emissions during establishment of eight alternative cellulosic bioenergy cropping systems in the North Central United States. *Global Change Biology Bioenergy*, **8**, 539–549.

Palmer MM, Forrester JA, Rothstein DE, Mladenoff DJ (2014) Conversion of open lands to short-rotation woody biomass crops: site variability affects nitrogen cycling and N_2O fluxes in the US Northern Lake States. *Global Change Biology Bioenergy*, **6**, 450–464.

Parkin TB, Kaspar TC (2006) Nitrous oxide emissions from corn-soybean systems in the Midwest. *Journal of Environmental Quality*, **35**, 1496–1506.

Pennock DJ, van Kessel C, Farrell RE, Sutherland RA (1992) Landscape-scale variations in denitrification. *Soil Science Society of America Journal*, **56**, 770–776.

Perez-Suarez M, Castellano MJ, Kolka RK, Asbjournsen HM (2014) Nitrogen and carbon dynamics in prairie vegetation strip across topographical gradients in mixed central Iowa Agroecosystems. *Agriculture, Ecosystem and Environment*, **188**, 1–11.

R Development Core Team (2012) R: A language and environment for statistical computing. R Foundation for Statistical Computing, Vienna, Austria. ISBN 3-900051-07-0. Available at: http://www.R-project.org/.

Ruan L, Robertson GP (2013) Initial nitrous oxide, carbon dioxide, and methane costs of converting conservation reserve program grassland to row crops under no-till vs. conventional tillage. *Global Change Biology*, **19**, 2478–2489.

Sharpley AN, Kleinman PJA, Heathwaite AL, Gburek WJ, Folmar GJ, Schimidt JP (2008) Phosphorus loss from an agricultural watershed as a function of storm size. *Journal of Environmental Quality*, **37**, 362–368.

Smith TE, Kolka RK, Zhou XB, Helmers MJ, Cruse RM, Tomer MD (2014) Effects of native perennial vegetation buffer strips on dissolved organic carbon in surface runoff from an agricultural landscape. *Biogeochemistry*, **120**, 121–132.

Strobl C, Malley J, Tutz G (2009) An introduction to recursive partitioning: rational, application, and characteristics of classification and regression trees, bagging, and random forests. *Psychological Methods*, **14**, 323–348.

Tilman D, Hill J, Lehman C (2006) Carbon-negative biofuels from low-input high diversity grassland biomass. *Science*, **314**, 1598–1600.

Tukey JW (1977) *Exploratory Data Analysis*. Addison-Wesley, Reading, MA, USA.

Vilain G, Garnier J, Tallec G, Cellier P (2010) Effect of slope position and land use on nitrous oxide (N_2O) emissions (Seine Basin, France). *Agriculture and Forest Meteorology*, **150**, 1192–1202.

Weier KL, Doran JW, Power JF, Walters DT (1993) Denitrification and the dinitrogen to nitrous oxide ratio as affected by soil water, available carbon and nitrate. *Soil Science Society of America Journal*, **57**, 66–72.

Zenone T, Chen J, Deal MW *et al.* (2011) CO_2 fluxes of transitional bioenergy crops: effect of land conversion during the first year of cultivation. *Global Change Biology Bioenergy*, **3**, 401–412.

Zhu Q, Schmidt JP, Buda AR, Bryant RB, Folmar GJ (2011) Nitrogen loss from a mixed land use watershed as influenced by hydrology and seasons. *Journal of Hydrology*, **405**, 307–315.

14

Demand for biomass to meet renewable energy targets in the United States: implications for land use

ANTHONY OLIVER[1] and MADHU KHANNA[2] (iD)

[1]*South Coast Air Quality Management District, 21865 Copley Drive, Diamond Bar, CA 91765, USA,* [2]*Department of Agricultural and Consumer Economics, University of Illinois at Urbana-Champaign, 1301 W. Gregory Drive, Urbana, IL 61801, USA*

Abstract

Renewable energy policies in the electricity and transportation sectors in the United States are expected to create demand for biomass and food crops (corn) that could divert land from food crop production. We develop a dynamic, open-economy, price-endogenous multi-market model of the US agricultural, electricity and transportation sectors to endogenously determine the quantity and mix of bioenergy likely to be required to meet the state Renewable Portfolio Standards (RPSs) and the federal Renewable Fuel Standard (RFS) if implemented independently or jointly (RFS & RPS) over the 2007–2030 period and their implications for the extent and spatial pattern of diversion of land from other uses for biomass feedstock production. We find that the demand for biomass ranges from 100 million metric tons (MMT) under the RPS alone to 310 MMT under the RFS & RPS; 70% of the biomass in the latter case can be met by crop and forest residues, while the rest can be met by devoting 3% of cropland to energy crop production with 80% of this being marginal land. Our findings show significant potential to meet current renewable energy goals by expanding high-yielding energy crop production on marginal land and using residues without conflicting with food crop production.

Keywords: bioelectricity, biofuels, dynamic optimization, partial-equilibrium model, Renewable Fuel Standard, Renewable Portfolio Standard, spatial analysis

Introduction

Growing interest in renewable energy for the electricity and the transportation sectors has led to state-level Renewable Portfolio Standards (RPSs) and the federal Renewable Fuel Standard (RFS) that mandate a share/quantity of demand being met by renewable energy sources. Bioenergy is expected to play a significant role in achieving these policies as it will be used to produce cellulosic biofuels and can also be used to generate bioelectricity (EIA, 2010a). Twenty-nine states have implemented RPSs that range from 10% to 40% of electricity being produced using renewable sources by 2030 (DSIRE, 2011). The RFS sets a target for 136 billion liters of biofuels of which at least 61 billion liters are to be from cellulosic biomass and at most 57 billion gallons from corn ethanol.

Biomass can be produced from a variety of different sources, including dedicated energy crops like miscanthus and switchgrass, woody biomass such as poplar, crop residues from corn and wheat and forest biomass. Biomass can be used to produce bioenergy in several ways. It can generate bioelectricity by co-firing it with coal in a coal-based power plant or combusting it in a

dedicated bioelectricity plant. Additionally, it can also generate bioelectricity as a coproduct during the process of conversion to cellulosic biofuels. The potential and cost of producing various feedstocks are likely to vary spatially due to differences in growing conditions; additionally, demand for bioelectricity may also vary spatially due to differences in the stringency of the RPSs across regions and the location of existing coal-based electricity plants that can potentially co-fire biomass with coal.

We examine the demand for biomass likely to be generated and the land required to meet this demand by the state RPSs and RFS policies over the 2007–2030 time period. We also examine the biomass price at which it will be feasible to do so and the implications of biomass production for food crop production as land is likely to be diverted from food to fuel production. Lastly, we examine the spatial pattern of biomass production under these policies. We compare the effects of implementing the RPSs and the RFS independently with those of implementing them jointly to analyze the effects of competing demands for biomass on biomass price, feedstock mix and the spatial pattern of feedstock production.

We undertake this analysis by extending a dynamic, price-endogenous, open-economy model of the electricity, transportation and agricultural sectors in the United States, the Biofuel and Environmental Policy Analysis

Correspondence: Madhu Khanna
email: khanna1@illinois.edu

Model (BEPAM-E) (Oliver & Khanna, 2017). The model endogenously determines the mix of renewable energy from different sources to meet the RPSs, the mix of feedstocks to produce bioelectricity and biofuels and the spatial location of their production. It also endogenously determines the allocation of available cropland and marginal land (in crop/pasture/fallow rotations) to be allocated to different uses (food, feed and fuel) and the prices and quantities of agricultural commodities, fuel, electricity and biomass. This paper extends the BEPAM used previously to analyze the effects of biofuel policies on the transportation sector (Huang *et al.*, 2013; Chen *et al.*, 2014) by including the US electricity sector and related fossil fuel sectors (coal and natural gas) that provide inputs to the electricity sector. The extended BEPAM-E (BEPAM with the electricity sector) incorporates the existing fossil and renewable electricity generating capacity in the United States as well as the potential for expansion in natural gas, wind and bioelectricity generation (Oliver & Khanna, 2017).

Early studies have examined the potential of co-firing at national levels (McCarl *et al.*, 2000) and regional levels: in Illinois (Khanna *et al.*, 2008; LaTourrette *et al.*, 2011) and Indiana (Brechbill *et al.*, 2011). Dumortier (2013) examines the supply of biomass from crop and forest resides and switchgrass at exogenously set biomass price and the availability of biomass for cellulosic biofuels assuming a binding constraint on demand for co-firing imposed by existing coal-based electricity plants. He shows that there would be shortfalls in biomass availability for cellulosic biofuel production in much of the Southeast. In contrast, our analysis endogenously determines the extent to which there will be an incentive to co-fire or produce bioelectricity in a biopower plant. Other studies have examined the implications of using a single feedstock, forest biomass, in the Southeast (Abt *et al.*, 2012) and the United States (Ince *et al.*, 2011) for producing bioelectricity for feedstock price.

A few studies have used the national scale Forest and Agricultural Sector Optimization Model (FASOM) to predict the mix of forest and agricultural biomass feedstock to meet exogenously given demand for bioelectricity (Latta *et al.*, 2013; White *et al.*, 2013). Palmer & Burtraw (2005) analyze the implications of a hypothetical federal RPS for the price of electricity assuming an exogenously given supply curve of biomass. Other studies have examined the feedstock price needed to induce the biomass supply needed to meet a 21-billion gallon cellulosic biofuel target (Langholtz *et al.*, 2014) and to meet exogenously given targets for both biofuel and bio-power targets (Langholtz *et al.*, 2012). Most recently, Sands *et al.* (2017) examine the land-use requirements for supplying about 250 million megawatt hours (M

MWh) of bioelectricity in 2030 using switchgrass as the only feedstock. They find that this will require 10–12 million hectares of land; two-thirds of this would be obtained from converting cropland to switchgrass production, and the rest would be obtained from marginal land and forest land.

We extend this literature by considering a broad range of feedstocks, including, crop and forest residues and energy crops. We endogenously determine the share of bioelectricity in meeting the RPS, and the mix, price, and location of biomass feedstocks to meet it.

There is a large literature examining the effect of the RFS on land use and food and fuel prices (Beach *et al.*, 2012; Chen *et al.*, 2014; Hudiburg *et al.*, 2016). These studies have assumed the RFS is implemented in isolation and not considered the implications of a concurrently implemented RPS with spatially varying targets for renewable electricity. We analyze the synergies and trade-offs in meeting goals of one renewable energy policy in the presence of the other and its spatially explicit implications for land use. We also extend previous work in Oliver & Khanna (2017) that examines the cost-effectiveness of GHG mitigation by jointly implemented RPSs and the RFS relative to a carbon tax policy. Here we compare biomass demand and land-use outcomes under the RPSs to those under the RFS and in the case with the two policies implemented concurrently.

Materials and methods

Model

Biofuel and Environmental Policy Analysis Model (BEPAM-E) is a nonlinear, dynamic, multi-sector, price-endogenous, open-economy, partial-equilibrium, mathematical programming model that simulates US agricultural, transportation fuel and electric power sectors including international trade with the rest of the world (ROW) (for a detailed description of model equations and data; see Oliver & Khanna, 2017). Market equilibrium is found by maximizing the sum of consumers' and producers' surpluses in the agricultural, transportation fuel and electric power sectors subject to various material balance constraints and technological constraints in a dynamic framework over the 2007–2030 period. BEPAM-E considers production of crop and biofuel feedstocks at the level of a Crop Reporting District (CRD) where crop production costs, yields and resource endowments are specified for each CRD and each crop. The 306 CRDs in the 48 contiguous US states are used as spatial units to model electricity generation from existing power plants by fuel type, while generation from new electricity capacity is considered at the level of twenty Electricity Market Regions (EMRs). The model endogenously determines the agriculture and transportation sector variables of food consumption, gasoline, diesel and biofuel consumption, imports of gasoline and sugarcane ethanol, mix of biofuels and regional land allocation among different food, feed and fuel crops and livestock activities over a given time horizon. It also endogenously determines electricity sector variables such as generation by energy type (coal, natural gas, oil, wind, co-fired biomass, dedicated biomass and coproduct), regional electricity consumption, inter-regional electricity transmission, bioelectricity feedstock transportation and GHG emissions.

A dynamic model allows us to distinguish among feedstocks that differ in their upfront costs and life span over which they yield returns. The model is run on an annual time scale using a rolling horizon approach in which decision makers are assumed to make land allocation plans for the next 10 years taking current prices, demand conditions, land availability in different categories and costs of technology as given and then to update their expectations about these variables every year. The model is solved iteratively; after solving each 10-year market equilibrium problem, we take the first-year solution values as 'realized', move the horizon 1 year forward and solve the updated model again. This structure enables us to incorporate technological change over time that lowers the costs of renewable technologies due to learning by doing across horizons (see Oliver & Khanna, 2017).

Electricity sector. Each of the twenty EMRs in the electricity sector has its own demand for electricity and potential to trade across geographically adjacent regions subject to constraints on transmission capacity. These EMRs are defined similarly to those in the EIA's National Energy Modelling System (NEMS) (EIA, 2011a), and for the most part an EMR consists of one or more states. A few EMRs in NEMS did not align with state boundaries or had regions smaller than a state. We redefined these to follow state boundaries to be able to link them consistently with the agricultural production regions defined at the CRD level. We allow for inter-regional electricity transmission between adjacent EMRs subject to transmission capacity constraints based on historically observed levels and a loss during transmission and distribution (EIA, 2011b).

Electricity demand functions are specified for each EMR as an aggregation of state-level, sector-specific (residential, commercial and industrial) demand functions. The annual demand for electricity is met by existing installed electricity generation capacity as well as expansion of generation capacity using natural gas, wind or biomass sources; the extent to which each type of source is used is endogenously determined based on relative costs. Fuel inputs for electricity generation include coal, natural gas, and fuel oil and biomass feedstocks from the agricultural and forest sectors.

Existing power plants are aggregated to the CRD level by energy type, which consists of coal, natural gas, oil, hydroelectric, nuclear, geothermal, municipal solid waste, solar, biomass, wind, and other. This power plant capacity is regionally heterogeneous in the energy type, nameplate capacity, and conversion efficiency. The decision can also be made to expand generation capacity that uses natural gas, wind, or biomass energy, while expansion from other sources is specified exogenously according to projections from the AEO (EIA, 2010a). Expansion of electricity generation capacity with each of these sources depends on the endogenously determined fuel cost per kilowatt hour (kWh), the conversion efficiency, and the levelized cost of generation which consists of the annualized fixed

cost and variable O&M costs per kWh. An exception is the cost (supply) of wind electricity generation in each EMR, which is modeled by an upward sloping function following Paul et al. (2009). These functions represent the marginal cost of generation from new wind turbine capacity that increases with the amount of wind-based electricity generated in a region as the availability of wind resources in the EMR diminishes.

Bioelectricity generation can be expanded along three pathways: co-firing at a coal power plant, firing at a new dedicated biomass power plant, and as a coproduct of cellulosic ethanol refining. Co-fired biomass is assumed to be converted using the same heat rate as the particular coal power plant, while generation from bio-power and coproduct sources is based on an assumed heat rate from the literature (Qin et al., 2006; Humbird et al., 2011). The heat rate determines the amount of feedstock required to generate a kWh of electricity and thus affects the relative price of bioelectricity; we examine the sensitivity of results to these assumptions. We assume a 10% limit on the mixture of biomass to coal, but also examine a more relaxed co-firing limit of 20% which falls within the range examined in other studies (Qin et al., 2006; Dumortier, 2013). We incorporate the cost of transporting biomass from a CRD producing it to other CRDs for co-firing with existing coal-based capacity in determining the location of biomass production for co-firing. As a result, the endogenous marginal cost of generation from co-firing is a function of the delivered price of biomass feedstock, the conversion efficiency of the coal power plant and biomass processing costs. The endogenous price of generation from a dedicated bio-power plant is similarly a function of the delivered price of biomass feedstock, the heat rate of the bio-power plant, the biomass processing cost, and the levelized cost of capacity net of fuel (for further details, see Oliver & Khanna, 2017). Transmission of electricity generated in an EMR to end-use consumers in other EMRs is subject to transmission capacity constraints, a transmission cost and some loss of electricity during transmission.

Agricultural sector. The agricultural sector includes fifteen conventional crops, eight livestock products, two perennial bioenergy crops, crop residues from the production of corn and wheat, forest residues, and coproducts from the production of corn ethanol and soybean oil (for further details, see Chen et al., 2014). In the crop and livestock markets, primary crop and livestock commodities are consumed either domestically or traded with the ROW (exported or imported). Primary crop commodities can also be processed or directly fed to various animal categories. Domestic and export demands and import supplies are incorporated by assuming linear price-responsive demand or supply functions. The commodity demand functions and export demand functions for tradable row crops and processed commodities are shifted upward over time at exogenously specified rates.

The simulation model incorporates CRD-specific data on costs of producing crops, livestock, biofuel feedstocks, yields of conventional and bioenergy crops, and land availability. Crops can be produced using alternative tillage, rotation, and irrigation practices. Yields and costs of production of crop residues and dedicated energy crops also differ across regions. We estimate the rotation-, tillage- and irrigation-specific costs of production in 2007 prices for each of the row crops and energy crops at a county level which are then aggregated to the CRD level for computational ease. Row crop yields increase over time as in Chen *et al.* (2014).

Biomass can be obtained from four different types of feedstocks: crop residues, energy crops, forest residues and pulpwood. The energy crops considered here are miscanthus and switchgrass, which can be produced on available cropland or cropland pasture. Production of dedicated energy crops is limited to the rain-fed regions which include the Plains, Midwest, South, and Atlantic, while conventional crops can be grown in the Western region as well. The supply of forest residues and pulpwood is modeled at the CRD level using available quantities obtained from the USDOE's Billion-Ton study on the annual quantity of county-level supplies at various biomass prices (Perlack & Stokes, 2011).

Five land types are included in the agricultural sector for each CRD: cropland, idle cropland, cropland pasture, pasture land and forestland pasture. As the demand for agricultural land increases, marginal lands (not currently utilized) can be converted to cropland, with the extent of conversion determined by variations in crop prices over time; agricultural land supply is, therefore, determined endogenously. We assume fixed cropland availability within the 10-year production planning period ahead. From the resulting multi-year equilibrium solution, we take the first-year values of the endogenous commodity prices and use them to construct a composite commodity price index that is used to adjust the assumption about cropland availability for the next 10-year planning period (see Chen *et al.*, 2014). To prevent unrealistic changes and extreme specialization in land use, we restrict CRD-specific row crop planting decisions to a convex combination (weighted average) of crop mixes in each CRD using methods described in Chen & Onal (2012). In the case of energy crops, we do not have historical crop mixes to constrain land-use change to their production. Energy crop production on cropland or cropland pasture occurs if the net discounted value of return from it is larger than the foregone returns from (or costs of conversion from) existing uses of these lands. To avoid the extreme changes in land use, between existing uses and energy crops, that could result from this we restrict the land allocated to perennial grasses (miscanthus and switchgrass) to less than 25% of total land availability in a CRD. This constraint is also intended to account for other factors such as inertia, transactions costs, risks and uncertainties that could limit the conversion of land to energy crops even if it were profitable to do so.

Transportation sector. Biofuel and Environmental Policy Analysis Model (BEPAM-E) includes linear demand curves for Vehicle Kilometers Travelled (VKT) with four types of vehicles, including conventional gasoline, flex fuel, gasoline hybrid and diesel vehicles (Chen *et al.*, 2014). The VKT production function considers the energy content of alternative fuels, fuel economy of each type of vehicle and the forthcoming Corporate Average Fuel Economy standards, and technological limits on blending gasoline and ethanol for each of these four types of vehicles, as specified by EIA (2010a). Demand curves are exogenously shifted for VKT with each type of vehicles over time as projected by the Annual Energy Outlook (AEO) (EIA,

2010a) to capture the growth in demand due to changes in vehicle fleet, income and population. Transportation fuel demand can be met through a mix of liquid fossil fuels and biofuels. The supply of transportation fossil fuels is represented with upward sloping linear supply functions for both gasoline and diesel produced the United States and ROW (Chen *et al.*, 2014). The supply of biofuels is derived from first-generation fuels, which include corn ethanol, biodiesel, and imported sugarcane ethanol, and from second-generation, cellulosic biofuels produced from the same types of feedstock used for bioelectricity generation (energy crops, crop residues and forest residues).

Policy. The federal RFS is implemented as a nested standard that sets a minimum annual requirement for cellulosic biofuels as projected annually by the AEO as a binding volumetric mandate (EIA, 2010a), but allows the mix of feedstocks used to meet them to be endogenously determined. There is an upper limit of 56 billion liters on corn ethanol after 2015 and a minimum requirement for 90 billion liters of advanced/cellulosic biofuels in 2030. As the RFS is implemented as a blend mandate, we estimate the blend rates that will achieve the volumetric targets each year 2007–2030 as in Chen *et al.* (2014) and impose those as binding annual blend mandates. As a federal mandate, the RFS provides flexibility in the amount of biofuels produced by any region and in the use of the lowest cost sources of feedstocks nationally, irrespective of which region they are located in.

This is in contrast to the state RPSs which will place demands on renewable energy resources for electricity generation that are region specific. Thus, the joint implementation of the state RPSs in the presence of the RFS has the potential to affect the mix of feedstocks and their regional production pattern to produce biofuels as well as the regional mix of renewable electricity generation.

The implementation of the RFS as a blend mandate imposes an implicit tax on fossil transportation fuels and an implicit subsidy on biofuels; the net impact on the blended fuel price could be positive or negative depending on the magnitudes of the implicit subsidy and tax (De Gorter & Just, 2009; Chen & Khanna, 2013). The RPS is also implemented as a blend mandate, where the percentage of electricity generated from renewable sources must be no less than the amount specified by the mandate. An RPS may raise or lower the price of electricity depending on the relative elasticity of the renewable and non-renewable fuel supply curves and the stringency of the RPS target (Fischer, 2010). Furthermore, the RPS can reduce the marginal cost of electricity from renewables over time through learning-by-doing.

We model the state RPSs as a constraint that requires eligible renewable generation consumed in each EMR, to be no less than the specified share by the RPS times the total electricity consumed in each EMR. The state-level RPSs are averaged at the EMR level using annual generation-weighted averages to obtain a regional RPS. The RPS proportions parameters are calculated from the Database of State Incentives for Renewables and Efficiency (DSIRE, 2011). In general, these parameters represent the RPS for an EMR and are proportions that increase over time until the target percentage is achieved in a target year.

Data

Electricity sector. Electricity generated at existing power plants is conditional on each plant's generation capacity. The existing capacity of each power plant at the CRD level is parameterized based on the Emission and Generation Integrated Database (EPA, 2010). The capacity by power plant type is aggregated by CRD. The capacity factor for each CRD is calculated based on the weighted average on the plant capacity factor of all power plants of a specific type in a given CRD (EPA, 2010). The fixed and variable O&M costs of existing power plants are obtained from a version of the NEMS model (UCS, 2011).

The costs of electricity generated from new sources (excluding co-firing) consist of a levelized cost and a fuel cost. We define the levelized cost as the annualized fixed cost [converted to a cost per kilowatt hour (kWh) based on the expected generation over the life of the plant] and a variable operating and maintenance (O&M) cost per kWh (EIA, 2010a). Additional cost of electricity generation includes the fuel cost per kWh and transmission losses, which are endogenously determined depending on the fuel price and the extent of transmission across different regions as described in Oliver & Khanna (2017). The cost of modifying a coal plant for co-firing is assumed to be obtained from EIA (2010a). Feedstock heat content, conversion efficiency and processing cost are described in Oliver & Khanna (2017). All feedstocks are assumed to have the same heat content (Haq, 2002). A constant cost of transportation per kilometer per ton of biomass from its harvest location to a potential co-firing or dedicated bio-power plant is assumed based on Searcy et al. (2007). Total transportation costs for biomass are calculated by multiplying this with the distance between centroid of the biomass-producing CRD and the biomass-consuming CRD. We also include an exogenously given annual demand for pellets for export.

Wind energy supply curves are specified for each EMR and are based on data from the NEMS model (EIA, 2011a). These data are projections of the amount wind capacity that is available by region at multiples of a base capacity cost. These capacity supply curves are converted to generation supply curves using a given capacity factor and calibrated using a base levelized cost per megawatt hour (EIA, 2010a). The relative cost of wind energy will be a factor in determining the regional share of bioelectricity; we thus test the sensitivity of our results to this assumption by examining a lower wind energy cost.

The natural gas supply function is a linear upward sloping function representing the national supply of natural gas for all sectors. It represents the national wellhead price of supplying natural gas. The supply function is calibrated annually for the year 2007–2011 with the observed national average wellhead price and production over this period and an assumed value for the natural gas supply elasticity (see Oliver & Khanna 2017) for details on parametric assumptions). The supply curve is assumed to be fixed during a rolling horizon but shifts inwards in each subsequent rolling horizon according to projections by the Annual Energy Outlook (EIA 2012). Natural gas used for power generation incurs a regional transportation and distribution cost. Therefore, the regional delivered natural gas price for the electricity sector is a function of the endogenously

determined national wellhead price plus the regional transportation and distribution cost. The demand for natural gas across sectors other than electricity is assumed to be exogenous. The supply of coal and fuel oil for electricity generation is assumed to be perfectly elastic at state-specific fixed prices; these prices are exogenously set annually, from observed data for 2007–2011 and from EIA growth rate projections for the following years (EIA, 2010b, 2012). The electricity demand functions are calibrated with parameters calculated from data on state electricity retail sales, retail electricity price and the price elasticity of demand for electricity as described in Oliver & Khanna (2017).

Agricultural sector. The agricultural sector consists of markets for primary and processed commodities and livestock products. Domestic and export demands for primary commodities, such as corn and soybeans, are determined in part by the demands for processed commodities obtained from them and by other uses (such as seed). We use two-year (2006–2007) average prices, consumption, exports and imports of crop and livestock commodities to calibrate the domestic demand, export demand and import supply functions for all commodities. Sources of data on prices, consumption, exports and imports are described in Chen et al. (2014). Domestic demands, export demands and import supplies are shifted upward over time at exogenously specified rates. The simulation model incorporates CRD-specific data on costs of producing crops, livestock, biofuel feedstocks, yields of conventional and bioenergy crops, and land availability. We estimate the rotation-, tillage- and irrigation-specific costs of production in 2007 prices for fifteen row crops (corn, soybeans, wheat, rice, sorghum, oats, barley, cotton, peanuts, potatoes, sugar beets, sugarcane, tobacco, rye and corn silage) and three perennial grasses (alfalfa, switchgrass and miscanthus) at county level. These are then aggregated to the CRD level for computational ease. Production of dedicated energy crops is limited to the rain-fed regions of the United States to the east of the 100th meridian. Domestically produced feedstocks used for biofuel production in the model include corn, corn stover, wheat straw, forest residues, miscanthus, switchgrass, waste grease, vegetable oils, DDGs and pulpwood.

County-specific corn stover and wheat straw availability are proportional to historically observed grain yields, and their harvest rates are limited to levels that depend on tillage practices to prevent soil degradation. The incremental costs of producing them include the cost of harvesting and replacement fertilizer application. Energy crops have the potential to be grown productively and at low cost on low-quality land (Khanna et al., 2011; Dwivedi et al., 2015). County-specific yields of miscanthus and switchgrass on two types of land qualities, high-quality cropland and low-quality cropland pasture, are simulated using the DayCent model (see Hudiburg et al., 2016). As the mix of feedstocks used for bioelectricity generation and biofuel production is sensitive to their yields, we examine a case where yields are lower than those simulated with the DayCent model. Their cost of production is estimated using methods described in Chen et al. (2014) and Dwivedi et al. (2015).

We obtain CRD-specific data on land availability for each of the five types of land (cropland, idle cropland, cropland pasture,

pasture land and forestland pasture) from USDA/NASS (2009). CRD-specific planted acres for 15 row crops are used to obtain the cropland available in 2007 and to obtain the historical and synthetic mixes of row crops. Cropland availability in each CRD is assumed to change in response to crop prices following acreage price elasticities estimated in Miao *et al.* (2016). Data on idle cropland, cropland pasture, pasture and forestland pasture for each CRD are obtained from USDA/NASS (2009). The analysis here assumes that land enrolled in CRP is preserved at 2008 levels and not used for conventional crop or bioenergy crop production as in Chen *et al.* (2014).

Transportation sector. Biofuel and Environmental Policy Analysis Model (BEPAM-E) includes linear demand curves for VKT with four types of vehicles, including conventional gasoline, flex fuel, gasoline hybrid and diesel vehicles, and a VKT production function for each type of vehicle. We exogenously shift demand curves for VMT with each type of vehicle over time as projected by the Annual Energy Outlook (EIA, 2010a,b) to capture the growth in demand due to changes in vehicle fleet, income and population. Gasoline and diesel supply curves are calibrated for 2007 using data on fuel consumption and production in the United States and for the rest of the world. Key assumptions about demand elasticity for VKT and supply elasticities of fuels are reported in Chen *et al.* (2014).

The biofuel sector includes several first- and second- generation biofuels. First-generation biofuels include domestically produced corn ethanol and imported sugarcane ethanol, soybean biodiesel, DDGS-derived corn oil and waste grease. Second-generation biofuels included here are cellulosic ethanol and biomass-to-liquid diesel produced using the Fischer–Tropsch process. The feedstock costs of biofuels consist of two components: a cost of producing the feedstock which includes costs of inputs and field operations, and a cost of land. Methods for estimating costs and the technological parameters for converting feedstock to different types of biofuel and the industrial costs of processing feedstocks and producing biofuels are described in Chen *et al.* (2014). These costs are assumed to decline due to learning-by-doing as cumulative production increases using an experience curve approach (see Oliver & Khanna, 2017).

Results

We first validate the model by examining the extent to which the simulated results deviate from the observed levels in the initial model year of 2007. We find that electricity generated from coal and natural gas sources and total electricity generation was deviated by less than 4% from the observed data. Average national electricity price deviates by 14%, and the national wellhead natural gas price deviates by 7% (for further details, see Oliver & Khanna, 2017).

For the purposes of this analysis, we consider three policy scenarios and compare them to a no-policy scenario and examine outcomes over the 2007–2030 period for the electricity, agricultural and transportation

sectors. Scenario 1 is the no-policy baseline scenario in which there is no renewable energy policy; we do, however, assume the presence of the Corporate Average Fuel Economy Standards, the excise tax on fuel and a low (3%) blend of ethanol for oxygenation. In Scenario 2, we constrain the eligible renewable generation consumed in each EMR to be no less than the regional RPS. In Scenario 3, we implement the RFS by imposing the annual volumetric targets for biofuels projected by the AEO as a binding mandate (EIA, 2010a). The RFS is implemented as a blend mandate with binding constraint on blend rates estimated to achieve the annual volumetric targets grow from 6% in 2007 to 24.5% in 2030 as in Chen *et al.* (2014). We find that this blend and level of biofuel consumption is achievable given the projections for flex-fuel vehicle demand by the Annual Energy Outlook (EIA, 2010a). By assuming the absence of demand-side constraints for blending biofuels, this analysis examines the supply-side implications of these renewable energy policies for biomass production. In Scenario 4, the RPS & RFS scenario, the RFS and state RPSs are implemented jointly as described above.

Effects of alternative policies on mix and level of electricity generation

The RPS increases renewable energy generation by 68% compared to the no-policy case, which increases the share of renewable energy in total generation from 10% to about 16% in 2030. It also reduces the generation from coal and natural gas by 5% and 13%, respectively. Nevertheless, total generation increases by 1.4% compared to the no-policy case. The amount of bioelectricity generated under the RPS is 140 M MWh; this requires about 87 MMT of biomass. The RFS has a negligible impact on renewable energy generation and increases renewable generation by 4% mainly due to coproduct electricity generation, which displaces natural gas-based electricity. The increased demand for biomass for cellulosic biofuel under the RFS raises the price of biomass and makes co-firing biomass with coal expensive. This results in a reduction in co-fired electricity generation and an increase in coal-based electricity generation. The effects of the joint implementation of the RFS & RPS on the electricity sector are very similar to those under the RFS alone, with total generation increasing by 2%, and the share of renewable energy increasing to about 16% (Table 1).

Under the no-policy scenario, bioelectricity and wind account for about 9% each of the total renewable electricity produced with the remaining 82% largely from hydroelectric sources and small quantities of solar, geothermal and other renewables. The implementation

Table 1 Effects of alternative policies on various energy sources in 2030

Energy source	No policy	RPS	RFS	RFS & RPS
		Percentage change relative to no-policy		
Total electricity generation (M MWh)	4372	1	1	2
Renewable-based (M MWh)	418	68	4	68
Share of renewables (%)	9.6	16	10	16
Fossil fuel-based (M MWh)	2998	−8	0	−7
Coal	1899	−5	1	−2
Natural gas	1099	−13	−2	−14
Sources of renewable electricity generation (%)				
Co-firing	9	16	3	9
Dedicated biomass	0	4	0	3
Coproduct	0	0	9	6
Total bioelectricity	9	20	12	17
Wind	9	31	9	34
Other	82	49	79	49
Total biofuels (billion liters)	19	19	147	146
Corn ethanol	15	15	57	57
Advanced biofuels	4	4	8	9
Cellulosic biofuels	0	0	82	81
Total biomass (M MT)	33	100	262	310
Biomass from agriculture (%)	0	55	85	75
Biomass from forest (%)	100	45	17	25
Biomass for bioelectricity (%)	60	87	3	17
Biomass for cellulosic biofuel (%)	0	0	95	79
Biomass for pellet exports (%)	0	13	2	4
Sources of feedstock (%)				
Corn stover	0	37	48	46
Wheat straw	0	7	9	11
Miscanthus	0	11	26	19
Switchgrass	0	0	3	0
Forest residues	100	44	17	14
Pulpwood	0	1	0	11

of the RPS would significantly expand the share of bio-electricity to 20% (with a 16% share of co-fired generation and 4% share of dedicated bioelectricity) and of wind generation to 31%. In contrast, the RFS would increase coproduct-based bioelectricity generation while reducing co-fired electricity generation. The RFS & RPS would result in a slightly lower share of bioelectricity (17%) and a higher share (34%) for wind generation than under the RPS (Table 2).

Effects of alternative policies on mix and volume of biofuels

Under the assumed targets of the RFS, the total volume of biofuels increases to 147 billion liters in 2030. Corn ethanol production is at the capped level of 57 billion liters. Additionally, the RFS induces 8 billion liters of advanced biofuels (imported sugarcane ethanol and renewable diesel) and 82 billion liters from cellulosic feedstocks. As the RPS creates no incentives for biofuel production, biofuel production under the RFS & RPS is close to that under the RFS alone (Table 1).

Feedstock production and mix

The demand for bioelectricity and biofuels creates a demand for biomass that increases from 33 million metric tons (MMT) under the no-policy scenario to 100 MMT under the RFS and to 262 MMT under the RFS. The combined demand for biomass under the RFS & RPS is 310 MMT in 2030. This is less than the sum of the demand for biomass under RPS and RFS individually because the higher price of biomass under the combined policy reduces the competitiveness of bioelectricity in favor of wind generation. Of the total biomass produced, 87% is used for bioelectricity under the RPS (and the rest exported as pellets), but 95% is used for biofuel under the RFS. With both RFS & RPS implemented jointly, about 80% would be used for cellulosic biofuels and 17% for bioelectricity.

Table 2 Prices in 2030

	No policy	RPS	RFS	RFS & RPS
Electricity price (average cents kWh^{-1})	11.5	11.2	11.3	11.0
Coal generation (max of all technologies)	5.3	5.2	5.3	5.1
Natural gas generation price	8.8	8.3	8.8	8.3
Co-fired electricity	6.0	7.8	7.1	9.2
Dedicated biomass	n/a	16.6	n/a	20.0
Wind	13.0	15.0	13.0	14.9
Natural gas price ($/MMBtu)	4.40	4.02	4.34	3.96
Farmgate biomass feedstock price ($/MT)	35.3	53.4	59.3	87.9
Land rental rate (average $/ha)	502	495	677	676

Of the total biomass produced under the various scenarios, a dominant share ranging from 55% to 85% is from agricultural feedstocks and the rest from forest residues and pulpwood. Crop and forest residues have the potential to account for almost 88% of the biomass under the RPS. This share decreases to about 70% under the RFS and under the RFS & RPS. Dedicated energy crops provide 11–29% of the biomass under the three scenarios with miscanthus having a dominant share among the energy crops produced.

Figure 1 shows the trend in the mix and quantity of biomass over the 2007–2030 period under the three alternative policy scenarios. Under the RPS (Fig. 1a), total biomass feedstock production increases over time, although the rate of increase slows over time; the growth in biomass feedstock demand corresponds to the increasing stringency of the RPSs over time. The mix of feedstock initially consists primarily of forest residues, but the proportion of corn stover, wheat straw and energy crops increases over time; by 2030, the share of feedstock from agriculture is 55% and from 44% from forest (Table 1). Corn stover makes up a large percentage of the total feedstock as it is relatively less costly to produce than other feedstocks in many areas, particularly those which have a high corn yield. Over 75% of the corn stover produced is utilized in the CRD in which it is produced and is therefore, not incurring an additional transportation costs, while the rest is consumed for co-firing in coal-based power plants in neighboring CRDs from the source of production. These results are in contrast to those obtained by Sands *et al.* (2017) who assume that all of the bioelectricity generated in 2030 will be produced using switchgrass only.

Under the RFS, biomass production increases more dramatically, particularly after 2021. The quantity of all feedstocks increases over time, but in particular, the share of energy crops increases significantly to 30% by 2030. The amount of crop residues is very similar over time.

Under the RFS & RPS scenario, pulpwood begins to be utilized for bioenergy, as forest residue production levels off after about 2025.

Spatial pattern of land use

The land required to produce these feedstocks for each scenario is listed in Table 3. Under the RPS, with much of the biomass feedstocks obtained from crop and forest residues, there is only a 0.7 million hectare diversion of land from cropland (0.4 million hectares) and from cropland pasture (0.2 million hectares) to produce dedicated energy crops. There is a marginal expansion of cropland from 120 to 120.5 million hectares and a very small reduction in land used for food production in 2030. This is significantly smaller than the land requirements in Sands *et al.* (2017) under the RPS scenario in 2030 in part due to the smaller level of bioelectricity generation determined here, the large share of crop residues in biomass supply and the relatively higher yield of miscanthus as compared to switchgrass.

The RFS and the RFS & RPS scenarios see a large increase in land use of about 4–5 million hectares relative to the no-policy scenario. Much of this increase in cropland is to produce energy crops as cropland pasture is converted to energy crop production to produce the additional feedstock required to meet the biofuel mandate. Land for food crop production under the RFS & RPS is only marginally lower than under the RFS alone, as a greater share of biomass is gathered from crop residues than under the RFS scenario only. This is likely due to the spatial distribution of the RPSs, which require biomass production in particular regions where crop residues are a lower cost source of biomass than energy crops.

Figure 2 shows the spatial pattern of biomass production in 2030 under the alternative policies. As shown in Fig. 2a, much of the production of biomass under the RPS occurs in areas where there are coal-based power plants that can co-fire biomass. The largest concentration of biomass feedstock production occurs in the Illinois and Missouri region at a level of about 14 M MT (Fig. 2a). Under the RFS & RPS scenario, production of agricultural and forest biomass for bioelectricity decreases relative to the RPS scenario (Fig. 2b). Much of the feedstock production for bioelectricity in upper

(a) Mix of biomass feedstocks under RPS

■ Corn residue ■ Straw ■ Miscanthus ■ Switchgrass ■ Forest Residue ■ Pulpwod

(b) Mix of biomass feedstocks under RFS

■ Corn residue ■ Straw ■ Miscanthus ■ Switchgrass ■ Forest Residue ■ Pulpwod

(c) Mix of biomass feedstocks under RFS & RPS

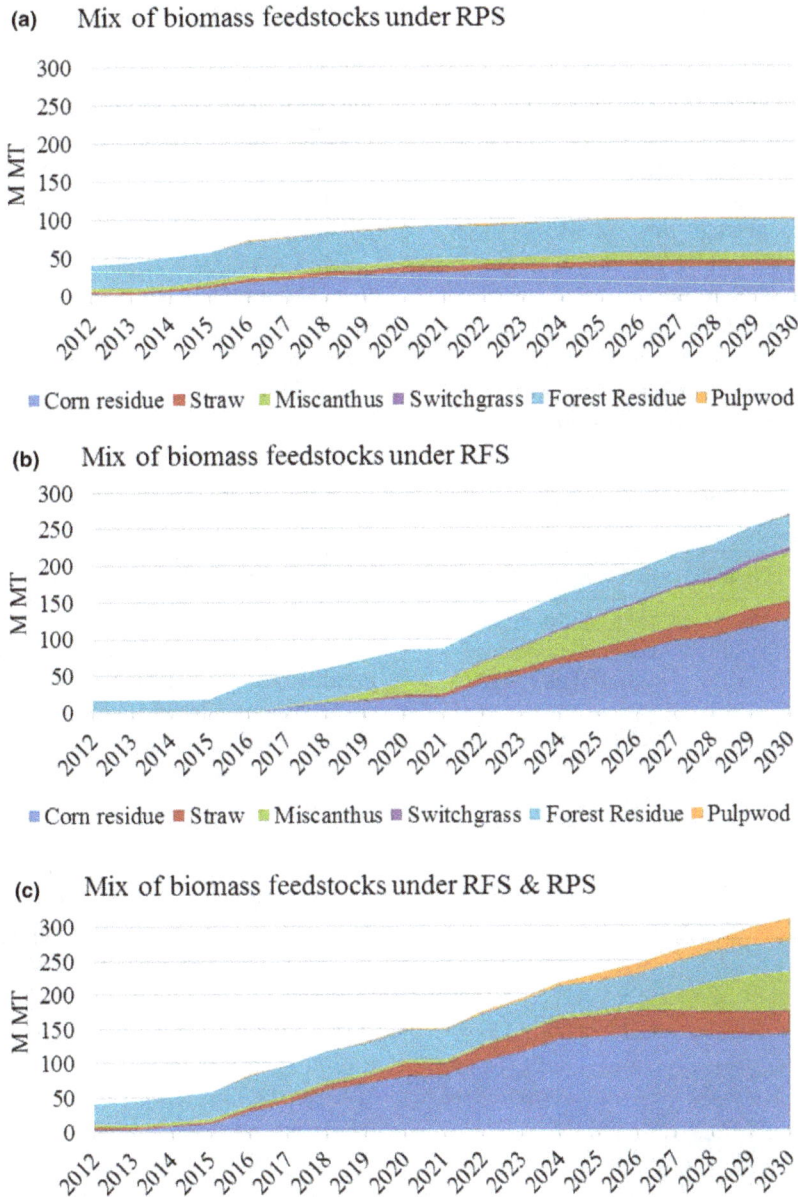

Fig. 1 Trends and mix of feedstock production under alternative policies. (a) Feedstocks under Renewable Portfolio Standard (RPS) (M MT). (b) Feedstocks under Renewable Fuel Standard (RFS) (M MT). (c) Feedstocks under RFS & RPS (M MT).

Midwest and Southeast that would occur under the RPS is entirely redirected toward biofuels under the RFS & RPS scenario as shown in Fig. 2d.

Under the RFS & RPS scenario, the intensity of biomass production increases in the rain-fed region of the United States (east of the 100th meridian) relative to the RPS alone. Production of biomass is particularly concentrated in Illinois, Tennessee, Missouri, Eastern Texas and the South Central region. Figure 2c, d shows biomass production for biofuels under the RFS and the RFS & RPS scenarios. Under the RFS & RPS scenario, crop residue production in the northern and north-western regions of the United States increases, while energy crop

production in the south and south central region is reduced relative to the RFS alone (Fig. 2d).

Effect of alternative policies on the prices

We now discuss the effects of the renewable energy policies considered here on the price of electricity, the price of biomass and the price of land. The price of electricity varies across EMRs and is determined by the demand and supply of electricity in the region. The price of biomass and the price of land vary across CRDs because land availability and land costs are defined at the CRD level. Price of food crops is determined at the national level given the national markets for agricultural commodities.

The farmgate price of biomass feedstock at the CRD level depends on the quantity of biomass production in the CRD, the CRD-specific biomass yields, land availability and costs of production (Fig. 3). In the RPS scenario, the price of biomass feedstock varies across the United States and has average price of $53/MT. The average price of biomass under the RFS is higher at $59/MT. The RFS & RPS results in an even larger increase in biomass price to $88/MT. Under each of these policies, there is considerable spatial variability in the price of biomass, as shown in Fig. 3. In the Midwest, where the greatest amount of feedstock is produced, the

price of biomass under the RPS ranges from about $46–$75/MT. In the Northeast, the price ranges from about $76–105/MT. The highest feedstock prices are found in the Southwest, but the quantity produced is relatively low. The range of biomass prices under the RFS & RPS scenario is much higher than under the RPS alone due to the increased level of production to meet the higher demand for biomass. In the Midwest prices range from about $76–105/MT, while in the Northeast prices range from about $106–135/MT. There is not much transportation of biomass across regions with new bioelectricity plants and biorefineries assumed to be established in CRDs with low costs of biomass production.

Bioelectricity prices also vary across regions, scenarios, and sources. Co-firing electricity tends to be less costly than electricity from a bio-power plant, which has a lower heat rate (conversion efficiency) of the bio-power plant than a co-fired process. Co-firing, however, is more expensive than coal-based generation, which costs on average about 5.3 cents kWh^{-1} (Table 2). These costs vary by region based on the conversion efficiency of existing coal-based plants and/or biomass transportation cost that depends on the distance travelled by the biomass. The average price of electricity from co-firing increases from 7.8 cents kWh^{-1} under the RPS to 9.2 cents kWh^{-1} under the RFS & RPS (Table 2). The average price of electricity from a bio-power plant increases from 16.6 cents kWh^{-1} under the RPS to 20 cents kWh^{-1} under the RFS & RPS. Regions such as CO, WY and IN, OH, WV, NJ have relatively lower

Table 3 Land-use implications of biomass production in 2030 (million hectares)

	No policy	RPS	RFS	RFS & RPS
Total cropland	120.0	120.5	125.0	124.0
Land under food crops	117.0	116.8	109.1	108.9
Land under corn for ethanol	3.0	3.0	11.6	11.6
Land under energy crops	0.0	0.7	4.3	3.5
From regular cropland	0.0	0.4	0.3	0.8
From marginal land	0.0	0.2	3.9	2.7
Land under corn stover	0.0	6.9	24.3	27.6
Land under wheat straw	0.0	2.7	10.6	16.1

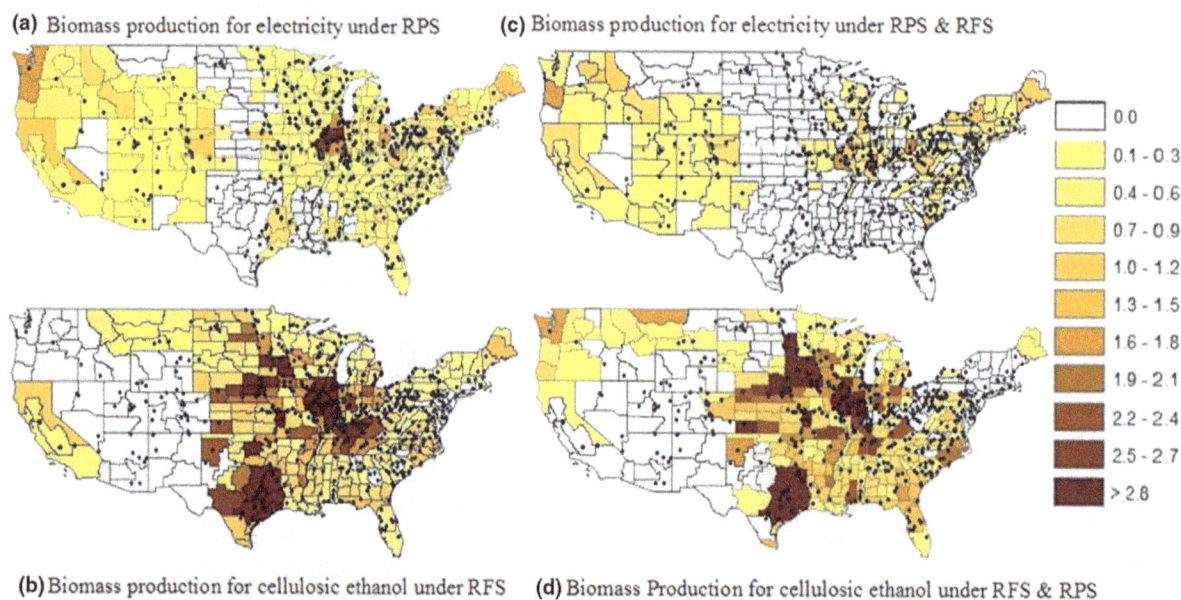

(a) Biomass production for electricity under RPS   **(c)** Biomass production for electricity under RPS & RFS

0.0
0.1 - 0.3
0.4 - 0.6
0.7 - 0.9
1.0 - 1.2
1.3 - 1.5
1.6 - 1.8
1.9 - 2.1
2.2 - 2.4
2.5 - 2.7
>2.8

(b) Biomass production for cellulosic ethanol under RFS   **(d)** Biomass Production for cellulosic ethanol under RFS & RPS

Fig. 2 Spatial pattern of biomass production in 2030 in million metric tons. Dots indicate existing coal power plant locations. (a) Biomass production for electricity under Renewable Portfolio Standard (RPS). (b) Biomass production for electricity under Renewable Fuel Standard (RFS) & RPS. (c) Biomass production for biofuel under RFS. (d) Biomass production for biofuel under RFS & RPS.

(a) Farmgate biomass price under RPS $/MT)
Dots indicate coal-fired power plant locations.

(b) Farmgate biomass price under RFS & RPS ($/MT)

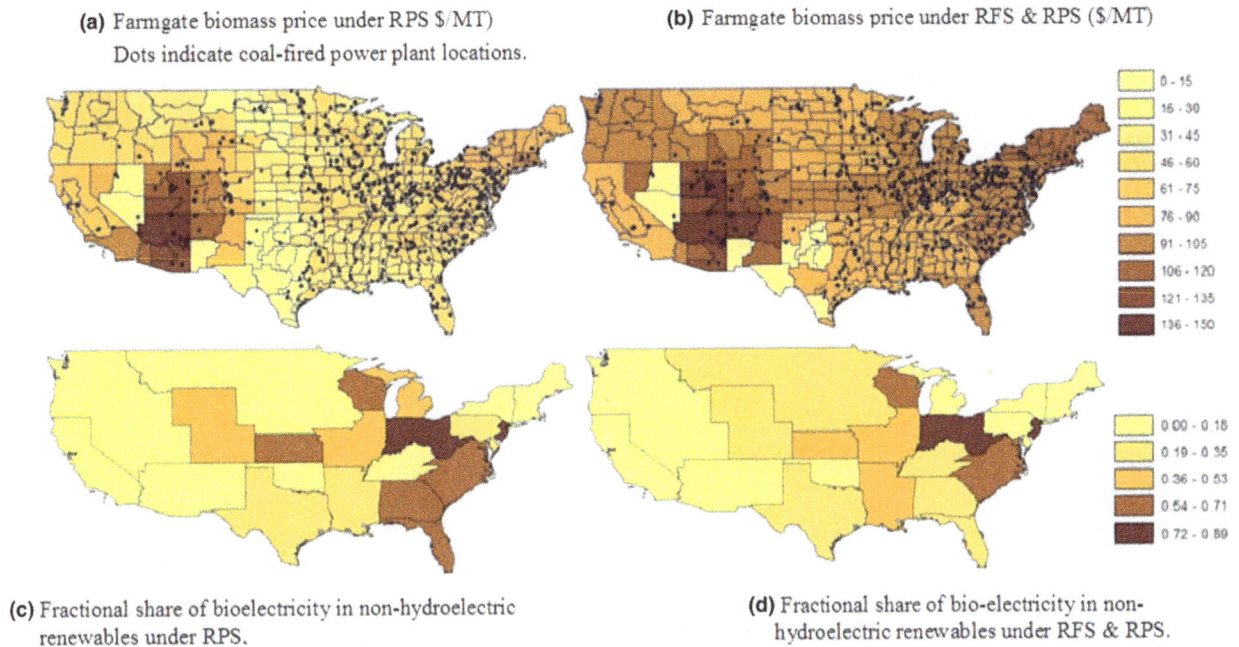

0 - 15
16 - 30
31 - 45
46 - 60
61 - 75
76 - 90
91 - 105
106 - 120
121 - 135
136 - 150

0.00 - 0.18
0.19 - 0.35
0.36 - 0.53
0.54 - 0.71
0.72 - 0.89

(c) Fractional share of bioelectricity in non-hydroelectric renewables under RPS.

(d) Fractional share of bio-electricity in non-hydroelectric renewables under RFS & RPS.

Fig. 3 Effect of alternative policies on regional biomass price and bioelectricity. (a) Farmgate biomass price under Renewable Portfolio Standard (RPS) ($/MT). (b) Farmgate biomass price under Renewable Fuel Standard (RFS) & RPS ($/MT). (c) Fractional share of bioelectricity in non-hydroelectric renewables under RPS. (d) Fractional share of bioelectricity in non-hydroelectric renewables under RFS & RPS. Dots indicate coal-fired power plant locations.

bioelectricity prices under both scenarios as they rely only on lower cost co-firing, while regions like CT, ME, MA, NH, RI, VT and NY have relatively high bioelectricity price under both scenarios, due to use of higher cost bio-power plants and relatively stringent RPSs. The increase in the average price of bioelectricity across regions with the combined RFS & RPSs implies an increase in the implicit subsidy that goes toward bioelectricity, which corresponds to an increase in the welfare cost of the policy (Oliver & Khanna, 2017).

We find that the increase in total electricity generation under the RPS leads to a marginal reduction in the average price of electricity. The reduction in coal and natural gas generation reduces the price of natural gas and the cost of fossil fuel-based electricity but increases the marginal cost of renewable electricity from bioelectricity and wind. The higher price of biomass under the RFS & RPS results in a higher average cost of bioelectricity, which increases from 16.6 cents kWh^{-1} under the RPS alone to 20 cents kWh^{-1} under the RFS & RPS.

We also find that the increased demand for biomass, particularly from energy crops, and the demand for corn for ethanol under the RFS and the RFS & RPS lead to an increase in the rental value of land. The land rental rate increases by 35% from $502 per hectare under the no-policy scenario to $677 per hectare under the RFS and the RFS & RPS scenarios.

Sensitivity analysis

We now examine the robustness of our findings about the effects of alternative policy scenarios on some key policy outcome variables of interest to parametric assumptions in the model. The policy outcome variables we examine are: bioelectricity generation, biomass quantity, average biomass price and land required for energy crop production. As compared to the benchmark case, we consider alternative parametric assumptions that: (i) increase the maximum biomass blend rate for co-firing to 20% instead of 10%, (ii) lower the rate of coproduct electricity production per liter of ethanol from a cellulosic ethanol refinery by 25% compared to the benchmark level based on (Humbird *et al.*, 2011), (iii) lower the marginal cost of wind generation by 25% than in the benchmark case (EIA, 2010a), (iv) raise the conversion efficiency of dedicated bio-power plants by 30% and (v) lower the yields of energy crops obtained from the simulation model DayCent by 20% (Dwivedi *et al.*, 2015).

In general, we find that the amount of bioelectricity generated decreases under the RFS & RPS scenario compared to the RPS scenario only, across all policy scenarios except for the case when the marginal cost of wind electricity is relatively low (Fig. 4a). This is consistent with the finding shown in Fig. 4d that the price of biomass under the RFS & RPS scenario is substantially higher ($61–$104/MT) than in the RPS scenario

($46–$59/MT). The extent to which this is the case is much lower in the scenario with the relatively low wind energy cost. In this case, the demand for bioelectricity under the RPS is less than a third of the demand in the benchmark case. This increases marginally under the RFS & RPS scenario due to the coproduct electricity generated. However, the overall demand for biomass is much lower in this scenario compared to the benchmark and other scenarios.

The relatively higher price of biomass under the RFS & RPS scenario accompanies the larger amount of biomass produced relative to the RPS scenario only. Across all the parametric assumptions, we find that the demand for biomass ranges between 52 and 168 MMT under the RPS and between 282 and 343 MMT under the RFS & RPS scenarios. Demand for biomass is lowest in the case with relatively low wind energy generation cost. Land requirement for biomass under the RFS & RPS is largely driven by the mandate for cellulosic biofuels and thus does not vary much with the variations in parametric assumptions considered here. Land requirements for biomass under the RPS are very small, ranging from 0.1 to 0.7 million hectares; it is much larger under the RFS & RPS and ranges from 2.9 to 3.8 million hectares. Land conversion to energy crops is lowest in the scenario with 25% lower energy crop yields because that makes energy crops less competitive with crop residues and forest biomass. Overall, we find that the qualitative direction of

our findings under the benchmark case is robust to assumed parameter values in the model.

Discussion

This paper examines the demand for biomass likely to be generated by existing renewable energy policies in the electricity and transportation sectors in the United States over the 2007–2030 period. We apply a dynamic, partial-equilibrium, open-economy, sector model (BEPAM-E) of the US agricultural, electricity and transportation sectors to endogenously determine the share of bioelectricity to meet the state RPSs and the allocation of biomass between electricity and cellulosic biofuel production to meet the RFS. We also examine the mix of biomass likely to be produced from agricultural and forestry sources and from residues vs. dedicated energy crops and its implications for the land requirements to meet the biomass demand.

We find that bioelectricity could provide 20% of the renewable electricity needed to meet state RPSs in 2030 and generate a total demand for biomass of 100 MMT (including pellets for export). In contrast, the quantity of biomass needed under the RFS in 2030 is 262 MMT. If implemented together, the two policies would induce a demand for 310 MMT, which is less than the sum of the demand that would be generated under each of the policies independently. This is due to the increase in price of biomass induced by the simultaneous implementation of

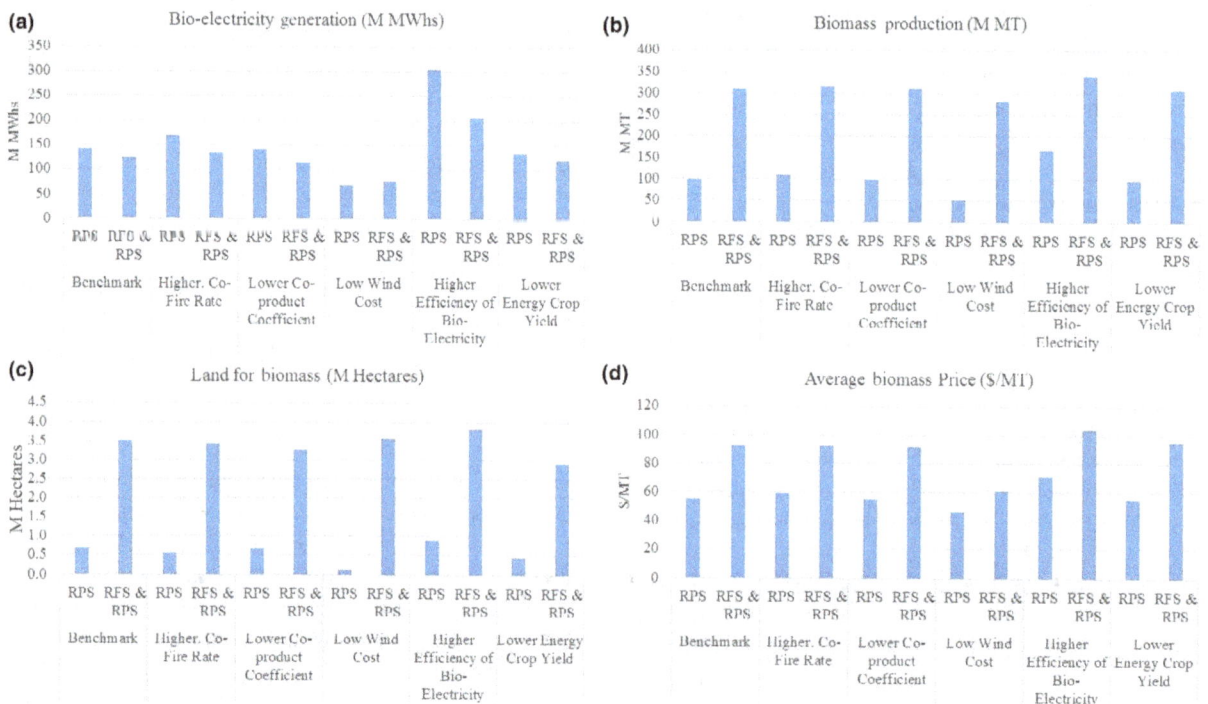

Fig. 4   Sensitivity analysis. (a) Bioelectricity generation (M MWh). (b) Biomass production (M MT). (c) Land for biomass (M hectares). (d) Average biomass price ($/MT).

the RFS & RPS relative to that under each of the policies independently; implementation of the RFS reduces the competitiveness of bioelectricity relative to wind energy generation.

We find that crop and forest residues could meet more than 70% of the demand for biomass under the policy scenarios considered here. The remaining supply of biomass can be achieved by converting 0.7 million hectares under the RPS and 4.3 million hectares of land under the RFS to high-yielding energy crops. Under the RFS & RPS scenario, 3.5 million hectares would need to convert to energy crop production; 80% of this land could be from marginal land in pasture/grazing/crop rotation, while the rest would be from regular cropland.

This shows the potential to meet a significant portion of the demand for biomass likely to be generated by existing renewable energy policies without diverting land from food crop production.

Acknowledgements

The authors are grateful for support provided by NIFA, USDA for this research.

References

Abt KL, Abt RC, Galik C (2012) Effect of bioenergy demands and supply response on markets, carbon, and land use. *Forest Science*, **58**, 523–539.

Beach RH, Zhang YW, Mccarl BA (2012) Modeling bioenergy, land use, and GHG emissions with FASOMGHG: model overview and analysis of storage cost implications. *Climate Change Economics*, **03**, 1250012.

Brechbill SC, Tyner WE, Ileleji KE (2011) The economics of biomass collection and transportation and its supply to Indiana cellulosic and electric utility facilities. *BioEnergy Research*, **4**, 141–152.

Chen X, Khanna M (2013) Food vs. fuel: the effect of biofuel policies. *American Journal of Agricultural Economics*, **95**, 289–295.

Chen X, Önal H (2012) Modeling agricultural supply response using mathematical programming and crop mixes. *American Journal of Agricultural Economics*, **94**, 674–686.

Chen X, Huang H, Khanna M, Önal H (2014) Alternative transportation fuel standards: welfare effects and climate benefits. *Journal of Environmental Economics and Management*, **67**, 241–257.

De Gorter H, Just DR (2009) The economics of a blend mandate for biofuels. *American Journal of Agricultural Economics*, **91**, 738–750.

DSIRE (2011) *Quantitative RPS data*. Database of State Incentives for Renewables and Efficiency, NC Clean Energy Technology Center. North Carolina State University.

Dumortier J (2013) Co-firing in coal power plants and its impact on biomass feedstock availability. *Energy Policy*, **60**, 396–405.

Dwivedi P, Wang W, Hudiburg T et al. (2015) Cost of abating greenhouse gas emissions with cellulosic ethanol. *Environmental Science & Technology*, **49**, 2512–2522.

EIA (2010a) *Annual Energy Outlook 2010: With Projections to 2035*. U.S. Energy Information Administration, Office of Integrated Analysis and Forecasting, U.S. Department of Energy, Washington, DC.

EIA (2010b) *State Energy Data System*. U.S. Department of Energy, Energy Information Administration, Washington, DC.

EIA (2011a) *The Electricity Market Module of the National Energy Modeling System Model Documentation Report*. U.S. Energy Information Administration, Office of Integrated Analysis and Forecasting, Washington, DC.

EIA (2011b) *Electricity Tends to Flow South in North America*. Energy Information Administration, Washington, DC.

EIA (2012) *Annual Energy Outlook 2012*. U.S. Department of Energy, Energy Information Administration. Office of Integrated Analysis and Forecasting. Washington, DC.

EPA (2010) *Emissions & Generation Integrated Database 2010*. U.S. Environmental Protection Agency, Washington, DC.

Fischer C (2010) Renewable Portfolio Standards: when do they lower energy prices? *Energy Journal*, **31**, 101–119.

Haq Z (2002) *Biomass for Electricity Generation*. U.S. Department of Energy, Energy Information Administration, Washington, DC.

Huang H, Khanna M, Önal H, Chen X (2013) Stacking low carbon policies on the renewable fuels standard: Economic and greenhouse gas implications. *Energy Policy*, **56**, 5–15.

Hudiburg TW, Wang W, Khanna M et al. (2016) Impacts of a 32-billion-gallon bioenergy landscape on land and fossil fuel use in the US. *Nature Energy*, **1**, 15005.

Humbird D, Davis R, Tao L et al. (2011) *Process Design and Economics for Biochemical Conversion of Lignocellulosic Biomass to Ethanol National Renewable Energy Laboratory Report NREL*. TP-5100-47764.

Ince PJ, Kramp AD, Skog KE, Yoo D, Sample VA (2011) Modeling future U.S. forest sector market and trade impacts of expansion in wood energy consumption. *Journal of Forest Economics*, **17**, 142–156.

Khanna M, Dhungana B, Clifton-Brown J (2008) Costs of producing miscanthus and switchgrass for bioenergy in Illinois. *Biomass and Bioenergy*, **32**, 482–493.

Khanna M, Chen X, Huang H, Önal H (2011) Supply of cellulosic biofuel feedstocks and regional production pattern. *American Journal of Agricultural Economics*, **93**, 473–480.

Langholtz M, Graham R, Eaton L, Perlack R, Hellwinkel C, De La Torre Ugarte DG (2012) Price projections of feedstocks for biofuels and biopower in the U.S. *Energy Policy*, **41**, 484–493.

Langholtz M, Eaton L, Turhollow A, Hilliard M (2014) 2013 feedstock supply and price projections and sensitivity analysis. *Biofuels, Bioproducts and Biorefining*, **8**, 594–607.

LaTourrette T, Ortiz DS, Hlavka E, Burger N, Cecchine G (2011) *Supplying Biomass to Power Plants: A Model of the Costs of Utilizing Agricultural Biomass in Cofired Power Plants*. RAND Corporation, Santa Monica, CA http://www.rand.org/pubs/technical_reports/TR876.html. Also available in print form.

Latta GS, Baker JS, Beach RH, Rose SK, McCarl BA (2013) A multi-sector intertemporal optimization approach to assess the GHG implications of U.S. forest and agricultural biomass electricity expansion. *Journal of Forest Economics*, **19**, 361–383.

McCarl BA, Adams DM, Alig RJ, Chmelik John T (2000) Competitiveness of biomass-fueled electrical power plants. *Annals of Operations Research*, **94**, 37–55.

Miao R, Khanna M, Huang H (2016) Responsiveness of crop yield and acreage to prices and climate. *American Journal of Agricultural Economics*, **98**, 191–211.

Oliver A, Khanna M (2017) What is the cost of a renewable energy based approach to greenhouse gas mitigation? *Land Economics*, **93**(3).

Palmer K, Burtraw D (2005) Cost-effectiveness of renewable electricity policies. *Energy Economics*, **27**, 873–894.

Paul A, Burtraw D, Palmer K (2009) *Haiku Documentation: RFF's Electricity Market Model Version 2.0*. Resources for the Future, Washington, DC.

Perlack RD, Stokes BJ (2011) *U.S. Billion-Ton Update: Biomass Supply for a Bioenergy and Bioproducts Industry*. Oak Ridge National Laboratory, Oak Ridge, TN.

Qin X, Mohan T, El-Halwagi M, Cornforth G, McCarl BA (2006) Switchgrass as an alternate feedstock for power generation: an integrated environmental, energy and economic life-cycle assessment. *Clean Technologies and Environmental Policy*, **8**, 233–249.

Sands RD, Malcolm SA, Suttles SA, Marshall E (2017) *Dedicated Energy Crops and Competition for Agricultural Land*. Economic Research Report Number 223, Economics Research Service United States Department of Agriculture, Washington, DC.

Searcy E, Flynn P, Ghafoori E, Kumar A (2007) The relative cost of biomass energy transport. In: *Applied Biochemistry and Biotechnology* (eds Mielenz JR, Klasson KT, Adney WS, McMillan JD), pp. 639–652. Humana Press.

UCS (2011) *A Bright Future for the Heartland: Technical Appendix*. Union of Concerned Scientists, Cambridge, MA.

USDA/NASS (2009) *U.S. & All States County Data - Crops*. U.S. Department of Agriculture, National Agricultural Statistical Service, Washington, DC. Available at: http://quickstats.nass.usda.gov/

White EM, Latta G, Alig RJ, Skog KE, Adams DM (2013) Biomass production from the U.S. forest and agriculture sectors in support of a renewable electricity standard. *Energy Policy*, **58**, 64–74.

PERMISSIONS

All chapters in this book were first published in GCB BIOENERGY, by John Wiley & Sons Ltd.; hereby published with permission under the Creative Commons Attribution License or equivalent. Every chapter published in this book has been scrutinized by our experts. Their significance has been extensively debated. The topics covered herein carry significant findings which will fuel the growth of the discipline. They may even be implemented as practical applications or may be referred to as a beginning point for another development.

The contributors of this book come from diverse backgrounds, making this book a truly international effort. This book will bring forth new frontiers with its revolutionizing research information and detailed analysis of the nascent developments around the world.

We would like to thank all the contributing authors for lending their expertise to make the book truly unique. They have played a crucial role in the development of this book. Without their invaluable contributions this book wouldn't have been possible. They have made vital efforts to compile up to date information on the varied aspects of this subject to make this book a valuable addition to the collection of many professionals and students.

This book was conceptualized with the vision of imparting up-to-date information and advanced data in this field. To ensure the same, a matchless editorial board was set up. Every individual on the board went through rigorous rounds of assessment to prove their worth. After which they invested a large part of their time researching and compiling the most relevant data for our readers.

The editorial board has been involved in producing this book since its inception. They have spent rigorous hours researching and exploring the diverse topics which have resulted in the successful publishing of this book. They have passed on their knowledge of decades through this book. To expedite this challenging task, the publisher supported the team at every step. A small team of assistant editors was also appointed to further simplify the editing procedure and attain best results for the readers.

Apart from the editorial board, the designing team has also invested a significant amount of their time in understanding the subject and creating the most relevant covers. They scrutinized every image to scout for the most suitable representation of the subject and create an appropriate cover for the book.

The publishing team has been an ardent support to the editorial, designing and production team. Their endless efforts to recruit the best for this project, has resulted in the accomplishment of this book. They are a veteran in the field of academics and their pool of knowledge is as vast as their experience in printing. Their expertise and guidance has proved useful at every step. Their uncompromising quality standards have made this book an exceptional effort. Their encouragement from time to time has been an inspiration for everyone.

The publisher and the editorial board hope that this book will prove to be a valuable piece of knowledge for researchers, students, practitioners and scholars across the globe.

LIST OF CONTRIBUTORS

Yong Chen
Texas A&M AgriLife Research (Texas A&M University System), PO Box 1658, 11708 Highway 70S, Vernon, TX 76384, USA
Department of Soil and Crop Sciences, Texas A&M University, 370 Olsen Blvd, TAMU MS 2474, College Station, TX 77843, USA

Srinivasulu Ale
Texas A&M AgriLife Research (Texas A&M University System), PO Box 1658, 11708 Highway 70S, Vernon, TX 76384, USA
Department of Biological and Agricultural Engineering, Texas A&M University, TAMU MS 2117, College Station, TX 77843, USA

Nithya Rajan
Department of Soil and Crop Sciences, Texas A&M University, 370 Olsen Blvd, TAMU MS 2474, College Station, TX 77843, USA

Clyde Munster
Department of Biological and Agricultural Engineering, Texas A&M University, TAMU MS 2117, College Station, TX 77843, USA

Janine Schweier
Chair of Forest Operations, Albert-Ludwigs-University Freiburg, Werthmannstraße 6, 79085 Freiburg, Germany,

Rüdiger Grote, Eugenio Díaz - Pinés, Saúl Molina-Herrera, Edwin Haas and Klaus Butterbach-Bahl
Karlsruhe Institute of Technology (KIT), Institute of Meteorology and Climate Research, Atmospheric Environmental Research, Kreuzeckbahnstraße 19, 82467 Garmisch-Partenkirchen, Germany

Andrea Ghirardo and Jörg-Peter Schnitzler
Helmholtz Zentrum München, Research Unit Environmental Simulation, Institute of Biochemical Pathology, Ingolstädter Landstraße 1, 85764 Neuherberg, Germany

Jürgen Kreuzwieser and Heinz Renne-Nberg
Chair of Tree Physiology, Albert-Ludwigs-University Freiburg, Georges-Köhler-Allee 53/54, 79110 Freiburg, Germany

Gero Becker
Chair of Forest Utilisation, Albert-Ludwigs-University Freiburg, Werthmannstraße 6, 79085 Freiburg, Germany

Carolyn Smyth and Werner A. Kurz
Natural Resources Canada, Canadian Forest Service, 506 Burnside Road West, Victoria, BC V8Z 1M5, Canada,

Greg Rampley, Tony C. Lempriére and Olaf Schwab
Natural Resources Canada, Canadian Forest Service, 580 Booth Street, Ottawa, ON K1A 0E4, Canada

Zhihong Song, Chengcheng Tao and Yangyang Fan
Key Laboratory of Plant Resofurces and Beijing Botanical Garden, Institute of Botany, Chinese Academy of Sciences, Beijing 100093, China
University of Chinese Academy of Sciences, Beijing 100049, China

Qin Xu and Cong Lin
Key Laboratory of Plant Resources and Beijing Botanical Garden, Institute of Botany, Chinese Academy of Sciences, Beijing 100093, China

Caiyun Zhu, Shilai Xing and Tao Sang
Key Laboratory of Plant Resources and Beijing Botanical Garden, Institute of Botany, Chinese Academy of Sciences, Beijing 100093, China
University of Chinese Academy of Sciences, Beijing 100049, China
State Key Laboratory of Systematic and Evolutionary Botany, Institute of Botany, Chinese Academy of Sciences, Beijing 100093, China

Wei Liu
State Key Laboratory of Systematic and Evolutionary Botany, Institute of Botany, Chinese Academy of Sciences, Beijing 100093, China

Juan Yan and Jianqiang L I
Key Laboratory of Plant Germplasm Enhancement and Speciality Agriculture, Wuhan Botanical Garden, Chinese Academy of Sciences, Wuhan 430074, China

Yudai Sumiyoshi, Susan E. Crow and Creighton M. Litton
Department of Natural Resources and Environmental Management, University of Hawaii Manoa, Honolulu, HI 96822, USA,

Jonathan L. Deenik, Brian Turano and Richard Ogoshi
Department of Tropical Plant and Soil Sciences, University of Hawaii Manoa, Honolulu, HI 96822, USA,

Andrew D. Taylor
Department of Biology, University of Hawaii Manoa, Honolulu, HI 96822, USA

Xiaoyin Sun, Ruifeng Shan, Jihua Pan, Xing Liu, Ruonan Deng and Junyao Song
Key Laboratory of Nansi Lake Westland Ecological Conservation & Environmental Protection (Shandong Province), College of Geography and Tourism, Qufu Normal University, Rizhao 276826, China

Xuhui L I
College of Environment and Planning, Henan University, Kaifeng 475004, China

Nathan M. Tarr and Matthew J. Rubino
North Carolina Cooperative Fish and Wildlife Research Unit, Department of Applied Ecology, North Carolina State University, Campus Box 7617, Raleigh, NC 27695, USA

Jennifer K. Costanza
Department of Forestry and Environmental Resources, North Carolina State University, 3041 Cornwallis Road, Research Triangle Park, NC 27709, USA

Alexa J. Mckerrow
U.S. Geological Survey, Core Science Analytics, Synthesis, and Libraries, Campus Box 7617, Raleigh, NC 27695, USA

Jaime A. Collazo
U.S. Geological Survey, North Carolina Cooperative Fish and Wildlife Research Unit, Department of Applied Ecology, North Carolina State University, Campus Box 7617, Raleigh, NC 27695, USA

Robert C. Abt
Department of Forestry and Environmental Resources, North Carolina State University, Campus Box 8008, Raleigh, NC 27695, USA

Ioannis Tsiropoulos, Ric Hoefnagels and Machteld Van Den Broek
Copernicus Institute of Sustainable Development, Utrecht University, Heidelberglaan 2, 3584 CS Utrecht, The Netherlands

Martin K. Patel
Energy Group, Institute for Environmental Sciences and Forel Institute, University of Geneva, Boulevard Carl-Vogt 66, 1205 Geneva, Switzerland

Andre P. C. Faaij
Energy and Sustainability Research Institute, University of Groningen, Nijenborg 4, 9747 AC Groningen, The Netherlands

Yongl I Zhao, Suma Basak, Marceline Egnin and Guohao He
College of Agriculture, Environment and Nutrition Sciences, Tuskegee University, Tuskegee, AL 36088, USA,

Christine E. Fleener, Erik J. Sacks
Department of Crop Sciences, University of Illinois, Urbana, IL 61801, USA

Channapatna S. Prakash
College of Arts and Sciences, Tuskegee University, Tuskegee, AL 36088, USA

Jule Schulze and Henning Nolzen
Department Ecological Modelling, UFZ – Helmholtz Centre for Environmental Research, Permoserstr. 15, 04318 Leipzig, Germany
Institute for Environmental System Research, Osnabrück University, Barbarastr. 12, 49076 Osnabrück, Germany,

Erik Gawel
Department of Economics, UFZ – Helmholtz Centre for Environmental Research, Permoserstr. 15, 04318 Leipzig, Germany,
Institute for Infrastructure and Resources Management, Leipzig University, Grimmaische Str. 12, 04109 Leipzig, Germany,

Hanna Weise
Institute of Biology, Biodiversity and Ecological Modelling, Freie Universität Berlin, Altensteinstr. 6, 14195 Berlin, Germany,

Karin Frank
Department Ecological Modelling, UFZ – Helmholtz Centre for Environmental Research, Permoserstr. 15, 04318 Leipzig, Germany

Institute for Environmental System Research, Osnabrück University, Barbarastr. 12, 49076 Osnabrück, Germany,
iDiv – German Centre for Biodiversity Research Halle-Jena-Leipzig, Deutscher Platz 5a, Leipzig, Germany

Richard R. Rodrigues
Interdisciplinary Ph.D. Program in Genetics, Bioinformatics, and Computational Biology, 1015 Life Science Circle, Virginia Tech, Blacksburg, VA 24061, USA

Jinyoung Moon and Bingyu Zhao
Department of Horticulture, 220 Ag Quad Lane, Virginia Tech, Blacksburg, VA 24061, USA

Mark A. Williams
Interdisciplinary Ph.D. Program in Genetics, Bioinformatics, and Computational Biology, 1015 Life Science Circle, Virginia Tech, Blacksburg, VA 24061, USA
Department of Horticulture, 220 Ag Quad Lane, Virginia Tech, Blacksburg, VA 24061, USA

Debasish Saha
Department of Ecosystem Science and Management, The Pennsylvania State University, University Park, PA 16802, USA

Department of Plant Science, The Pennsylvania State University, University Park, PA 16802, USA

Benjamin M. Rau and Paul R. Adler
Pasture Systems and Watershed Management Research Unit, USDA Agricultural Research Service, University Park, PA 16802, USA

Jason P. Kaye
Department of Ecosystem Science and Management, The Pennsylvania State University, University Park, PA 16802, USA

Felipe Montes and Armen R. Kemanian
Department of Plant Science, The Pennsylvania State University, University Park, PA 16802, USA

Anthony Oliver
South Coast Air Quality Management District, 21865 Copley Drive, Diamond Bar, CA 91765, USA

Madhu Khanna
Department of Agricultural and Consumer Economics, University of Illinois at Urbana-Champaign, 1301 W. Gregory Drive, Urbana, IL 61801, USA

Index